SOLID-STATE IMAGING WITH CHARGE-COUPLED DEVICES

SOLID-STATE SCIENCE AND TECHNOLOGY LIBRARY

Volume 1

Editorial Advisory Board

Aims and Scope of the Series

The aim of this series is to present monographs on semiconductor materials processing and device technology, discussing theory formation and experimental characterization of solid-state devices in relation to their application in electronic systems, their manufacturing, their reliability, and their limitations (fundamental or technology dependent). This area is highly interdisciplinary and embraces the cross-section of physics of condensed matter, materials science and electrical engineering.

Undisputedly during the second half of this century world society is rapidly changing owing to the revolutionary impact of new solid-state based concepts. Underlying this spectacular product development is a steady progress in solid-state electronics, an area of applied physics exploiting basic physical concepts established during the first half of this century. Since their invention, transistors of various types and their corresponding integrated circuits (ICs) have been widely exploited covering progress in such areas as microminiaturization, megabit complexity, gigabit speed, accurate data conversion and/or high power applications. In addition, a growing number of devices are being developed exploiting the interaction between electrons and radiation, heat, pressure, etc., preferably by merging with ICs.

Possible themes are (sub)micron structures and nanostructures (applying thin layers, multi-layers and multi-dimensional configurations); micro-optic and micro-(electro)mechanical devices; high-temperature superconducting devices; high-speed and high-frequency electronic devices; sensors and actuators; and integrated opto-electronic devices (glass-fibre communications, optical recording and storage, flat-panel displays).

The texts will be of a level suitable for graduate students, researchers in the above fields, practitioners, engineers, consultants, etc., with an emphasis on readability, clarity, relevance and applicability.

Solid-State Imaging with Charge-Coupled Devices

by

Albert J. P. Theuwissen

Philips Imaging Technology,
Eindhoven, The Netherlands

KLUWER ACADEMIC PUBLISHERS

DORDRECHT / BOSTON / LONDON

Library of Congress Cataloging-in-Publication Data

```
Theuwissen, Albert J. P.
   Solid-state imaging with charge-coupled devices / by Albert J.P.
 Theuwissen.
     p.   cm. -- (Solid-state science and technology library)
   Includes index.
   ISBN 0-7923-3456-6 (acid free)
   1. Charge-coupled devices.  2. Imaging systems.   I. Title.
 II. Series.
 TK7871.99.C45T48   1995
 621.36'7--dc20                                        95-11945
```

ISBN 0-7923-3456-6

Published by Kluwer Academic Publishers,
P.O. Box 17, 3300 AA Dordrecht, The Netherlands.

Kluwer Academic Publishers incorporates
the publishing programmes of
D. Reidel, Martinus Nijhoff, Dr W. Junk and MTP Press.

Sold and distributed in the U.S.A. and Canada
by Kluwer Academic Publishers,
101 Philip Drive, Norwell, MA 02061, U.S.A.

In all other countries, sold and distributed
by Kluwer Academic Publishers Group,
P.O. Box 322, 3300 AH Dordrecht, The Netherlands.

First published 1995
Reprinted with corrections 1996

Printed on acid-free paper

Printed in the Netherlands

to Elke, Lien, Kim
to Marie-Jeanne

CONTENTS

Symbol list — xiii

List of figures — xix

Preface — xxv

Chapter 0 : Introduction — 1

Chapter 1 : Fundamentals of charge-coupled devices — 7
1.1. An ideal MOS capacitor — 7
    1.1.1. MOS capacitance in accumulation — 8
    1.1.2. MOS capacitance in deep depletion — 10
    1.1.3. MOS capacitance in inversion — 13
    1.1.4. MOS capacitance in weak inversion — 16
1.2. A real MOS-capacitor in a charge-coupled device — 18
1.3. Charge transfer — 25
    1.3.1. Thermal diffusion — 28
    1.3.2. Self-induced drift — 29
    1.3.3. Fringing fields — 31
1.4. Charge transfer (in)efficiency — 36
1.5. Buried channel CCD — 39
    1.5.1. From SCCD to BCCD — 40
    1.5.2. Fringing field and transfer time — 42
    1.5.3. Charge-handling capability — 45
1.6. One-dimensional potential analysis — 48
1.7. Conclusions — 51

Chapter 2 : Into, through and out of a charge-coupled device — 53
2.1. Transport systems — 54
    2.1.1. Four-phase system — 54
    2.1.2. Three-phase system — 57
    2.1.3. Two-phase system — 58
    2.1.4. One-and-a-half phase system — 62
    2.1.5. Virtual-phase system — 63
    2.1.6. Ripple clock — 63
2.2. Channel definition — 66
2.3. Input structures — 69
    2.3.1. Diode cut-off — 70

CONTENTS

   2.3.2. Fill-and-spill   73
  2.4. Output structures   75
   2.4.1. Floating diffusion with reset   76
   2.4.2. Floating gate without reset   79
  2.5. Conclusions   83

Chapter 3 : A real CCD delay line   85
  3.1. Effect of transport inefficiency   87
  3.2. Effect of dark current   92
  3.3. Effect of dark-current nonuniformities   94
  3.4. Effect of noise   97
   3.4.1. Shot noise   97
   3.4.2. Trapping noise   98
   3.4.3. kTC noise   99
   3.4.4. Amplifier noise   99
  3.5. Sampling of an electrical signal   100
  3.6. Conclusions   106

Chapter 4 : Solid-state imaging at a glance   109
  4.1. Photon sensing   109
  4.2. Imager configurations   112
   4.2.1. Linear imagers   112
   4.2.2. Array imagers   114
    Frame-transfer CCD   114
    Interline-transfer CCD   117
    Frame-interline-transfer CCD   119
    MOS-XY imager   120
    Charge-injection device   123
    Overview array imagers   125
  4.3. Conclusions   128

Chapter 5 : Fundamentals of solid-state imaging   131
  5.1. Absorption of photons   131
  5.2. Collection of generated carriers   134
  5.3. Spectral response   136
  5.4. Quantum efficiency   139
  5.5. Resolution   142
   5.5.1. Diffusion MTF   143
   5.5.2. Transport MTF   145
   5.5.3. Geometric MTF   148
  5.6. Aliasing and Moiré effects   152
  5.7. Conclusions   154

Chapter 6 : Solid-state imaging for television applications	157
6.1. Scanning modes	159
6.1.1. Interlaced scanning	161
Frame-transfer CCD	161
(Frame-)interline-transfer CCD	161
6.1.2. Progressive scanning	162
6.2. Color imaging	165
6.2.1. Stripe filters	166
6.2.2. Mosaic filters	168
6.2.3. Primary or complementary, stripe or mosaic	170
6.2.4. Color separation by a prism	171
6.2.5. Color imaging with linear arrays	173
6.3. Blooming and antiblooming	176
6.3.1. Lateral or horizontal antiblooming	177
6.3.2. Charge-pumped or clocked antiblooming	178
6.3.3. Vertical antiblooming	179
6.4. Charge reset or electronic shutter	183
6.4.1. Charge reset in frame-transfer CCDs	184
6.4.2. Charge reset in (frame-) interline-transfer CCDs	184
6.4.3. Charge reset in MOS-XY and CID imagers	188
6.5. Conclusions	190
Chapter 7 : Advanced imaging : light sensitivity	193
7.1. Increasing the light sensitivity	193
7.2 Aperture ratio of the pixels	195
7.2.1. Microlenses	196
7.2.2. Photoconversion top layer	201
7.2.3. Optimizing the vertical CCDs	202
7.3. Light transmission of multilayered structure	205
7.3.1. Adapting the optical thicknesses	205
7.3.2. Minimizing the number of gates	207
7.3.3. Transparent conductive gates	208
7.4. Back-side illumination	211
7.5. Pixels with an amplification function	211
7.5.1. Static-induction transistor	212
7.5.2. Charge-modulation device	213
7.5.3. Bulk charge-modulated device	213
7.5.4. Amplified MOS intelligent imager	214
7.5.5. Base-stored image sensor	215
7.6. Conclusions	217

CONTENTS

Chapter 8 : Advanced imaging : noise and smear   219
  8.1. Decreasing noise levels   220
  8.2. Technology-related noise   221
    8.2.1. Point defects   221
    8.2.2. Column defects   222
    8.2.3. Transfer noise   222
    8.2.4. Striations   222
    8.2.5. Pixel nonuniformities   222
    8.2.6. Dark-current shot noise   223
  8.3. Output-amplifier noise   224
    8.3.1. Thermal noise   225
    8.3.2. 1/f noise   227
    8.3.3. Reset noise   227
    8.3.4. Elimination of the reset noise   228
    8.3.5. New output-amplifier architectures   231
  8.4. Output-amplifier sensitivity   234
  8.5. Smear   236
    8.5.1. Smear in frame-transfer CCDs   236
    8.5.2. Smear in (frame-)interline-transfer CCDs   237
    8.5.3. Smear compensation techniques   239
    8.5.4. Smear in MOS-XY and CID imagers   240
    8.5.5. State of the art in smear suppression   242
  8.6. Conclusions   244

Chapter 9 : Advanced imaging : device architectures   247
  9.1. Increasing horizontal pixel density   248
    9.1.1. Design adaptation   253
    9.1.2. Gate tapering   253
    9.1.3. Extra channel implantations   254
    9.1.4. Compound channels   257
    9.1.5. Combined techniques   258
  9.2. New clocking schemes for vertical registers   258
    9.2.1. Accordion CCD   259
    9.2.2. Charge-sweep device (CSD)   262
    9.2.3. High-speed clocks for vertical CCDs   262
    9.2.4. Dynamic pixel management   264
  9.3. Electronic still picture   267
  9.4. Time-delay and integrating CCD   274
  9.5. Conclusions   276

Chapter 10 : Nonconsumer imaging   279
  10.1. Scientific imagers   280
    10.1.1. Large-area devices   280

10.1.2. Buttable devices   281
10.1.3. Notch CCD   282
10.1.4. Skipper CCD   283
10.1.5. Pinned-phase CCDs   285
        Open-phase pinned CCDs   287
        Multi-phase pinning CCDs   288
        Dynamic pinning   290
10.2. Smart image sensors   293
        10.2.1. ASIC Vision   293
        10.2.2. Three-dimensional integrated image sensor   294
        10.2.3. Foveated-retina sensor   296
10.3. Nonvisible imaging   297
        10.3.1. Infrared imaging   298
        10.3.2. UV imaging   302
        10.3.3. X-ray imaging   306
              X-ray imaging for spectroscopic purposes   306
              X-ray imaging for medical purposes   308
10.4. High-speed imagers   310
        10.4.1. Imagers with multiple outputs   310
        10.4.2. Gallium-arsenide CCD Imagers   311
10.5. Contact-type linear imagers   311
10.6. Conclusions   313

APPENDIX 1 : How CCD imagers are made   317
A1.1. Substrate preparation   320
A1.2. Implantation of the p well   321
A1.3. Drive-in of the p well   322
A1.4. Implantation of the CCD channel   323
A1.5. Drive-in of the channel implant   324
A1.6. Implantation of the channel stoppers   325
A1.7. Gate oxidation   326
A1.8. Deposition of the first poly-Si layer   327
A1.9. Definition of the first poly-Si layer   328
A1.10. Interpoly isolation   329
A1.11. Deposition of the second poly-Si layer   330
A1.12. Definition of the second poly-Si layer   331
A1.13. Interpoly isolation   332
A1.14. Deposition of the third poly-Si layer   333
A1.15. Definition of the third poly-Si layer   334
A1.16. Source and drain implantation   335
A1.17. Back-end isolation   336
A1.18. Contact-hole definition   337
A1.19. Deposition of the metal layer   338

CONTENTS

A1.20. Definition of the metallization pattern   339
A1.21. Scratch-protection deposition   340
A1.22. Deposition of the first color filter layer   341
A1.23. Lift-off of the first color filter   342
A1.24. Deposition of the second filter layer   343
A1.25. Lift-off of the second color filter   344
A1.26. Deposition of planarization layers   345
A1.27. Etch-back of the planarization layers   346
A1.28. Deposition of the lens material   347
A1.29. Reflow of the microlenses   348

APPENDIX 2 : How to interpret CCD artifacts   349
A2.1. Reference picture   349
A2.2. Smear   350
A2.3. Charge reset   350
A2.4. Blooming   351
A2.5. Integration-time shortening   351
A2.6. Column defects   353
A2.7. Cover glass damage   354
A2.8. Damage by electrostatic discharge   355
A2.9. Gate dielectric damage   355
A2.10. White point defects   356
A2.11. Fixed-pattern noise   357
A2.12. Other noise sources   358

APPENDIX 3 : How to compromise on CCD specifications   359
A3.1. Sensitivity and quantum efficiency   359
A3.2. Resolution   359
A3.3. Horizontal modulation transfer function   360
A3.4. Vertical modulation transfer function   360
   A3.4.1. Frame-transfer CCD   360
   A3.4.2. Frame-interline-transfer and interline-transfer CCD   362
A3.5. Antiblooming   363
A3.6. Smear   364
A3.7. Cost price   365

References   367

Index   381

SYMBOL LIST

A	constant	[-]
A_o	AC component of the optical input	[-]
A_{cell}	area of a CCD cell	[μm^2]
A_{SF}	gain of the source follower stage	[-]
C_D	depletion capacitance per unit area	[F/cm^2]
C_G	input capacitance of source follower stage	[fF]
C_{FD}	floating diffusion capacitance	[fF]
C_{ox}	oxide capacitance per unit area	[F/cm^2]
C_{ox1}	floating-gate capacitance per unit area	[F/cm^2]
C_{ox2}	interpoly capacitance per unit area	[F/cm^2]
C_p	parasitic output capacitance	[fF]
C'_p	reduced parasitic output capacitance	[fF]
C_1	quantum efficiency coefficient	[/eV]
CL	CCD cell length	[μm]
c	velocity of light	[$3*10^8$ m/sec]
D	defect density	[/cm^2]
D_{eff}	effective diffusion constant	[cm^2/sec]
D_n	diffusion constant of electrons	[cm^2/sec]
D_p	diffusion constant of holes	[cm^2/sec]
$d\phi$	flux decrement	[W/m^2]
E_f	fringing field	[V/cm]
$E_{f,min}$	minimum value of the fringing field	[V/cm]
E_G	semiconductor bandgap	[eV]
E_{ox}	electrical field in the gate dielectric	[V/cm]
E_s	self-induced	[V/cm]
f	frequency of the signal transferred by a CCD	[Hz]
f'	frequency of the output signal	[Hz]
f_c	clock frequency	[Hz]
f_{in}	frequency of the input signal	[Hz]
f_N	Nyquist frequency	[Hz]
f_{out}	frequency of the output signal	[Hz]
f_s	sampling frequency or CCD clocking frequency	[Hz]
f_{sig}	frequency of an input signal	[Hz]
$g(x)$	generation velocity	[/cm^2.sec]
$H(z)$	ideal transfer function of a CCD in the z-domain	[-]
$H(\omega)$	ideal transfer function of a CCD in the frequency domain	[-]
$H_{in}(x)$	optical input signal	[-]
H_o	DC component of optical input signal	[-]

$H_{out}(x_0)$	output signal at location x_0	[-]
$H_{em}(z)$	real transfer function of a CCD in the z-domain	[-]
$H_{em}(\omega)$	real transfer function of a CCD in the frequency domain	[-]
h	Plank's constant	$[6.63*10^{-34}$ J.sec]
I	bias current	$[\mu A]$
I_{out}	CCD output current	$[A/cm^2]$
J_d	current density by thermal diffusion	$[A/cm^2]$
J_{dark}	dark-current density	$[A/cm^2]$
$J_{dark,*}$	dark-current density at temperature *	$[A/cm^2]$
J^n_{dark}	dark-current density at CCD cell number n	$[A/cm^2]$
J_f	current density by fringing field	$[A/cm^2]$
J_s	current density by self-induced field	$[A/cm^2]$
$J(y,t)$	current density by charge packet movement	$[A/cm^2]$
K	body factor	$[V^{0.5}]$
k	Bolzmann's constant	$[1.38*10^{-23}$ J/K]
L	length of a single CCD gate	$[\mu m]$
L	length of the gate of an MOS transistor	$[\mu m]$
L'	length of a small part of a CCD gate	$[\mu m]$
L_K	diffusion MTF parameter	$[\mu m^2]$
L_n	diffusion length	$[\mu m]$
l	number of charge packets in a unit cell	[-]
M	parameter in the MTF expression	[-]
MTF	modulation transfer function	[%]
MTF_D	diffusion modulation transfer function	[%]
MTF_G	geometric modulation transfer function	[%]
MTF_e	transport modulation transfer function	[%]
m	number of phases	[-]
N	number of CCD cells in a CCD shift register	[-]
N	equivalent number of noise electrons	[-]
N_A	doping level of the p-type silicon substrate	$[/cm^3]$
N_D	doping level of the buried-channel implant	$[/cm^3]$
NED	noise equivalent density	$[e^2/Hz]$
N_{ss}	number of surface states	$[/cm^2.eV]$
N(x)	dopant concentration	$[/cm^3]$
n	random integer	[-]
n_i	intrinsic carrier concentration	$[1.4*10^{10}/cm^3$ at RT]
n_{CCD}	equivalent number of noise electrons generated in the CCD	[-]
n_{kTC}	number of kTC-noise electrons	[-]
$n_{kTC,RT}$	number of kTC-noise electrons at RT (= room temperature)	[-]
n_{SF}	equivalent number of noise electrons in the output amplifier	[-]
n_{shot}	equivalent number of shot-noise electrons	[-]
$n_{tr,BCCD}$	number of trapping-noise electrons in a BCCD	[-]

$n_{tr,SCCD}$	number of trapping-noise electrons in a SCCD	[-]
P	pitch of the CCD gate	[μm]
P_{pix}	pixel pitch	[μm]
Q	density of charge	[C/cm^2]
Q_{acc}	density of charge in the accumulation layer	[C/cm^2]
Q_{CCD}	total charge contained by a CCD	[C]
Q_d	density of fixed charge in the depletion layer	[C/cm^2]
Q_{dark}	density of dark-current-generated charge	[C/cm^2]
Q^N_{dark}	density of dark-current-generated charge in a CCD with N cells	[C/cm^2]
Q^n_{dark}	density of dark-current-generated charge in CCD cell number n	[C/cm^2]
Q^1_{dark}	density of dark-current-generated charge in one CCD cell	[C/cm^2]
Q_{in}	signal charge fed into a CCD shift register	[C/cm^2]
Q_k	charge available at stage k of a CCD shift register	[C/cm^2]
Q_{k+1}	charge available at stage k+1 of a CCD shift register	[C/cm^2]
Q_n	density of mobile charge in the inversion layer	[C/cm^2]
$Q_{n,sat}$	density of mobile charge at saturation in the inversion layer	[C/cm^2]
Q_{out}	charge fed to the output amplifier	[C/cm^2]
Q_{ox}	density of oxide charge	[C/cm^2]
Q_{Si}	density of total charge in the silicon	[C/cm^2]
q	elementary charge	[1.6*10^{-19} C]
R	spectral response	[A/W]
RT	room temperature	[300 K]
r	polar coordinate	[-]
S	equivalent number of signal electrons	[-]
Sm	amount of smear	[%]
Sm_a	amount of smear generated after the integration	[%]
Sm_b	amount of smear generated before the integration	[%]
T	absolute temperature	[K]
T_{int}	integration time	[msec]
$T_{int,M}$	maximum integration time	[msec]
T_s	sampling period	[nsec]
T_{tr}	transport time	[μsec]
t_{ox}	thickness of the gate oxide	[nm]
t_{Si}	thickness of the silicon substrate	[μm]
tt	transfer time	[nsec]
tt_{opt}	transfer time at a depth x'$_{opt}$	[nsec]
u	cortical coordinate	[-]
V_B	bulk voltage	[V]
V_D	diode voltage	[V]

SYMBOL LIST

V_{DC}	DC bias at a CCD gate	[V]
V_{DD}	positive supply voltage	[V]
V_{in}	AC input signal	[V]
V_{FB}	flat-band voltage	[V]
V_G	gate voltage	[V]
V_{G1}	gate voltage to compensate for Φ_{MS}	[V]
V_{G2}	gate voltage to compensate for Q_{ox}	[V]
V_{out}	output voltage at the output pin	[V]
V^*_{out}	output voltage at the floating diffusion or floating gate	[V]
V_s	source voltage	[V]
V_{SB}	source-bulk bias	[V]
V_T	threshold voltage	[V]
V_{T0}	threshold voltage at zero source-bulk bias	[V]
v	cortical coordinate	[-]
W	width of the gate of a MOS transistor	[μm]
W'	width of a small part of a CCD gate	[μm]
w	cortical coordinate	[-]
x	depth coordinate ($x=0$ at SiO_2-Si interface)	[μm]
x_d	depletion-layer width	[nm]
x_n	junction depth of the buried-channel implant	[μm]
x^*	penetration depth	[μm]
x'	depth coordinate ($x'=0$ at gate-SiO_2 interface)	[μm]
x'_{opt}	optimal value of x'	[μm]
z	polar coordinate	[-]
α	absorption coefficient	[/cm]
γ	center of gravity	[μm]
ΔV	voltage swing on the CCD gates during the charge transfer	[V]
ΔV_{out}	voltage swing at the output pin	[V]
ΔV^*_{out}	voltage swing at the floating diffusion or floating gate	[V]
ΔV_R	voltage swing through reset-clock feedthrough	[V]
Δx	pixel width	[μm]
$\Delta \Phi_R$	reset-pulse amplitude	[V]
$\Delta \Phi_S$	surface-potential change needed to fill an empty well	[V]
$\Delta \Psi$	ideal phase shift introduced by a CCD transport	[rad]
δ	angle between the optical axis and ray	[rad]
ϵ	transfer inefficiency of a single CCD transport	[-]
ϵ_m	transfer inefficiency of a complete CCD cell with m gates	[-]
ϵ_{ox}	permittivity of silicon oxide	[$3.38*10^{-13}$ F/cm]
ϵ_{Si}	permittivity of silicon	[$10.7*10^{-13}$ F/cm]
η	quantum efficiency	[%]
η_{bulk}	collection efficiency in the bulk	[%]
η_c	overall collection efficiency	[%]
η_{dl}	collection efficiency in the depletion layer	[%]

θ	polar coordinate	[rad]
λ	wavelength of the light	[nm]
λ_c	cutoff wavelength	[nm]
μ_n	mobility of electrons	[cm²/V.sec]
ν	photon frequency	[eV]
π	constant	[3.14]
$\rho(x,y)$	space-charge density	[C/cm²]
τ_{empty}	time constant for emptying surface states	[nsec]
τ_{fill}	time constant for filling surface states	[nsec]
Φ_{CH}	channel potential	[V]
Φ_{DC}	DC-biased CCD clock	[-]
Φ_F	Fermi potential	[V]
Φ_M	work function of the MOS-gate material	[V]
Φ_{MS}	work-function difference	[V]
Φ_{ox}	potential across the gate oxide	[V]
Φ_R	reset pulse	[-]
Φ_S	surface potential	[V]
Φ_{Si}	work function of silicon	[V]
Φ_{Ssi}	surface potential in strong-inversion mode	[V]
Φ_{Sdd}	surface potential in deep-depletion mode	[V]
$\Phi_{1,2,..}$	CCD clocks	[-]
ϕ_0	incoming photon flux	[W/cm²]
$\phi(x)$	photon flux at depth x	[W/cm²]
ω	angular frequency of the signal transferred by a CCD	[Hz]
φ	angle of the cone generated by the incoming rays	[rad]

LIST OF FIGURES

1.1.	MOS capacitance in accumulation	9
1.2.	MOS capacitance in deep depletion	11
1.3.	MOS capacitance in inversion	14
1.4.	MOS capacitance in weak inversion	17
1.5.	Surface potential versus effective gate voltage	21
1.6.	Surface potential as a function of the charge content	22
1.7.	Influence of different parameters on surface potential	24
1.8.	Charge transport in a CCD	26
1.9.	Effect of fringing fields on the surface potential	32
1.10.	Influence of fringing fields on the transport efficiency	33
1.11.	Transfer efficiency as a function of time	34
1.12.	Transfer efficiency as a function of gate length	35
1.13.	Effect of surface states on the CCD transport	37
1.14.	Fringing fields at various depths in the substrate	40
1.15.	Surface and channel potential of SCCD and BCCD	41
1.16.	Channel potential as a function of the effective gate voltage	43
1.17.	Two-dimensional device setup	44
1.18.	Transfer time as a function of depth	45
1.19.	Location of the charge storage in SCCD and BCCD	46
2.1.	Cross section of a CCD	53
2.2.	Cross section of a four-phase CCD	55
2.3.	Timing diagram of a four-phase CCD	56
2.4.	Antiparallel pulses for a four-phase CCD	56
2.5.	Cross section of a three-phase CCD	57
2.6.	Timing diagram of a three-phase CCD	58
2.7.	Cross section of a two-phase CCD	59
2.8.	Timing diagram of a two-phase CCD	60
2.9.	Cross section of a two-phase implanted CCD	60
2.10.	Timing diagram of a two-phase CCD	61
2.11.	Cross section of a one-and-a-half phase CCD	62
2.12.	Timing diagram of a one-and-a-half phase CCD	63
2.13.	Cross section of a virtual-phase CCD	64
2.14.	Timing diagram of a virtual-phase CCD	64
2.15.	Cross section of a ripple clocked CCD	65
2.16.	Timing diagram of a ripple clocked CCD	66
2.17.	Top view of a CCD delay line	67
2.18.	CCD channel definition	68

2.19.   Basic input configuration   69
2.20.   Diode cut-off technique   71
2.21.   Timing diagram of the diode cut-off method   72
2.22.   Fill-and-spill technique   73
2.23.   Timing diagram of the fill-and-spill method   74
2.24.   Floating-diffusion output with reset   77
2.25.   Timing diagram of the floating-diffusion output   78
2.26.   Electrical representation of the output node   78
2.27.   Definitions of the different voltage variations   79
2.28.   Floating gate without reset   80
2.29.   Timing diagram of the floating-gate readout   81
2.30.   Electrical equivalent of the floating-gate readout   82

3.1.   Illustration of a CCD delay line   85
3.2.   Modulation transfer function as a function of frequency   90
3.3.   Phase shift as a function of frequency   92
3.4.   Illustration of the sampling process   101
3.5.   Illustration of aliasing   102
3.6.   The fundamental sampling theory   103
3.7.   Summarization of the theory for a band-limited signal   104
3.8.   Summarization of the theory for a band-unlimited signal   105

4.1.   Photon-converting sites   111
4.2.   Readout structures   112
4.3.   Device concepts of linear image sensors   113
4.4.   Device architecture of a frame-transfer CCD   115
4.5.   Working principle of a frame-transfer CCD   116
4.6.   Device architecture of an interline-transfer CCD   117
4.7.   Working principle of an interline-transfer CCD   118
4.8.   Device architecture of a frame-interline-transfer CCD   119
4.9.   Working principle of a frame-interline-transfer CCD   120
4.10.   Device architecture of an MOS-XY image sensor   121
4.11.   Working principle of an MOS-XY image sensor   122
4.12.   Device architecture of an MOS-XY imager with CCD readout   123
4.13.   Device architecture of a charge-injection device   124
4.14.   Working principle of a charge-injection device   125

5.1.   Illustration of the generation of electron-hole pairs   132
5.2.   Absorption coefficient as a function of wavelength   133
5.3.   Cross section of an MOS capacitor   135
5.4.   Spectral response as a function of wavelength   138
5.5.   Relation between output voltage and integration time   140
5.6.   Spectral response of a linear image sensor   141

5.7.	Degradation of the MTF due to misdiffusion	144
5.8.	Diffusion MTF as a function of spatial frequency	145
5.9.	MTF degradation due to transfer inefficiency	147
5.10.	Representation of a single pixel row	149
5.11.	Geometric MTF as a function of signal frequency	150
5.12.	Illustration of the origin of the Moiré effect	153
6.1.	Various display modes	160
6.2.	Interlacing with a frame-transfer CCD	162
6.3.	Frame-integration mode interlacing with interline-transfer CCD	163
6.4.	Field-integration mode interlacing with interline-transfer CCD	164
6.5.	Progressive and interlaced scanning with interline-transfer CCD	165
6.6.	Color filter arrangement in stripes	166
6.7.	Color filters in a mosaic configuration	169
6.8.	Color separation by means of a prism	172
6.9.	Spatial offset of three imagers to increase the resolution	173
6.10.	Color imaging with linear image sensors	174
6.11.	The origin of the blooming effect	176
6.12.	Lateral or horizontal antiblooming	177
6.13.	Antiblooming by a dynamic charge-pumping mechanism	179
6.14.	Antiblooming by a vertical n-p-n photodiode	180
6.15.	Antiblooming by a vertical n-p-n frame-transfer cell	181
6.16.	Basic principle of integration-time shortening	183
6.17.	Charge reset in a frame-transfer pixel	185
6.18.	Charge reset in a frame-interline-transfer CCD with a drain	186
6.19.	Charge reset in a frame-interline-transfer CCD	187
6.20.	Charge reset in an MOS-XY imager	189
7.1.	Cross section of an image cell with microlenses	196
7.2.	Different types of microlenses	197
7.3.	Effect of the F number of the main lens on the microlenses	199
7.4.	Quantative illustration of the effect of the F number	200
7.5.	Effect of nonperpendicular light rays on the microlenses	200
7.6.	Quantative data on the effect of nonperpendicular light rays	201
7.7.	Cross section of an image cell with photoconversion top layer	202
7.8.	Charge-sweep device	203
7.9.	Trench charge-coupled device	204
7.10.	Influence of the layer stack of the multilayered structure	206
7.11.	Cross section of virtual-phase CCD and its photoresponse	207
7.12.	Blue response enhancement by light windows	208
7.13.	Cross-gate CCD	209
7.14.	Photoresponse of a CCD with an ITO gate	210
7.15.	Cross section of back-side thinned CCD and its photoresponse	212

7.16.   Equivalent circuit for a static-induction transistor imager   213
7.17.   Equivalent circuit for a charge-modulation imager   214
7.18.   Equivalent circuit for a bulk charge-modulation imager   214
7.19.   Equivalent circuit for an amplified MOS intelligent imager   215
7.20.   Equivalent circuit for a base-stored image sensor   216

8.1.   Illustration of the different noise sources of a CCD imager   220
8.2.   Shielding interface states by a shallow p^+ layer   223
8.3.   Commonly used output stage for CCD imagers   225
8.4.   Influence of the W and L of the first driver transistor   226
8.5.   Influence of the bias current through the driver transistor   227
8.6.   Noise reduction by means of correlated-double sampling   228
8.7.   Noise reduction by means of a clamping circuit   229
8.8.   Noise reduction by means of a delay line   230
8.9.   Comparison between video-preprocessing circuits   231
8.10.   Output stage including a JFET as first follower   232
8.11.   Output stage including a MOSFET as first follower   233
8.12.   Different output-stage configurations   235
8.13.   Definition of smear measurement   236
8.14.   Origin of smear in a (frame-) interline-transfer CCD   238
8.15.   Techniques to reduce the smear   238
8.16.   Origin of smear in an MOS-XY imager   241
8.17.   Introduction of the transversal signal line device   242
8.18.   Comparison of smear performance for various image sensor types   243

9.1.   Limiting factor in defining the horizontal resolution   249
9.2.   Charge-multiplexing structure   250
9.3.   Timing diagram needed to perform the charge-multiplexing   251
9.4.   A layout concept of the transfer region between two CCDs   254
9.5.   Design optimization by gate tapering   255
9.6.   Extra implants in the charge-multiplexing structure   256
9.7.   Basic idea of the compound-channel construction   257
9.8.   Working principle of the accordion CCD   260
9.9.   Driving the accordion CCD by two digital shift registers   261
9.10.   Speeding up the shift registers of an interline-transfer CCD   263
9.11.   Speeding up the vertical frame shift of a frame-transfer CCD   265
9.12.   Working principle of dynamic pixel management   266
9.13.   Two MOS image sensors in a still-picture camera   268
9.14.   Device architecture of a frame-transfer CCD dedicated for ESP   270
9.15.   Device architecture of a frame-transfer CCD dedicated for ESP   271
9.16.   Device architecture of a multiple-frame-interline-transfer CCD   272
9.17.   An optimized interline-transfer CCD for ESP applications   273
9.18.   Working principle of a time-delay and integration CCD   275

10.1. Four buttable sensors forming a two-dimensional array 281
10.2. Basic configuration of a notch CCD 283
10.3. Schematic diagram of the skipper CCD 285
10.4. Photon transfer curve of a CCD provided with a skipper output 286
10.5. Basic working principle of pinned-phase CCDs 287
10.6. Cross section of an open-phase pinned CCD 288
10.7. Potential plots of an open-phase pinned CCD 289
10.8. Cross section of a multi-phase pinned CCD 290
10.9. Dark current of CCDs with and without pinning 291
10.10. The basis of ASIC vision 294
10.11. Architecture of a 3-D image sensor 295
10.12. Schematic representation of the foveated-retina sensor 297
10.13. Penetration depth of silicon for different wavelengths 299
10.14. Cross section of a Schottky-barrier infrared detector 300
10.15. Cross section of a detector with optical cavity 300
10.16. Responsivity of a PtSi Schottky-barrier infrared imager 301
10.17. Photoresponse of a virtual-phase CCD covered with a phosphor 303
10.18. Cross section of a deep-depletion CCD 304
10.19. Cross section of a visible, UV and IR detector 305
10.20. Quantum efficiency of a visible, UV and IR detector 306
10.21. Image sensor coupled to an imaging tube by a fiber-optic plate 309
10.22. System concept using a contact-type linear imager 312
10.23. Equivalent circuit of a contact-type linear imager 313
10.24. Design concept of a contact-type linear imager 314

A1.1. Illustration of a full-frame imager used as technology vehicle 318
A1.2. Coding of the various regions of the CCD 319
A1.3. Substrate preparation 320
A1.4. Implantation of the p well 321
A1.5. Drive-in of the p well 322
A1.6. Implantation of the CCD channel 323
A1.7. Drive-in of the channel implant 324
A1.8. Implantation of the channel stoppers 325
A1.9. Gate oxidation 326
A1.10. Deposition of the first poly-Si layer 327
A1.11. Definition of the first poly-Si layer 328
A1.12. Interpoly isolation 329
A1.13. Deposition of the second poly-Si layer 330
A1.14. Definition of the second poly-Si layer 331
A1.15. Interpoly isolation 332
A1.16. Deposition of the third poly-Si layer 333
A1.17. Definition of the third poly-Si layer 334
A1.18. Source and drain implantation 335

A1.19.   Back-end isolation   336
A1.20.   Contact-hole definition   337
A1.21.   Deposition of the metal layer   338
A1.22.   Definition of the metallization pattern   339
A1.23.   Scratch-protection deposition   340
A1.24.   Deposition of the first color filter layer   341
A1.25.   Lift-off of the first color filter   342
A1.26.   Deposition of the second color filter layer   343
A1.27.   Lift-off of the second color filter   344
A1.28.   Deposition of planarization layers   345
A1.29.   Etch-back of the planarization layers   346
A1.30.   Deposition of the lens material   347
A1.31.   Reflow of the microlenses   348

A2.1.   Reference picture   349
A2.2.   Smear induced by a highlight   350
A2.3.   The effect of charge reset on the smear   351
A2.4.   Blooming effects caused by local overexposures   352
A2.5.   Unsharply displayed objects which move too fast   352
A2.6.   Sharply displayed moving objects   353
A2.7.   Column defects caused by incorrect charge transport   354
A2.8.   Damage of the cover glass   355
A2.9.   Damage caused by electrostatic discharge   355
A2.10.   Leakers caused by local defects in the gate dielectric   356
A2.11.   White point defects generated by a dominant contaminant   357
A2.12.   Fixed-pattern noise   358
A2.13.   Amplifier noise, photon shot noise and other noise sources   358

A3.1.   Effect of the gate bias on the active pixel area   361
A3.2.   Dependence of the pixel configuration on the integration mode   363
A3.3.   Effect of the substrate bias on the potential profiles   364
A3.4.   Yield versus chip area   365

PREFACE

The idea of writing a book on solid-state imaging with charge-coupled devices occurred to the author after a couple of teaching sessions in which he described the basic materials to an audience of university students. It soon became apparent that no up-to-date reference books existed which dealt with charge-coupled devices in general or with solid-state imaging in particular. At first the questions were posed by the students, and these were followed by questions from customers concerned with charge-coupled imaging devices, and finally questions from the scientific world in general convinced the author to start writing the book you are now reading.

The following is an overview of the circles interested in the publication of a reference work on the subject under discussion and the audience for whom this book has been compiled :

- the educational world : charge-coupled devices constitute the cornerstone of the present-day semiconductor technology, but until now there has not been a really up-to-date textbook on this subject and none at all on solid-state imaging. In the world of education, at various types of public and private establishments, CCD imagers are frequently the main theme of colloquia, tutorials, workshops, summer schools, etc. This book will, it is hoped, prove suitable for use as a textbook for all such institutions and projects;
- the industrial world : many engineers, technicians and developers concerned with solid-state imaging feel the need for a reference book or tutorial guide in which the fundamentals and the applications of charge-coupled devices are expounded;
- the scientific world : over the last 10 years or so, solid-state imagers have been replacing the old familiar imaging tubes. Immense progress is being made in the basic fabrication technology, in the device physics and in the applications of charge-coupled devices. Now, for the first time, a book has been written which gives an overview of all the different developments in the specific field of solid-state imaging using CCDs.

Starting from a knowledge of the various backgrounds and interests of potential readers, the contents of the book, though as widely based as possible, are restricted to the scope of the subject. As far as possible, all the basics and copious details of solid-state imaging with charge-coupled devices are presented. Absolutely complete coverage of the subject is, of course, out of the question because imaging technology has still not reached the end of its development. On the other hand, all the material between the covers of this book is based on published work, making the overview as comprehensive as possible and guaranteeing that the volume is not based selectively on a single technology.

Since the mid-seventies, the author has been active in the field of solid-state imaging with charge-coupled devices. In the course of his work toward an M. Sc. degree he developed the driving electronics and video-processing hardware for a two-dimensional imaging system based on a linear array comprising 256 elements with a mechanical scanning system in the other direction. His involvement in CCDs continued during his work toward the Ph.D. degree. His research subject was the incorporation of indium-tin oxide gates in the CCD technology. His research in CCD hardware and technology was done under supervision of Prof. Dr. Gilbert Declerck at the ESAT laboratory of the University of Leuven, Belgium. The author gratefully acknowledges Prof. Dr. Declerck's scholarship and enthusiasm and also thanks him for the technical discussions they had during the period when they worked together. Some ideas generated in this early CCD period can be found in the present volume, particularly its first part.

In the mid-eighties, the author moved on from the university to industry : the Philips Research Laboratories in Eindhoven, the Netherlands. There he was enabled by Max Collet, department head in the Microcircuits Division, to continue his CCD career as designer and later as project leader in CCD imaging projects for broadcast applications. The author is grateful to Max Collet for his invaluable technical and scientific support, which led ultimately to the designing of the first commercially available HDTV image sensor.

In the early nineties, the author was appointed department head in the Microcircuits Division of which he was a member. CCD imaging is still the chief concern of his R&D group. At this point the author takes the opportunity to thank all his former and present group members for helping to build the foundations on which new imaging developments are based.

As already mentioned, this volume is largely based on publications, papers and conference proceedings. This imposes the author the pleasant obligation to thank members of the CCD world as a whole for their contribution to this volume. Without their ideas, their research and their output, the book would never have been possible.

Special thanks are also due to :
   - those who provided the photographic material : N. Teranishi, K. Yonemoto, Y. Kitamura, T. Kuroda, N. Tsubouchi, H. Peek, P. Johnson, H. Akimoto, I. Debusschere, H. Wildenberg, Y. Okada, T. Nakamura, J. Nakamura, L. Brissot, H. Matsuzawa, A. Okano, S. Smith, B. Doody, H. Titus, W. McColgin;

- C. Schaeffer and E. Tienkamp for reading the manuscript;
- H. Peek for his comments on the appendix dealing with the CCD technology : this part of the book is based on his daily work;
- W. Toren and S. Bazelmans for their contribution on the CCD artifacts;
- D.R. Welsh of Philips Translation Services, for linguistic correction of the manuscript;
- K. Nederveen of Kluwer Academic, for his kind cooperation;
- the "unknown" reviewers for their review and their kind comments;

Finally, the author gratefully acknowledges the forbearance of his family - wife and three daughters - who allowed him to sit at his PC for hours, evenings, nights and even weekends in order to complete this book.

Chapter 0

INTRODUCTION

The year 1969, the year of the moon landing, the year of Woodstock, was also the year when the first paper about Bucket-Brigade Devices was published. Of course the first two events received (and still receive) much more publicity, but the effect on society might be equal for all three of the milestones.

It is very doubtful whether Sangster and Teer, the inventors of the Bucket Brigade Device, and Boyle and Smith, the inventors of the Charge-Coupled Device, could ever have anticipated the impact of their pioneering work. In 1969 F. Sangster and K. Teer of the Philips Research Labs, the Netherlands, published their findings on a completely new type of device : the Bucket-Brigade Device or BBD (Sangster 69). Their original application was as an analog delay line, but Sangster and Teer soon realized that their invention was also capable of being used as a solid-state image sensor. Basically, the BBD is a charge-transfer device in which charge packets are transported from one transistor to another. In 1970, W. Boyle and G. Smith of Bell Labs, the United States, improved the charge-transfer concept by introducing a transport mechanism from one capacitor to another one (Boyle 70). The Charge-Coupled Device or CCD was born ! The charge-coupled devices completely covered the same application areas as the bucket-brigade devices : analog delay lines, (programmable) analog filters, analog memories, digital memories and image sensors. Not surprisingly that CCDs and BBDs had so much in common, the BBD was in fact a two-phase CCD.

In the decade after the first announcements about CCDs there followed many papers about the device's physics, technology and applications. Beside the paperwork in the scientific world, however, few devices came on the commercial market. Knowledge about the physics and the ideas about the applications were available, but the know-how about the production of the charge-coupled devices was not. The CCDs had to wait until the digital world pushed semiconductor technology so that CCDs could be produced in a similar fabrication environment. But this also meant that while CCDs could be made with a reasonable yield and acceptable price, the digital devices were able to take over some of the CCD's applications. Delay lines, filters and memories could be made cheaper and bigger (in bit size) in digital circuitry. Nevertheless, solid-state imaging remained the exclusive application field of charge-coupled devices.

Once CCDs entered the consumer market as the "eyes" of the camera-recorder combination (camcorder), they ruled over the imaging world. The classical imaging tube disappeared completely, first from the consumer scene and a couple of years later from the professional broadcasting scene. Today the shift towards solid-state imaging is taking place in the medical world.

Compared to the classical imaging tubes, CCDs have many advantages, ranging from size, weight, cost, power consumption, stability, reliability, image lag, burn-in, etc. The smaller size of the CCDs, as an example, opened new picture shooting possibilities. For instance, in capturing sports events, the very small, high-quality cameras can be placed in outstanding locations :
- right above the basket in basketball,
- inside, at the front, or at the rear of automobiles in races,
- on the helmet of motocrossers,
- right above or parallel to the bar in high-jump and pole-vaulting competitions,
- in the back of the goal in soccer games,
- right above the net in volleyball,
- just in the middle, close to the net, of a tennis court.

These kinds of picture-shooting locations were not possible with cameras using the classical imaging tubes.

The other characteristics of CCDs also open up interesting new applications. All together, today's imaging world would look completely different without charge-coupled devices. Nevertheless, new applications are still possible as device improvements and extra features create new opportunities.

As knowledge of the device's physics in the seventies increased, a few books were published about the charge-coupled devices and their applications. In the last decade, CCDs have been the subject of only single or few chapters of a book : no complete work has been devoted to this material. At the same time the growth of the imaging application field pushed the technology further and totally new concepts and device architectures were invented and developed. This book is the first completely devoted to solid-state imaging with these charge-coupled devices. It tries to give an overview of the state-of-the-art in this field. In the meantime, it should fill the gap of more than 10 years of technological progress in research on and development of charge-coupled devices.

This book contains four main parts. The first part deals with the basics of charge-coupled devices in general. The second explains the imaging concepts in close relation to the classical television world. The third part goes into detail about new developments in the solid-state imaging world, and the fourth part rounds off the discussion with the remaining subjects which were not covered in the previous sections. Although not explicitly mentioned in the rest of the book, the different parts form more or less stand-alone sections and it should be possible to read them independently of each other. Every section also "closes" with a short summary : "Worth memorizing". Once these concluding remarks are read, the section can be skipped and the next can be started without missing the main messages from the previous section.

Chapter 1 explains the fundamentals of the charge-coupled device for those readers who are not yet familiar with CCDs. The starting point is the Metal-Oxide Semiconductor (MOS) capacitor which is studied for different bias conditions. After a single MOS capacitor, the next step is a series of MOS capacitors so that charge transfer can take place. An in-depth study of the various transport mechanisms is made in order to understand the limitations of CCDs and to introduce the buried-channel charge-coupled device. The latter is a technology with a very high transport speed and is almost exclusively used in today's imagers.

Chapter 2 deals with the different ways of introducing charge into a CCD, how to transport it through the channel, and how to get the charge packet out of the CCD. Quite a lot of time is spent on different clocking schemes and their effect on transport. All have pros and cons, and most find their way into the imaging application field. Output structures are crucial in the study of charge-coupled devices because they convert the charge packet, transported through the CCD channel, into a measurable quantity : an electric voltage or an electric current. In practical imaging applications the charge packets are rather small. This places some stringent requirements on the output stages as far as the conversion factor, noise behavior, and bandwidth are concerned. Although not directly linked to imaging, input structures to introduce an electric signal into the CCD are described to complete the overall study. The theory behind the input structures gives some extra insight into the creation of charge packets in general.

Chapter 3 describes a real delay line. This vehicle is chosen to discuss the limitations of charge-coupled devices. What is the effect of the finite value of the transport efficiency on the electrical signal passed through the CCD ? (Answer : an amplitude decrease and a phase shift.) What is the effect of the dark current ? (Answer : a limitation to the dynamic range.) What is the effect of the dark-current nonuniformities ? (Answer : the introduction of fixed-pattern noise.) What is the effect of noise ? (Answer : a noisy output signal.) What is the effect of sampling on the electrical signal ? (Answer : if the Nyquist criterion is not fulfilled, low-frequency Moiré patterns are introduced.)

Chapter 4 is the first chapter dealing with solid-state imaging. Readers who are already acquainted with charge-coupled devices or who are not interested in the fundamentals of these devices can start their reading here. The chapter gives a very short overview of photon sensing in general and of the various imager configurations (ranging from linear arrays to two-dimensional arrays) ending with an overview and comparison of the various architectures used to build two-dimensional imaging arrays. The latter item includes not only CCDs but also MOS-XY addressable and charge-injection devices.

Chapter 5 deals with the fundamentals of solid-state imaging, irrespective of the imager's configuration. Starting with the absorption of the photons and the collection of the generated electrons, the spectral response of the solid-state imaging device is defined. One parameter, commonly used to compare different imagers with each other, is the quantum efficiency. This is the ratio of the number of electrons collected by the imager to the number of impinging photons. Special attention is paid to the dependence of the quantum efficiency on the wavelength of the incoming light. Resolution is a characteristic which has to do with the ability of the sensor to resolve spatial information. The description of resolution in this chapter reveals its dependence on the diffusion of generated electrons, on the transport characteristics of the CCD, and finally on the geometry of the individual pixels. The chapter concludes with the effect of the optical sampling action of the imager and the generation of Moiré patterns if the Nyquist theorem is not obeyed.

Chapter 6 is completely devoted to solid-state imagers in connection with television applications. What kind of design issues have to be faced when the CCD imager has to fit the television standard and the consumer's needs ? The most typical requirement is the interlaced scanning of a TV. For the sake of completeness the alternative scanning mode is also described : the progressive method. Interlaced scanning depends on the device architecture. All the various combinations of imager architecture and scanning modes are described in this chapter. Also important to television applications is color imaging. Different kinds of color filter arrangements are compared with each other, and with the color-splitting technique using an optical prism. The terms blooming and antiblooming are characteristics describing the ability of the sensors to withstand overexposure. In close relation to antiblooming are the options of charge reset, electronic shutter or integration-time shortening. Antiblooming and charge reset depend very much on sensor architecture. The chapter compares different types with each other, and solid-state imagers other than the CCD are also considered.

Chapter 7 is the first of three chapters dealing with new developments in the imaging field, so-called advanced imaging. In this chapter special attention is paid to light sensitivity. How can CCD imagers be made more light sensitive ? For instance, on optimizing the aperture ratio of the pixels (with microlenses, by introducing new photoconverting layers or with special designs) or optimizing the light transmission of the multilayered structure, will improve the imager response. The light sensitivity can also be increased by back-side illumination or by the introduction of an amplification function inside each pixel. All methods are discussed in this chapter, together with their advantages and disadvantages.

Chapter 8 covers recent developments in the fields of noise and smear (which is a spurious signal). A CCD contains several noise sources which have not only different origins but also different temperature dependencies or spectra. Several types of noise sources in the technology or amplifier are described. For the latter, several techniques are reported which diminish the output amplifier noise or compensate for it. Totally new readout methods appear, which outperform the former output amplifiers with their limits as to noise performance. Although not directly related to noise, the sensitivity or conversion factor of the output amplifier is also studied here. The chapter concludes with an overview of the smear issue. How is smear introduced in different architectures and how is it solved ?

Chapter 9 concludes the advanced imaging section with special attention to device architectures. It starts with the increase of the horizontal resolution of the imager and consequent very high pixel rates at the output amplifier. This problem can be circumvented by the introduction of several parallel output registers. The description of this multiplexing method together with its consequences on the design of the device is a study object intended for real CCD freaks. After the increase in horizontal resolution, new developments in the vertical clocking are reported. The section on high-speed clocks gives some important information needed in the design phase of very fast and large devices such as those for the HDTV application area. More details of the electronic still-picture application and time-delay integration devices conclude this chapter.

Chapter 10 deals with all important material not covered in the previous chapters. For example, a special group of devices, called scientific imagers which are characterized by an extremely large number of pixels. Consequently, their read-out time is very large and the dark current plays an important role. To keep dark current and dark-current nonuniformities low, the devices can be cooled or operated in a pinned mode. Not only for its benefits, but also from the physical point of view, pinning the CCD is quite interesting.

As a second example of a separate group of devices, CCD imagers might be specially designed to fulfill some typical image-processing functions. A few are discussed to give some insight into this very challenging and rapidly evolving field. A third example beyond the realm of traditional imaging are imaging applications outside the visible spectrum. Infrared imaging, UV imaging and X-ray imaging cannot be done in a straightforward way with CCDs. Special precautions have to be taken, for instance, introducing new materials to make the CCD sensitive to the wavelength of interest, to cover the device with light-emitting phosphors (when hit by UV photons or X-rays), or to illuminate the CCD from the back. The chapter concludes with a short overview of high-speed devices and contact-type image sensors.

Appendix 1 guides the reader through the complete semiconductor process so as to fabricate CCD solid-state imagers. A hypothetical structure is chosen and a step-by-step explanation describes how a new imaging device is fabricated.

Appendix 2 contains photographs of several artifacts possible with a charge-coupled device if it is not operated correctly, if it contains technology-related noise sources or if other noise sources dominate the video signal.

Appendix 3 gives an overview of the interlinkage between different imager's specification items. If the main emphasis is placed on one item, how will this influence the other items ? Resolution, light sensitivity, dynamic range, chip size, cost, etc., are all parameters which influence each other.

Chapter 1

FUNDAMENTALS OF CHARGE-COUPLED DEVICES

The basic concept of Charge-Coupled Devices (CCDs) is a simple series connection of Metal-Oxide-Semiconductor capacitors (MOS capacitors). The individual capacitors are physically located very close to each other. The CCD is a type of charge storage and transport device : charge carriers are stored on the MOS capacitors and transported. To operate the CCD, digital pulses are applied to the top plates of the MOS structures. The charge packets can be transported from one capacitor to its neighbor capacitor. This transport of isolated charge packets can be effected in an almost perfect way, without any noticeable deterioration of the charge content. If the chain of MOS capacitors is closed with an output node and an appropriate output amplifier, the charges forming part of a moving charge packet can be translated into a voltage and measured at the outside of the device.
The way the charges are loaded into the charge-coupled device is application-dependent. In the case of filters, delay lines, and memories, the CCD is provided with an electrical input stage. In the case of solid-state imagers, the impinging photons generate charge carriers which are collected and transported to the output amplifier.

The basics of charge-coupled devices - the storage of charge carriers on the capacitor and the charge transfer or transport - are subjects discussed in this chapter. The charge storage is described starting from the fundamental physics of the MOS capacitor. The extension from a single capacitor to a group of four introduces the transport mechanism. Various charge-transfer mechanisms are studied which all apply to real CCDs. Special attention is paid to incomplete charge transfer.

In a separate paragraph an almost ideal charge-transfer device is introduced : the buried-channel CCD. The most interesting features of the buried-channel charge-coupled device are described : the fringing fields, the transfer time and the charge-handling capability. These aspects are not all advantages compared to the surface-channel CCD. Nevertheless, buried-channel CCDs are exclusively used in today's commercially available imagers.

The last section of this chapter contains an analytical expression for the electrostatic potential analysis of all types of CCD.

1.1. An ideal MOS capacitor

The basic working principle of a charge-coupled device rests on the theory of Metal-Oxide-Semiconductor capacitors. The physical behavior of an MOS capacitor is totally different from that of a simple dual-plate capacitor. Depending on the gate electrode voltage or gate voltage (which is the voltage applied to the top or metal plate of the MOS capacitor), the structure is forced into accumulation or into depletion. (The exact definitions of the accumulation and depletion modes will be given later in this section.) The depletion state is further influenced by the availability of free charge carriers in the semiconducting bulk.

To examine details of the working principle of charge-coupled devices, a fundamental understanding of MOS capacitance is essential. In this section, the MOS capacitor is described in accumulation, in deep depletion, in weak inversion, and in strong inversion.

1.1.1. MOS CAPACITANCE IN ACCUMULATION

The case when an MOS capacitance is biased into accumulation is the only situation in which the MOS capacitance resembles that of a dual-plate capacitor. A cross section of a typical MOS structure is shown in Figure 1.1a. In this example, the gate or top plate of the capacitance is located at the left side of a dielectric. A typical material used in CCD technology to construct the gate is polycrystalline silicon. The dielectric can be made from silicon oxide, silicon nitride, or a combination of both. In the case shown in Figure 1.1, only silicon oxide is used. The thickness of this layer is typically 100 nm. The semiconducting substrate is p-type silicon, e.g. boron-doped, and its majority carriers are holes. Typical doping levels for the substrate are in the range of 10^{14} ... 10^{15} ions/cm^3.

If the substrate is grounded (as is the case in Figure 1.1a), the MOS capacitor will operate in accumulation for negative gate voltages V_G, e.g. $V_G = -10$ V. (For this simplified discussion, the work-function difference between gate and substrate material, and the oxide charge in the dielectric will be neglected. In this way the flat-band voltage is automatically set to zero. In the next section a definition of these parameters and their influence on the behavior of the MOS capacitance will be given.)

As shown in Figure 1.1a, the negative voltage on the gate is compensated by the availability of free holes in the p-type silicon substrate. These free holes are represented by Q_{acc}.

In the accumulation situation the potential across the oxide, Φ_{ox}, equals the negative gate voltage V_G. This potential dependence is illustrated in Figure 1.1b. (Note the nonconventional way of pointing positive values of the electrostatic potential Φ downwards. The reason for this notation will become clear later in this section.)

The electric field E_{ox} in the dielectric layer, which is generated by the potential Φ_{ox}, is given by (in one dimension x) :

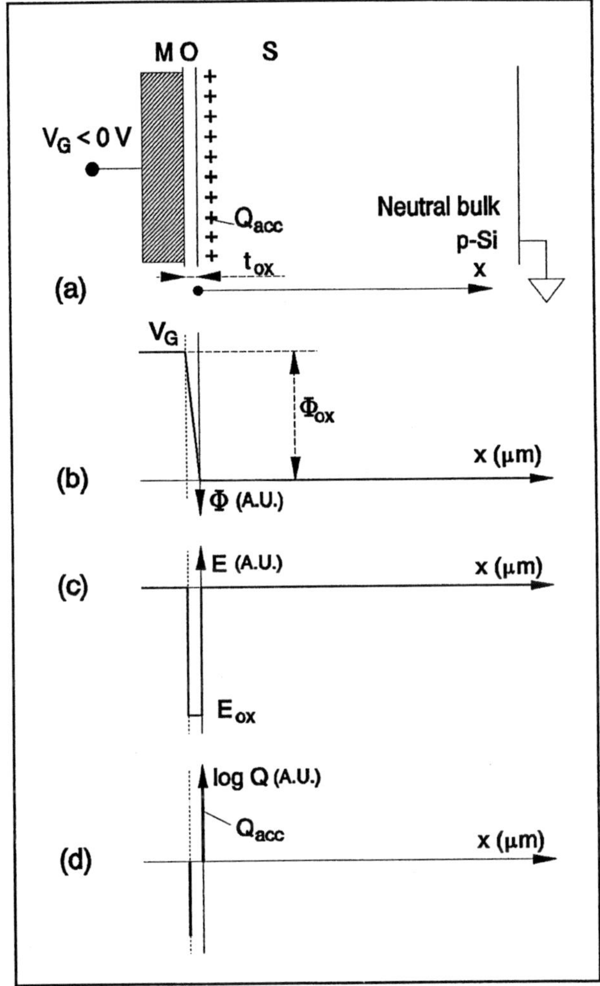

FIGURE 1.1. The MOS capacitance (a) in accumulation with its electrostatic potential (b), electric field (c), and charge distribution (d) diagram.

$$E_{ox} = -\frac{d\Phi_{ox}}{dx} \qquad [1.1]$$

and with :

   - t_{ox} = 100 nm, being the thickness of the SiO_2 dielectric layer,

   - V_G = -10 V,

this yields E_{ox} = 1 MV/cm.

A representation of the electrical field through the structure is illustrated in Figure 1.1c.

The charge distribution is schematically shown in Figure 1.1d. The two clouds of charges, one on the gate electrode and the other on the substrate electrode, are sketched as two lines because the charges are located in a very thin layer.

The accumulated charge Q_{acc} held by the electric field E_{ox} is given by Gauss' law :

$$Q_{acc} = \epsilon_{ox} \cdot \frac{d\Phi_{ox}}{dx} = \epsilon_{ox} \cdot \frac{dV_G}{dx} , \qquad [1.2]$$

in which ϵ_{ox} is the permittivity for silicon oxide ($\epsilon_{ox} = 3.38*10^{-13}$ F/cm). In this example $Q_{acc} = 3.0*10^{-6}$ C/cm^2. If this number is divided by q, the elementary charge (being equal to $1.6*10^{-19}$ C), it corresponds to $1.9*10^{13}$ electrons/cm^2 or $1.9*10^5$ electrons/μm^2.

1.1.2. MOS CAPACITANCE IN DEEP DEPLETION

By means of a positive gate voltage, the free charges or holes in the silicon substrate are pushed away from the gate and gate dielectric. Only negatively charged ions, which are fixed in the silicon crystal, are left behind. This situation is defined as deep depletion. The structure has no free charges available in the p-type bulk to compensate the charges applied to the gate. In this deep-depletion mode the MOS capacitance will tend to overcome this effect (= lack of free carriers) by depleting the top part of the semiconducting bulk. The substrate makes negative dopant ions free by pushing the majority charge carriers or holes away from the gate dielectric under the influence of the positive gate voltage. The depletion mode can only be maintained as long as no free electrons are available to compensate the charges available on the gate electrode. For this reason the deep-depletion mode is a nonequilibrium situation. (Or by simple definition : free charges on the gate electrode are not compensated by free charges in the silicon bulk, but by means of fixed ions.)

The deep-depletion situation, completely different from that of the dual-plate capacitor, is shown in Figure 1.2. The same types of drawings as in Figure 1.1 are repeated again : a cross section of the structure (a), its electrostatic potential (b), electric field (c), and charge distribution (d).

In fact, the voltage available on the gate is distributed across the gate dielectric and the depleted part of the substrate, called depletion region :

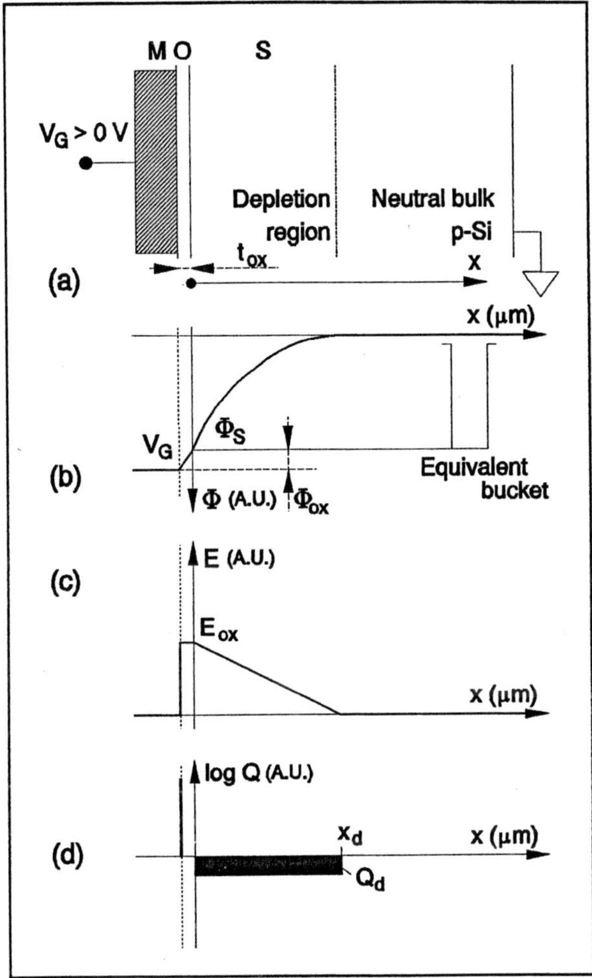

FIGURE 1.2. *The MOS capacitance (a) in deep depletion with its electrostatic potential (b), electric field (c), and charge distribution (d) diagram.*

$$V_G = \Phi_{ox} + \Phi_S , \qquad [1.3]$$

with :

- Φ_{ox} : representing the electrical voltage across the silicon oxide,
- Φ_S : denoting the potential at the silicon surface. If the silicon substrate is grounded, Φ_S is equal to the voltage across the depletion region.

According to the classical one-sided diffused junction theory, the width of the depletion layer x_d is given by (Carr 72):

$$X_d = \sqrt{\frac{2.\epsilon_{Si}.\Phi_S}{q.N_A}} \,,\qquad [1.4]$$

where ϵ_{Si} is the permittivity for silicon ($\epsilon_{Si} = 10.7 * 10^{-13}$ F/cm) and N_A the doping concentration of the p-type bulk.

The amount of fixed charge of negative acceptor ions Q_d per unit area which is needed to compensate the gate voltage is given by :

$$Q_d = q.N_A.x_d = \sqrt{2.\epsilon_{Si}.q.N_A.\Phi_S} \,.\qquad [1.5]$$

The charge distribution for the structure biased in deep depletion is depicted in Figure 1.2d.
The electric field, which is constant in the oxide, decreases to zero at the edge of the depletion width. The field in the dielectric is given by :

$$E_{ox} = \frac{Q_d}{\epsilon_{ox}} \,.\qquad [1.6]$$

This electric field in turn is generated by a voltage drop Φ_{ox} across the oxide. This last parameter can be calculated with :

$$\Phi_{ox} = t_{ox}.E_{ox} \,,\qquad [1.7]$$

which, after substituting relation [1.6] for E_{ox}, yields :

$$\Phi_{ox} = \frac{t_{ox}}{\epsilon_{ox}}.Q_d = \frac{Q_d}{C_{ox}} \,,\qquad [1.8]$$

where C_{ox} is the oxide capacitance per unit area.

With the foregoing theory and by substituting [1.8] and [1.5] in [1.3], the relation between the gate voltage V_G and the surface potential Φ_S is found, namely :

$$V_G = \frac{1}{C_{ox}}.\sqrt{2.\epsilon_{Si}.q.N_A.\Phi_S} + \Phi_S \,.\qquad [1.9]$$

From [1.4] x_d is found to be equal to 3.3 μm. The electrostatic potential at the dielectric-substrate interface, further indicated as surface potential Φ_S, under the given conditions (V_G = 10 V, N_A = 1.0*10^{15}/cm^3) can be calculated to be 8.5 V.

Note the parabolic characteristic for the electrostatic potential in the silicon bulk. According to :

$$\Phi = \int E \cdot dx \qquad [1.10]$$

and with E linearly decreasing to zero, the potential curve in the silicon is quadratic.

In the same illustration in which the electrical potential in Figure 1.2b is depicted, a potential well or bucket is drawn which has a "depth" up to 8.5 V (with respect to the bulk potential), the surface potential. This analogy between the electrostatic surface potential and a bucket will clearly illustrate the overall working principle of charge-coupled devices. At this moment the bucket remains empty, but in a later stage of this study, the bucket will also be filled.

Observe, by means of relation [1.9], that the depth of the potential well, represented by the surface potential Φ_S, can easily be changed by the external gate potential, but also by the parameters of the structure itself : the thickness of the gate dielectric (t_{ox}), the nature of the dielectric material (ϵ_{ox}), and the doping level of the substrate (N_A).

1.1.3. MOS CAPACITANCE IN INVERSION

If enough free electrons are made available or provided (through thermal generation, for instance) to the MOS capacitor in deep depletion, the structure will move toward an equilibrium phase. Not all negative ions are now needed to compensate the charge on the gate. The free electrons take over their job. The situation is shown in Figure 1.3. The channel at the interface contains such an amount electrons, that the depletion region is reduced to a minimum. The free or mobile charge (electrons) per unit area in the inversion layer is denoted by Q_n.

Deeper in the bulk of the silicon the substrate becomes p-type again and neutral. In the transition region between the inversion layer at the SiO$_2$-Si interface and the neutral bulk, the depletion layer still exists, but no longer with the original width. Under strong inversion conditions (when by definition the minority-carrier concentration Q_n in the inversion layer equals the majority-carrier concentration in the bulk), the surface potential Φ_S is by definition (Nicollian 82) :

$$\Phi_S = 2 \cdot \Phi_F , \qquad [1.11]$$

where Φ_F represents the Fermi potential of the bulk material, given by :

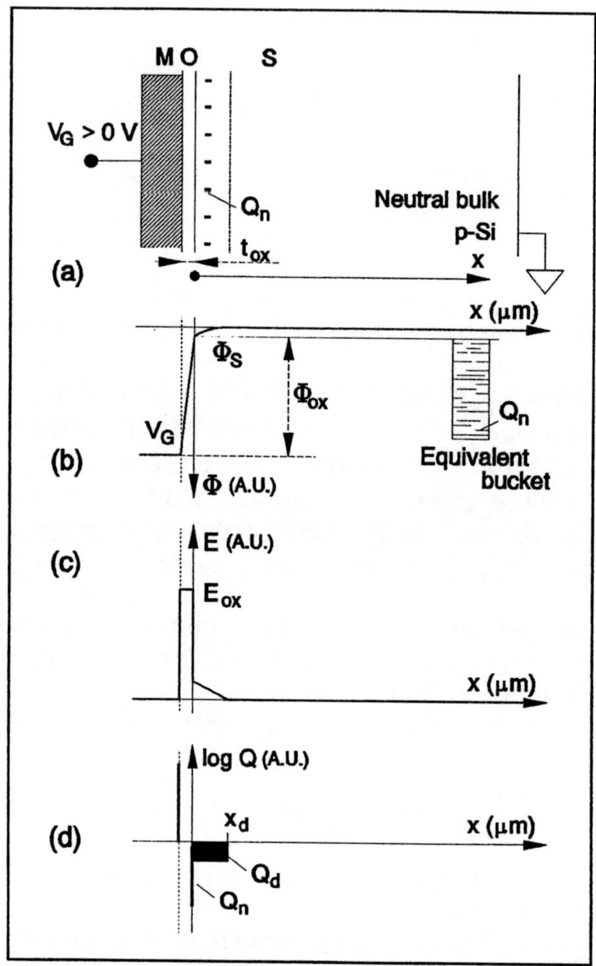

FIGURE 1.3. *The MOS capacitance (a) in inversion with its electrostatic potential (b), electric field (c), and charge distribution (d) diagram.*

$$\Phi_F = \frac{k.T}{q} \cdot \ln\frac{N_A}{n_i} \qquad [1.12]$$

with :

- k : Bolzmann's constant (= $1.38*10^{-23}$ J/K);
- T : absolute temperature (K);
- n_i : the intrinsic carrier concentration (= $1.4*10^{10}/cm^3$ at room temperature).

Filling in the parameters appropriate to the structure under study, $\Phi_F = 0.3$ V, $\Phi_s = 0.6$ V, and the depletion layer width $x_d = 0.9$ μm, calculated using [1.4].

The total charge density in the silicon, Q_{Si}, consists of a mobile charge of electrons Q_n in the inversion layer and the depletion charge Q_d of fixed ions :

$$Q_{Si} = Q_n + Q_d \, .$$

[1.13]

This charge requires an electric field in the oxide of :

$$E_{ox} = \frac{Q_{Si}}{\varepsilon_{ox}} \, .$$

[1.14]

The electric field and the potential distribution are also depicted in Figure 1.3. For the voltage drop across the oxide, the same applies as mentioned above in [1.7] :

$$\Phi_{ox} = t_{ox} . E_{ox} = \frac{Q_{Si}}{C_{ox}} = \frac{Q_n + Q_d}{C_{ox}} \, .$$

[1.15]

The gate voltage V_G necessary to bring the structure near to strong inversion is called the threshold voltage V_T. This voltage is the gate potential needed to effect strong inversion $2.\Phi_F$ plus the voltage across the oxide Φ_{ox}, which is necessary to generate the electric field E_{ox} holding the charge Q_{Si} in the silicon. However, at the onset of strong inversion the mobile charge Q_n is still negligible in relation to the depletion charge Q_d and therefore the total gate voltage at the onset of strong inversion, V_T, can be written as :

$$V_T = \Phi_s + \Phi_{ox} = \Phi_s + \frac{Q_d}{C_{ox}} = 2.\Phi_F + \frac{Q_d}{C_{ox}} \, .$$

[1.16]

According to the data of the Figures 1.1 through 1.3, V_T for this example is equal to 1.0 V.

If the gate voltage V_G is increased above V_T, the potential Φ_{ox} also increases. Large amounts of electrons become available in the inversion layer and Q_n increases considerably. The depletion charge Q_d on the contrary only increases slightly, as suggested by the equation [1.5].
It can be concluded that the difference in voltage $(V_G - V_T)$ is mainly used to increase the mobile charge in the inversion layer Q_n, which is given by :

$$Q_n = C_{ox} . (V_G - V_T) \, .$$

[1.17]

Coming back to the analogy between the MOS-capacitance and the bucket, note that in Figure 1.3b the potential well or the bucket is filled up to the level of the surface potential $\Phi_s = 2.\Phi_F = 0.6$ V. This is the voltage for which the structure behaves in the inversion mode under equilibrium conditions. To continue the analogy between the potential well and a bucket, the former can be filled with electrons and the latter with water. To "contain" electrons in the potential well in analogy to water in a bucket, it is necessary to point the positive direction of the electrostatic potential downward, which is contrary to normally used conventions.

1.1.4. MOS CAPACITANCE IN WEAK INVERSION

If, under the nonequilibrium conditions of deep depletion, as shown in Figure 1.2, the MOS capacitor described above is provided with "only a few" electrons, the structure will be unable to operate in strong inversion. The amount of free minority charges is not enough to build an inversion channel and these charges will be used immediately by the structure to compensate some of the charges on the gate and to decrease the width of the depletion layer by a corresponding amount. This situation is shown in Figure 1.4. The various illustrations are in the same order as before.

In the example shown, it is assumed that the charge packet lowers the surface potential to a level of 5.8 V. The exact relations with which to calculate this value of the surface potential will be deduced in the next section.
With this value for Φ_s, the depletion layer width can be calculated with the foregoing relation [1.4] as being decreased to 2.7 μm. The compensating charges in the structure are now partly built up by the free electrons Q_n available at the SiO_2-Si interface. The remaining quantity needed to compensate the charges on the gate is generated by the negative ions Q_d in the depletion layer. As such, the electric field and the charge distribution are also illustrated in Figure 1.4.

The curve illustrating the electrostatic potential shows a surface potential Φ_s of 5.8 V. This reduction in surface potential compared to the deep-depletion situation is due to the presence of minority charge carriers, i.e. in the situation under study the free electrons. This is also schematically shown by the bucket which is partly filled by these electrons. Any additional electron in this bucket will "higher" the water level in the bucket, will further decrease the surface potential and consequently will decrease the depletion region width.
Note that in the weak inversion mode, the MOS capacitor is also behaving in a non-equilibrium situation. Any additional minority carrier becoming available to the structure will immediately be added to the mobile charge content in the inversion layer. Consequently, the depletion layer width will be also adapted, because fewer ionized doping atoms are needed to compensate the gate voltage.

As mentioned earlier, a charge-coupled device is a simple series connection of MOS capacitors. To operate the CCD the MOS capacitors are pulsed into deep depletion. By supplying minority carriers to the structure, the capacitor changes from a deep-depletion state towards a weak-inversion mode. Leaving a non-equilib-

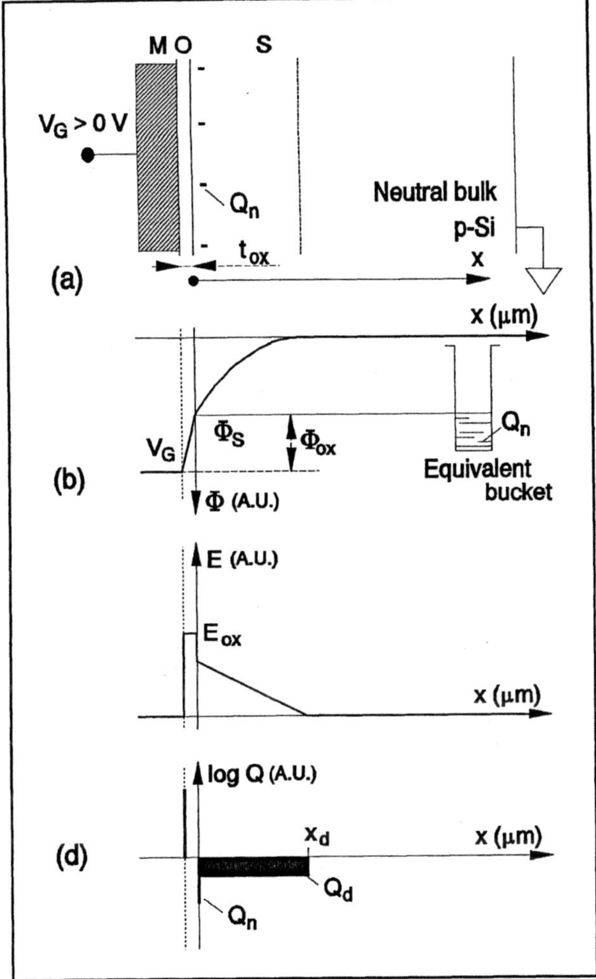

FIGURE 1.4. The MOS capacitance (a) in weak inversion with
its electrostatic potential (b), electric field (c), and charge
distribution (d) diagram.

rium phase, the device will only be brought into full equilibrium by a completely
filled inversion layer.
The step from the ideal situation, as described in this section, toward reality is made
in the next section.

WORTH MEMORIZING

The charges applied to the gate of an MOS capacitor have to be compensated. This can be done by means of :
  - **majority carriers if the capacitor is biased into accumulation;**
  - **ionized doping atoms if the capacitor is biased into deep depletion and no minority carriers are available;**
  - **an inversion layer of minority carriers if the capacitor is biased towards depletion and enough minority carriers are available;**
  - **a combination of a small number of minority carriers and ionized doping atoms, if the capacitor is biased towards deep depletion and only a small amount of minority carriers are available.**

The fundamental working principle of charge-coupled devices or CCDs is based on this last situation : the MOS capacitance biased into deep depletion and behaving in a nonequilibrium situation.

1.2. A real MOS capacitor in a charge-coupled device

In the previous section the discussion dealt with the characteristics of an ideal MOS capacitor. In the construction of charge-coupled devices, real MOS capacitors differ slightly from those discussed in the foregoing theoretical study. For instance, to obtain no potential drop at all in the silicon and a net charge density of zero, a gate voltage different from zero is needed. This is a result of the difference between the polycrystalline-silicon gate material and the monocrystalline-silicon bulk work functions.

The work function of a material is the energy needed to remove an electron at the Fermi level from the material into a vacuum. Thus, the gate voltage necessary to counterbalance the work-function difference Φ_{MS} between the metal and the silicon bulk is given by :

$$V_{G1} = \Phi_{MS} = \Phi_M - \Phi_{Si} \qquad [1.18]$$

with Φ_M and Φ_{Si} respectively the work function of the gate material and that of the silicon. For an n-type doped polysilicon gate on a p-type silicon bulk, Φ_{MS} is about 0.1 V.

The gate voltage necessary to achieve flat bands in the silicon (or to counterbalance all internal effects which influence the effective gate voltage) is also modified by the oxide charge Q_{ox}. This charge consists of fixed oxide charges and mobile oxide charges (alkali ions). The value of Q_{ox} depends on the crystal orientation of the silicon bulk material, but is also very sensitive to the processing conditions. This charge is always positive and is usually represented at the silicon-silicon oxide interface. To achieve a flat-band situation, the gate voltage must be sufficiently negative to cause an equal charge $-Q_{ox}$ to appear on the gate and is thus given by :

$$V_{G2} = - \frac{Q_{ox}}{C_{ox}} \; . \tag{1.19}$$

With both the work-function difference and oxide charge present, the gate voltage to cause flat-band condition (V_{FB}) is obtained as :

$$V_G = V_{FB} = V_{G1} + V_{G2} = \Phi_{MS} - \frac{Q_{ox}}{C_{ox}} \; . \tag{1.20}$$

The threshold voltage V_T, needed to bring the real MOS capacitance near to strong inversion, consists of the sum of the flat-band voltage V_{FB} and the threshold voltage of the ideal MOS capacitor, and is given by the combination of [1.16] and [1.20] :

$$V_T = \Phi_{MS} + 2.\Phi_F - \frac{Q_{ox}}{C_{ox}} + \frac{Q_d}{C_{ox}} \; . \tag{1.21}$$

The main parameter to control the value of V_T is the oxide thickness t_{ox} through C_{ox}, and the doping level of the substrate N_A through Q_d. The oxide charge Q_{ox} is usually reduced as much as possible by specialized processing techniques.

A charge-coupled device is no more than a series connection of simple MOS capacitors. To store "information" charges in such a CCD, the MOS capacitors are sequentially driven into deep depletion by digital pulses on the gates of the MOS structures. But when the device is forced into deep depletion, free electrons will be supplied to the capacitors - behaving in a nonequilibrium mode - through a process of thermal generation of minority carriers. So the clock voltages on the gates may only be applied for a time shorter than the time needed to build up the equilibrium inversion layer by thermal generation, otherwise the thermally-generated minority carriers would add up to the information charges which have to be stored in and transferred through the CCD. This would cause the signal to be deformed. Under these conditions of thermal nonequilibrium, an expression for the surface potential Φ_S as a function of the minority-carrier charge density in deep depletion can be deduced. Starting from the knowledge that the gate voltage, corrected for the work-function difference ($V_G - \Phi_{MS}$), is divided between the oxide (Φ_{ox}) and the silicon (Φ_S), the following expression can be written :

$$V_G - \Phi_{MS} = \Phi_{ox} + \Phi_S \; . \tag{1.22}$$

Together with [1.7] and

$$E_{ox} = \frac{Q_{ox} + Q_{Si}}{C_{ox} \cdot t_{ox}} \qquad [1.23]$$

for the nonideal MOS capacitor, the relation [1.22] can be rewritten as :

$$V_G - \Phi_{MS} = \frac{Q_{ox}}{C_{ox}} + \frac{Q_{Si}}{C_{ox}} + \Phi_S \qquad [1.24]$$

with Q_{Si} the charge in the silicon, built up from the mobile charge in the channel Q_n and the fixed doping ions in the silicon Q_d. In combination with [1.5] and [1.20], this yields :

$$V_G - V_{FB} = \frac{Q_n}{C_{ox}} + \frac{\sqrt{2 \cdot \epsilon_{Si} \cdot q \cdot N_A \cdot \Phi_S}}{C_{ox}} + \Phi_S . \qquad [1.25]$$

This equation is solved to find Φ_s as a function of the minority-carrier density Q_n :

$$\Phi_S = V_G - V_{FB} + \frac{Q_n}{C_{ox}} - \frac{\epsilon_{Si} \cdot q \cdot N_A}{C_{ox}^2} \cdot \left[1 - \sqrt{1 - \frac{2 \cdot C_{ox}^2 \cdot \left(V_G - V_{FB} + \frac{Q_n}{C_{ox}} \right)}{\epsilon_{Si} \cdot q \cdot N_A}} \right] . \qquad [1.26]$$

As can be seen from the above-mentioned expression [1.26], the surface potential Φ_s is a function of the gate voltage V_G, the oxide thickness t_{ox}, the doping concentration of the substrate N_A, and the number of free charge carriers Q_n.

The relation between the surface potential and the effective gate voltage $V_G - V_{FB}$ is shown in Figure 1.5a with the substrate doping as a parameter and in Figure 1.5b with the gate dielectric thickness as a parameter. In both cases the charge content of the potential well, Q_n is zero. Note that with this boundary condition the surface potential is almost equal to the effective gate voltage. In both figures the curve $\Phi_s = V_G$ is shown as a dotted line.

The maximum amount of charge that can be stored in the potential well is the charge found at thermal equilibrium. This charge content denoted by $Q_{n,sat}$, is given by [1.17] :

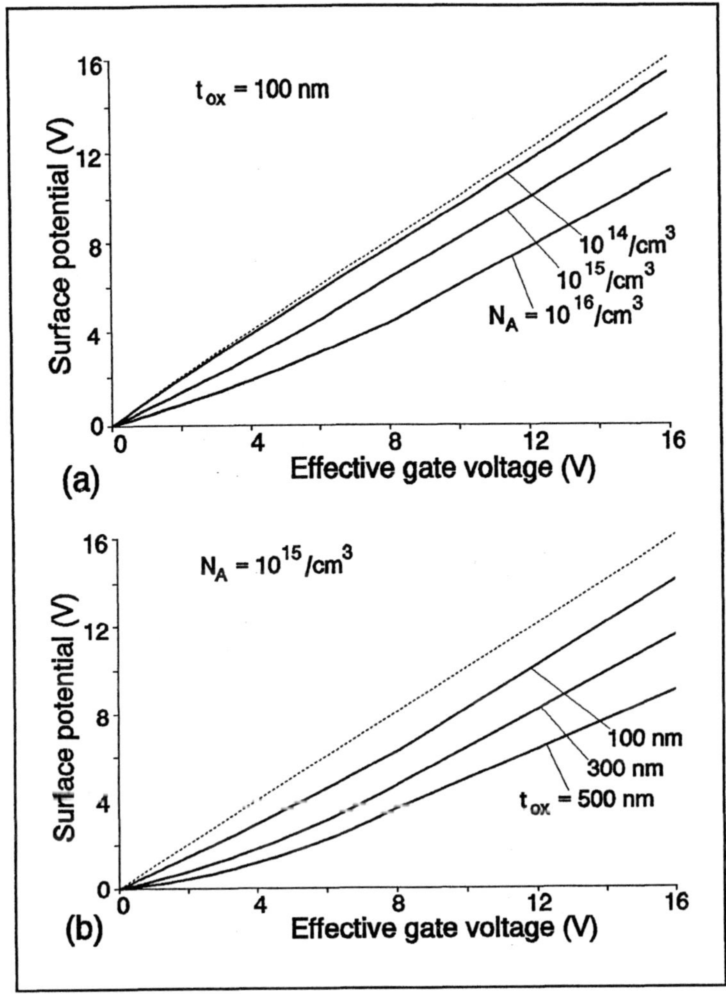

FIGURE 1.5. *Surface potential versus the effective gate voltage, parameters are the substrate doping (a) and the oxide thickness (b). Reference line :* $\Phi_S = V_G$.

$$Q_{n,sat} = C_{ox} \cdot (V_G - V_T) \, . \qquad [1.27]$$

Substituting C_{ox} in [1.26] by its value derived from [1.27] yields :

$$\Phi_S = V_G - V_{FB} - \frac{Q_n \cdot (V_G - V_T)}{Q_{n,sat}} -$$

$$\frac{\varepsilon_{Si} \cdot q \cdot N_A}{C_{ox}^2} \cdot \left[1 - \sqrt{1 - \frac{2 \cdot C_{ox}^2 \cdot (V_G - V_T)}{\varepsilon_{Si} \cdot q \cdot N_A} \cdot \left(1 - \frac{Q_n}{Q_{n,sat}} \cdot \frac{V_G - V_T}{V_G - V_{FB}} \right)} \right] . \qquad [1.28]$$

The dependence of the surface potential Φ_S on the relative amount of charge $Q_n/Q_{n,sat}$ or on the normalized charge content is shown in Figure 1.6. (Values for the various parameters from [1.28] are unaltered.) The surface potential is nearly a linear function of the minority-carrier concentration.

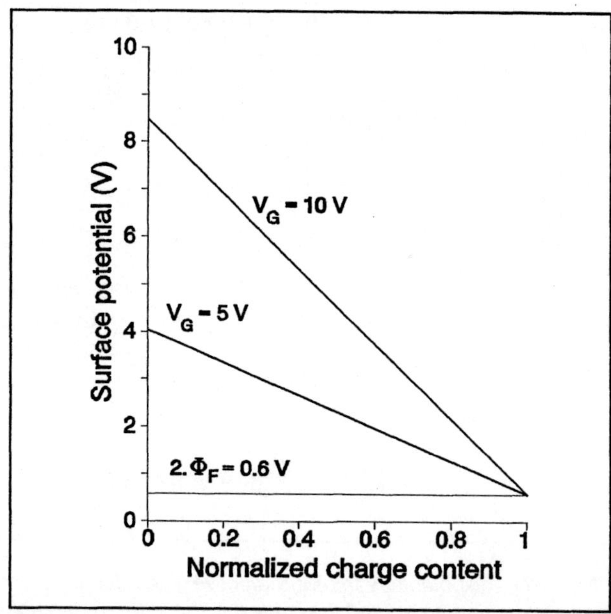

FIGURE 1.6. Surface potential as a function of the (normalized) charge content in the potential well for two values of the gate voltage.

Defining two new parameters :
  - Φ_{Sdd}　: the surface potential in deep depletion ($Q_n = 0$),
  - Φ_{Ssi}　: the surface potential in strong inversion ($Q_n = Q_{n,sat}$),

the relation [1.28] between the surface potential Φ_S and Q_n can be linearized by the following formula :

$$\Phi_S = \Phi_{Sdd} + (\Phi_{Ssi} - \Phi_{Sdd}) \cdot \frac{Q_n}{Q_{n,sat}} \qquad [1.29]$$

The influence of the minority-carrier density on the surface potential, illustrated in Figure 1.5 and given by the previous relations [1.28] and [1.29], causes a disturbance of the surface-potential distribution under the gates of a CCD. This has an important consequence on the charge transport which is described in the next section on charge transfer.

As already mentioned, the free charges can be stored in a potential well located at the dielectric-substrate interface. To understand the physics of charge-coupled devices it is very helpful to use the analogy with buckets. The size in width and depth of the potential wells or buckets can be very easily changed by external factors or by the design of the MOS structures themselves. Some of these possibilities are illustrated in Figure 1.7, which is also an overview of the theory explained up to now in this section. The figure illustrates the influence of four different parameters : the substrate-doping level N_A, the gate voltage V_G, the thickness of the gate dielectric t_{ox}, and the charge content of the bucket itself Q_n :
- influence of the doping level of the substrate material : the bucket is deeper for lower doping levels of the substrate (Figure 1.7a);
- influence of the gate voltage : the depth of the bucket can be changed in a continuous manner simply by changing the gate voltage. Higher gate voltages, deeper potential wells in the bulk, and more charges can be stored or the storage capacity becomes higher (Figure 1.7b);
- thickness of the gate dielectric : thicker oxides make the buckets shallower. The same dependence of the potential depth on the dielectric material is also illustrated(Figure 1.7c). The permittivity of the gate dielectric used has a direct influence on the potential well depth;
- charge content of the bucket : the remaining bucket depth is, of course, determined by the level to which it is already filled (Figure 1.7d).

If electrons are released in the middle (between the two potential wells sketched for each situation) of all four structures shown in Figure 1.7, they will preferably move to the highest potential or the deepest potential well in the neighborhood. Consequently, the electrons will flow to the lowest substrate dope, to the highest gate voltage, to the thinnest gate dielectric, and to the bucket which is filled to its lowest level.

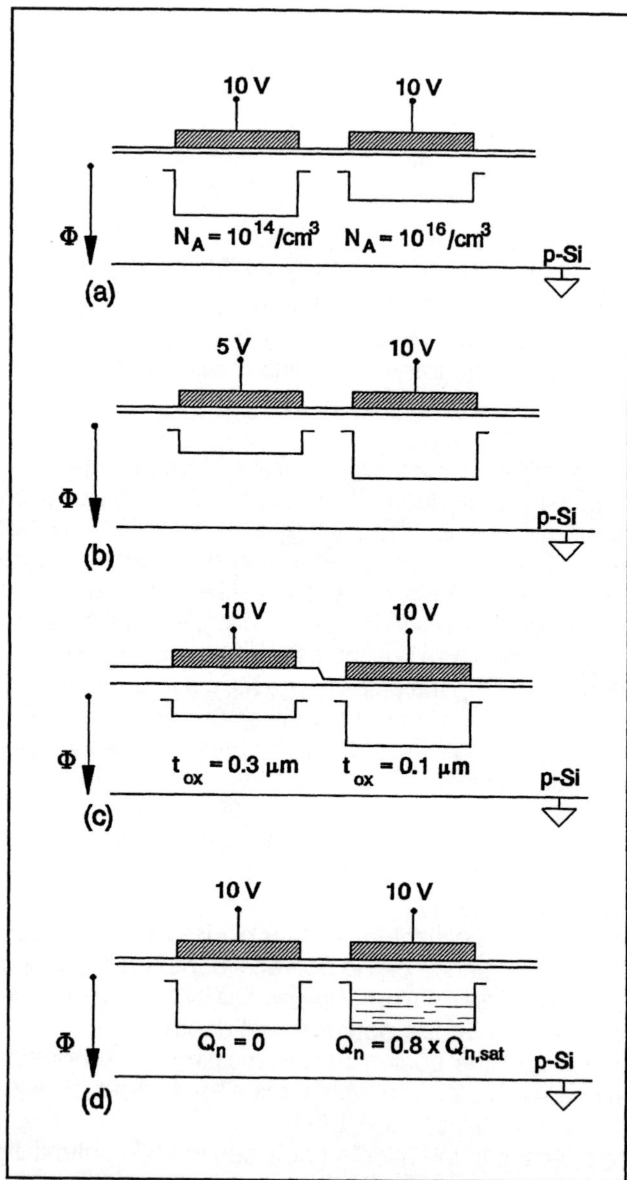

FIGURE 1.7. *Influence of different parameters on the surface potential : the substrate doping (a), the gate voltage (b), the gate dielectric (c), and the free-charge content (d).*

WORTH MEMORIZING

The behavior of a potential well (which might be filled with minority carriers) of a real MOS capacitor is quite comparable to a bucket (which might be filled with water). The shape of the potential well can be controlled by means of different parameters (intrinsic or extrinsic to the MOS structure). The doping concentration of the silicon, the thickness of the gate dielectric, and the gate dielectric material are parameters available to the designer of the MOS capacitor to shape the potential well. The user of the device can define the depth of the potential well via the charge content and the externally applied gate voltage.

1.3. Charge transfer

In this section on charge transfer details necessary in order to understand the charge transport in a charge-coupled device will be presented. After the previous discussion concerning the storage of charge packets in a potential well, the next logical step in CCD physics is the transport of the charges. The aim is to move the charge packets of minority carriers through the silicon from one gate to another by shaping and reshaping the form of the potential wells. Some possibilities for influencing the shape of the potential wells have already been shown in Figure 1.7. If, however, a dynamic action is needed, only the option of influencing the external potential well by means of external parameters or more specifically by changing the gate voltages can be applied.

Figure 1.8 explains the transfer of a charge packet from one gate to another (Beynon 80). In the small structure under consideration, only four gates on top of the gate dielectric are shown. To start with, the gates are biased to 0 V, except the second one which is biased at 10 V. From the previous section, it is known that a potential well is formed underneath the positively-biased gate (if the silicon substrate is p-type doped). Electrons, if available, will gather in this potential well. This situation is depicted in Figure 1.8a. At the onset of the charge transfer, shown in Figure 1.8b, the third gate in the row is set to 10 V, and an extra potential well is created underneath the third gate. If the two positively-biased gates are closely spaced, the individual potential wells under these gates will mix, forming just a single wide bucket. As stated before, the electrons will flow to the deepest point in the buckets and, in the enlarged bucket from Figure 1.8b, the charge packet will be redistributed across the whole width of the potential well. The ultimate result is shown in Figure 1.8c, where the packet of electrons is spread out across the second and third gates.

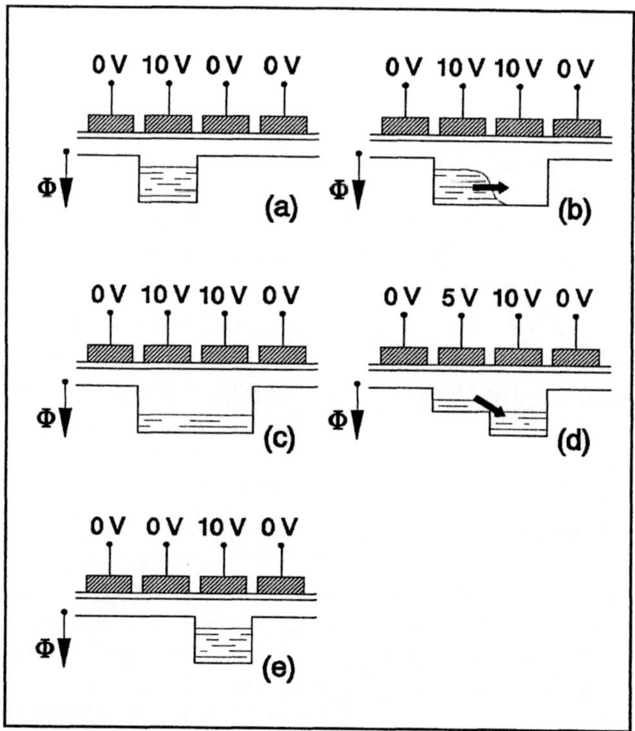

FIGURE 1.8. Illustration of the charge transport in a CCD. The charge packet of minority carriers is moved through the silicon by means of digital pulses on the CCD gates.

The next step is to push the charge packet away from the second gate. This is done in the way opposite to that used previously to attract the electrons. If the gate voltage on the second gate is lowered from 10 V to 0 V, the charges are pushed further. This process is illustrated in Figure 1.8d where the second gate is biased at an intermediate level of 5V. Proceeding from a high voltage to a low voltage, the bottom of the bucket is raised, and the electrons "try" to stay at the highest potential or in the deepest bucket. Due to the lowering voltage on the second gate the electrons are forced to flow underneath the third gate. The end result is shown Figure 1.8e, where the width of the potential well is reduced again to the dimension of a single gate. The potential well is moved from one gate to its neighbor, and consequently the charge packet of minority carriers is transferred from the second gate underneath the third one.

Note that, during this complete process of charge transport, the gate voltages on the first and fourth gate did not change. This is quite important in order to keep the charge packet under consideration isolated from the outside world in all situations.

Otherwise, it could interfere with other charge packets or it can even get lost in the device to which this small structure belongs.

The charge transfer from one gate to another is shown fairly schematically in Figure 1.8. In the remaining part of this section, the same charge transfer process will be explained and described in a more analytical way.
The movement of free carriers, in this case electrons, along the surface of the charge-coupled device is driven by three different mechanisms, all of which can operate and be described quite independently of each other (Carnes 72) :
- the thermal diffusion of charge carriers : even in the absence of any electric field, a charge packet which has the possibility to redistribute its local gradient toward an equilibrium situation will do so by means of thermal diffusion;
- self-induced fields : if a gradient exists in charge concentration, charges of the same type will repel each other and reshuffle the concentration of charge carriers so that the gradient becomes zero;
- fringing fields : charges will be forced to move due to the existence of electric fields generated by the voltages on the gates. These electric fields, which are gate-voltage, and also gate-geometry dependant, are called fringing fields. More details of the fringing fields are given further on in this section.

From the analytical point of view the charge transport of the minority carriers through a CCD channel can be described by means of the current density and continuity equations.
The first can be written as :

$$J(y,t) = J_d + J_s + J_f ,$$
[1.30]

where J(y,t) is the current density created by the movement of the charge packet, as a function of place and time. (The direction along the silicon surface in which the charges move is denoted as the positive y direction.) J_d, J_s, and J_f are the current densities of respectively the thermal diffused part, the part due to the self-induced fields, and the part resulting from the fringing fields. These three components are given by the classical expressions for diffusion current and drift current generated through the action of an electrical field (Carnes 72) :

$$J_d = q.D_n.\frac{\partial Q_n(y,t)}{\partial y}$$
[1.31]

with D_n the diffusion constant of electrons and $Q_n(y,t)$ the charge distribution as a function of place and time,

$$J_s = Q_n . \mu_n . E_s$$
[1.32]

with μ_n the mobility of the electrons and E_s the self-induced electric field,

$$J_f = Q_n \cdot \mu_n \cdot E_f \qquad\qquad [1.33]$$

with E_f the fringing field.

Charge carriers which are transferred from one position to another define the current between these two positions. The continuity equation connects the amount of charge to be transferred with the current density (Carnes 72) :

$$\frac{\partial Q_n(y,t)}{\partial t} = \frac{\partial J(y,t)}{\partial y} \qquad\qquad [1.34]$$

This relation is only valid if no extra charges are added or subtracted from the original charge packet. Or, the recombination and the generation of minority carriers is not considered. If the transport of the charge packet is fast enough, the rate of carrier generation and recombination is very small and negligible relative to the total amount of charge carriers in a charge packet.
It is very complicated to find an analytical solution for the complete problem of charge transfer and several numerical techniques have been presented in the literature. To facilitate discussion and study the various mechanisms in further detail, however, it is of interest to consider the three mechanisms separately.

1.3.1. THERMAL DIFFUSION

As already described, even in the absence of any electric field, a charge packet which has the possibility to redistribute its local gradient toward an equilibrium situation, will do so by means of thermal diffusion. (The possibility to redistribute the gradient can be created by an adaptation of gate voltages in a charge-coupled device.) This effect can be studied simply by introducing the current-density relation [1.31] into the continuity equation [1.34], which results in :

$$\frac{\partial Q_n(y,t)}{\partial t} = \frac{\partial}{\partial y}\left(q.D_n.\frac{\partial Q_n(y,t)}{\partial y}\right) \qquad\qquad [1.35]$$

The solution to this partial differential equation is given by :

$$Q_n(t) = \frac{8}{\pi^2}.Q_n(0).e^{-\frac{\pi^2.D_n.t}{4.L^2}} \qquad\qquad [1.36]$$

In this expression the parameter L represents the length of a single CCD gate, which is its dimension in the direction of charge transport.

The diffusion constant D_n can be related to the mobility of the charge carriers μ_n through :

$$D_n = \frac{k.T}{q}.\mu_n \ . \tag{1.37}$$

As can be seen from the analytical solution of the problem presented by [1.36], the remaining charge packet at time point t, $Q_n(t)$, decreases exponentially in value from its starting value $Q_n(0)$. With this exponential decay, only a first small part of the charge packet will move relatively fast, but the remaining packet needs some time to be transported to the next stage in the CCD.

Note that the time constant of the process is inversely proportional to the diffusion constant D_n and proportional to L^2.

In conclusion, the transport mechanism based on thermal diffusion is enhanced if the diffusion constant D_n is increased and if the gate length L is made as small as possible. Charge-coupled devices which make use of electrons as minority carriers (on a p-type substrate) will transport faster than CCDs transferring holes, because D_n is about three times as high as D_p, the diffusion constant of holes.

1.3.2. SELF-INDUCED DRIFT

In general, a gradient in charge concentration built up by charges of the same type will repel the charges to reshuffle their concentration so that the gradient will become zero. This reordering will take place through the electric field generated by the gradient in charge distribution. In the case of the self-induced fields alone, the continuity equation [1.34] can be written as :

$$\frac{\partial Q_n(y,t)}{\partial t} = \frac{\partial}{\partial y}(Q_n.\mu_n.E_s) \ . \tag{1.38}$$

As can be seen from Figure 1.8b, the presence of a charge gradient underneath two neighboring gates at the same potential causes a change in surface potential Φ_s under the gates and consequently produces an electric field E_s. An accurate determination of the surface potential Φ_s should be based on the solution of the two-dimensional Poisson equation with space-charge density $\rho(x,y)$, but the problem studied in this section can be dealt with sufficiently accurately using a one-dimensional Poisson equation. This permits the determination of $\Phi_s(y)$ as a function of $Q_n(y)$ along the direction of transport or CCD channel, and the self-induced field is then found from :

$$E_S(y) = \frac{\partial \Phi_s}{\partial y} = \frac{\partial \Phi_s}{\partial Q_n}.\frac{\partial Q_n}{\partial y} \ . \qquad\qquad [1.39]$$

The relation between the surface potential Φ_s and the inversion charge Q_n is :

$$\Phi_S = \Phi_{Sdd} + \frac{Q_n}{C_{ox}+C_D} \ , \qquad\qquad [1.40]$$

with Φ_{Sdd} the surface potential for an empty well, C_{ox} the oxide capacitance, and C_D the depletion capacitance.

Combining the foregoing relations [1.38], [1.39], and [1.40], yields :

$$\frac{\partial Q_n(y,t)}{\partial t} = \frac{\partial}{\partial y}\left(\frac{Q_n.\beta}{C_{ox}+C_D}\right).D_n.\frac{\partial Q_n(y,t)}{\partial y} \qquad\qquad [1.41]$$

with :

$$\beta = \frac{q}{k.T} \ . \qquad\qquad [1.42]$$

Compared to the theory of the transport of charges by thermal diffusion, the process involving the self-induced fields looks similar and both, thermal-diffusion and self-induced drift processes, can be combined using a single, effective diffusion constant D_{eff} given by :

$$D_{eff} = \left(\frac{Q_n.\beta}{C_{ox}+C_D}+1\right).D_n = \left(\Delta\Phi_S.\beta.\frac{Q_n}{Q_{n,sat}}+1\right).D_n \qquad\qquad [1.43]$$

in which the maximum amount of free charge $Q_{n,sat}$ is rendered by [1.27].
The parameter $\Delta\Phi_s$ represents the change in surface potential needed to fill up an empty well completely and is given by :

$$\Delta\Phi_S = \frac{Q_{n,sat}}{C_{ox}+C_D} \ . \qquad\qquad [1.44]$$

Note that the effective diffusion constant depends on the concentration of minority carriers and is, through this parameter, a function of time and position along the channel.

Due to the presence of the self-induced fields, the total diffusion of charge carriers during the transport phase has become much faster compared to the case of thermal diffusion only. Especially if the charge packet is greater than 1 % of the saturation level (e.g. $Q_n > 0.01\ Q_{n,sat}$), the drift current J_s is much greater than the thermal-diffusion current J_d (e.g. $J_s = 200\ J_d$). But in the case of the charge packet becoming really small (e.g. $Q_n < 0.01\ Q_{n,sat}$), the transport process is completely determined by the thermal diffusion of the minority carriers, and becomes slow again.

So the drift current is important only during the initial period of the charge transfer as long as Q_n is of sufficiently great value compared to $Q_{n,sat}$. The transfer efficiency of the transport is only high if enough time is available to effect the CCD transport and to allow the thermal diffusion to transport also the latest part of the charge packet. This theory might set an upper limit for the transport frequency. However, if the CCD needs to be operated with a high transport efficiency or in a high speed mode, a significant improvement in transfer efficiency can be achieved by the introduction of a "fat zero". With this effect, the zero charge content is represented by a certain amount of stored charge (e.g. 20 % of $Q_{n,sat}$) and no longer by absolute zero (see also 1.4).

1.3.3. FRINGING FIELDS

A practical device can be operated much faster than described by the previous theory of thermal diffusion and self-induced drift. This is due to the forcing of the transport by electric fields generated by the voltages on the gates. These fringing fields at the Si-SiO$_2$ interface, along the direction of charge flow, arise from the two-dimensional nature of the structure. This means that the surface potential is not only set by the gate voltage on top of the dielectric but also influenced by its neighboring gates.

The occurrence of the fringing fields is schematically shown in Figure 1.9. In Figure 1.9a the surface potential is depicted as a series of rectangular buckets, while a more realistic curve for the surface potential is shown in Figure 1.9b. The last situation is only valid if the spacing between two adjacent gates is small enough (e.g. comparable to the oxide thickness) and if the length L of the gates is not too long. If they are, for instance, greater in length than 10 μm, the potential surface becomes more like the one shown in Figure 1.9a.

When the surface potential is completely flat underneath the CCD gates, the fringing field E_f are locally zero. But when the surface potential has a certain gradient, the fringing field E_f differs from zero. An indication of the location of the minimum fringing field $E_{f,min}$ is included in Figure 1.9.

An empirical approximation for the minimum value of the fringing field $E_{f,min}$ at the middle of the transferring electrode in a CCD cell built with three gates (or a three-phase CCD structure as shown in Figure 1.9b) is found to be equal to (Beynon 80) :

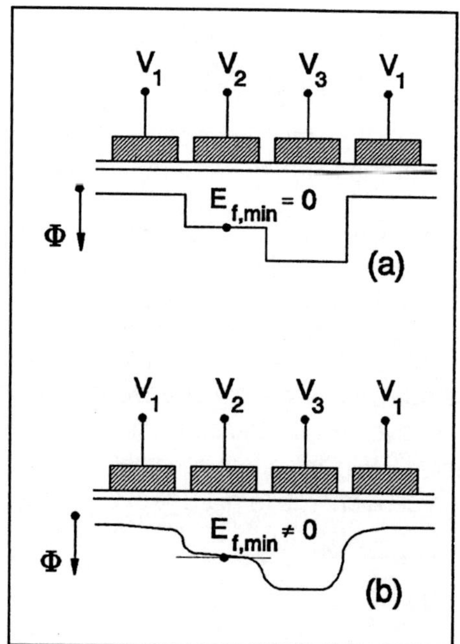

FIGURE 1.9. Schematic illustration of the surface potential without (a) and with (b) fringing fields ($V_1 < V_2 < V_3$).

$$E_{f,min} = A.\frac{t_{ox}}{L^2}.\frac{\Delta V}{2}.\left[\frac{5.x_d/L}{(5.x_d/L)+1}\right]^4 \qquad [1.45]$$

with :
    - A   : a constant equal to 6.5,
    - ΔV : the voltage swing on the gate during the charge transfer.
This relation [1.45] is only valid if the gap between two gates is small enough, e.g. comparable to the thickness of the dielectric. In the opposite case, there will even be a potential barrier between the two gates. Smooth transport of charges can be prevented by this barrier.

The minimum fringing field can be influenced by the substrate doping : at doping levels higher than $10^{15}/cm^3$ the fringing field drops off due to the decreasing x_d/L ratio. A smaller electrode length gives a much higher fringing field and a significant improvement in transfer efficiency. Other parameters through which a CCD designer can increase the fringing-field drift are the oxide thickness t_{ox}, the pulse voltages on the gates ΔV and, the substrate bias through x_d.

The relative importance of the various transport mechanisms is illustrated by Figure 1.10 (based on Carnes 72), where the transport efficiency is shown as a function of the time available for the charge transfer. The transport efficiency is defined as the part of a charge packet transported through the CCD cell, normalized to the original charge packet that had to be transported.

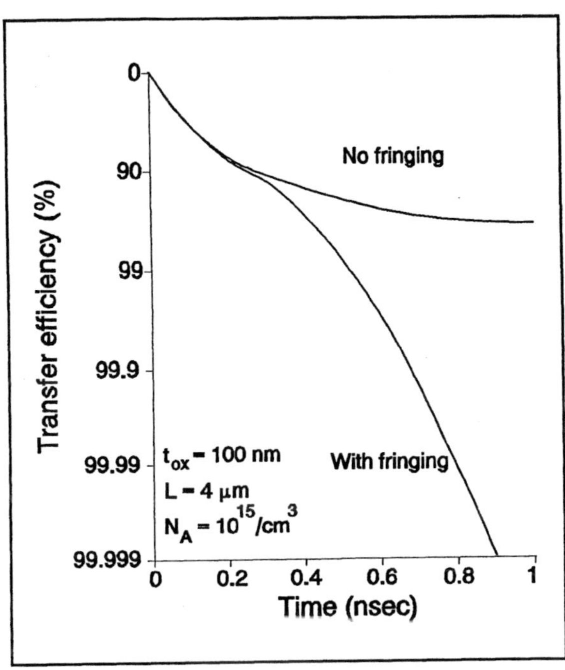

FIGURE 1.10. Influence of the fringing fields on the transport efficiency as a function of the transport time.

The two curves of Figure 1.10 (obtained with computer simulations) compare the transport of the same charge packet with and without the influence of the fringing fields. Note that the two curves coincide for very short periods of time. This means that during the initial phase of the charge transport (immediately after the positive bias of the receiving gate), the movement of the charges is primarily driven by the self-induced drift. After 200 psec, in this typical case, the fringing fields are dominant and are responsible for the transport of the remaining 90 % of the charge packet that still has to move. From the curve it can be deduced that the time needed to allow the transport to take place up to 99.99 % is about 800 psec.

As mentioned earlier, the fringing fields can be influenced by the substrate doping. When the substrate doping is lower, the thickness of the depletion layer is greater and the fringing fields also increase. This is illustrated in Figure 1.11 (based on Carnes 72), which shows the transport efficiency as a function of transport time,

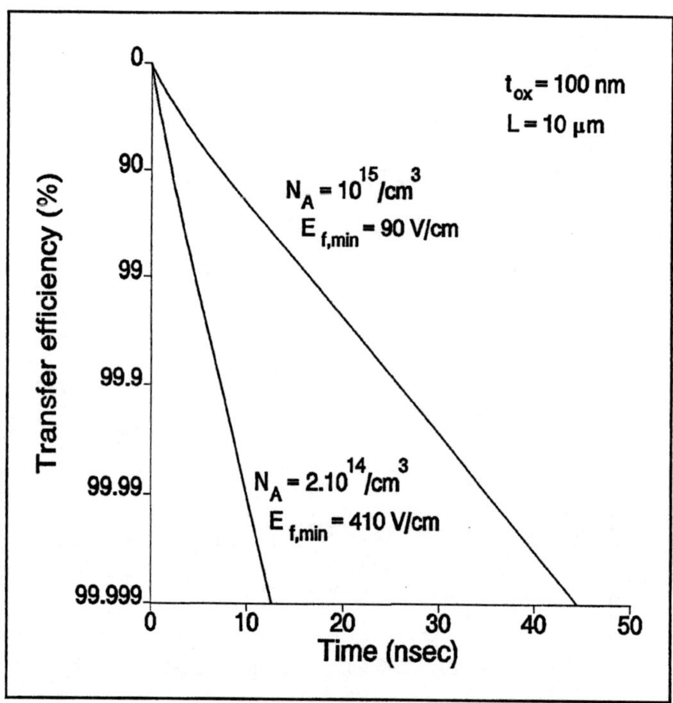

FIGURE 1.11. The transfer efficiency as a function of the transfer time
for two values of the substrate doping.

with the substrate doping or fringing field value as a parameter. To obtain a transport efficiency of 99.99 %, the CCD on the $10^{15}/cm^3$ doped substrate is a factor 3 to 4 slower than the device build on the $2*10^{14}/cm^3$ doped material.

Figure 1.12 (based on Carnes 72) summarizes the results obtained : it shows the transport time - needed to move a charge packet from one gate to its neighbor with a transport efficiency of 99.99 % - as a function of the gate length. The dashed line in Figure 1.12, represents the theoretical situation of only thermal diffusion. This phenomenon is always active, despite self-induced or fringing fields, which means that in all situations considered, the worst case transport time can be found on the thermal diffusion line. On the other hand, the figure clearly illustrates the improvement in transfer speed by using low doped substrate material to maximize the fringing fields. Above the dashed curve relating to the thermal diffusion, the three curves for different substrate dopings have only a theoretical value, because in reality the transport of charges will proceed at least as fast as determined by the thermal process.

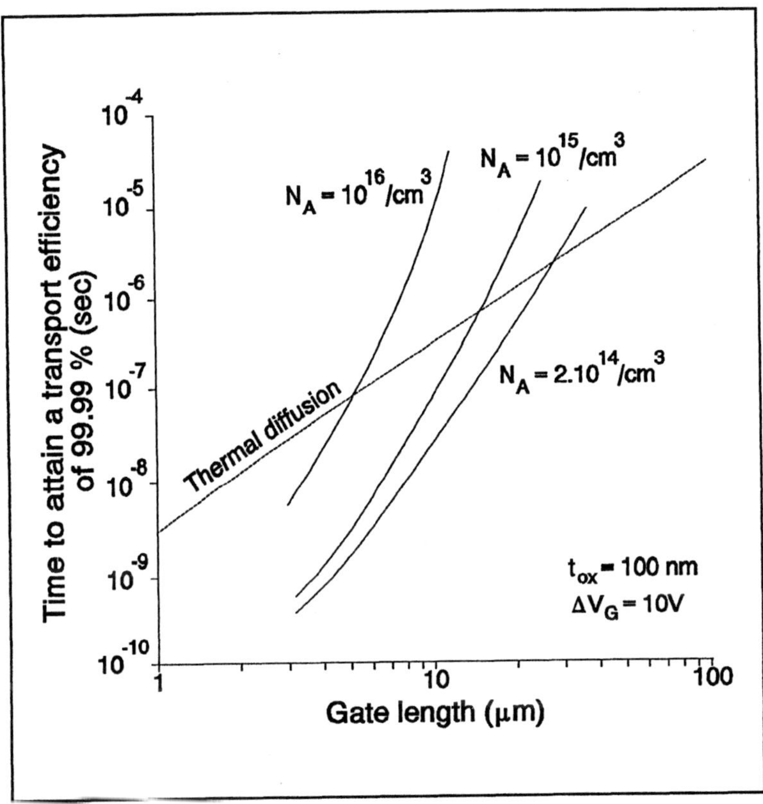

FIGURE 1.12. Transport time, as a function of gate length, to reach an efficiency of 99.99 %, with the substrate doping as parameter, and the thermal diffusion as theoretical limit.

WORTH MEMORIZING

The transport of a charge packet from one gate to another through a charge-coupled device is based on three different mechanisms :
- thermal diffusion, which is characterized by a very slow transport and is due to the existing concentration gradient, followed by;
- self-induced drift to redistribute the charge content and which is responsible for a fairly fast transfer of the very first charge carriers;
- fringing-field induced drift, which effects of the transport of the remaining part of the charge packet.

Compared to the fringing-field induced transfer, thermal diffusion and self-induced drift are slow processes. Otherwise, to operate charge-coupled devices at high speeds, care has to be taken during the design

and operation of the devices to derive maximum benefit from the presence of the fringing fields.

1.4. Charge-transfer (in)efficiency

Up to now the charge-transfer process has been studied for the ideal case. If the time needed for a complete charge transfer is available, transport will take place without any inefficiency. In a real CCD structure, however, the transport will hardly ever be ideal, and the charge transport will be degraded by a certain level of inefficiency. The charge packet leaving a "gate position" will not be the same as the one which arrived one clock cycle earlier. The difference between the two can be negative or positive : the amount of charge carriers in a packet can be increased or decreased. In general, during the movement of a charge packet from one gate toward the next one, the transport efficiency will be degraded by :
   - the limited time available to perform the charge transport;
   - charge trapping by the surface states.

Things that can happen when the transport time becomes too short have been extensively described in the previous section. Complete charge transfer under the influence of thermal diffusion, driven by self-induced fields or even by fringing fields needs a finite time. But an additional effect which makes the charge transport no longer perfect is the trapping of charges from the charge packet in surface states. (Surface states are crystalline defects at the transition between the monocrystalline-silicon bulk and the amorphous silicon oxide. Surface states generate discrete energy levels lying within the forbidden bandgap of silicon.) In this situation the electrons are "falling" from the conduction band into the lower energy levels associated with the surface states. This effect of catching electrons from the charge packet and storing them in surface states is illustrated in Figure 1.13. Once the electrons are trapped in these surface states, they will not recombine because there are no holes in which are needed to do so.

On the other hand, however, these electrons do not remain trapped in the surface states but after a while are released again. When they become free again, they will move toward a potential well but, as illustrated in Figure 1.13, freed electrons can easily move to a potential containing a charge packet different from the charge packet where the freed electron originally came from. This charge packet might already have moved further on.

The processes of trapping and "detrapping" the electrons into the surface states are not related to each other. This effect can be translated into different time constants for the filling and the emptying of the traps (Tompsett 73). The time constant associated with the filling of the surface states, τ_{fill}, depends on the surface concentration of the carriers. A typical value can be $\tau_{fill} = 10^{-9}$ sec. But, something

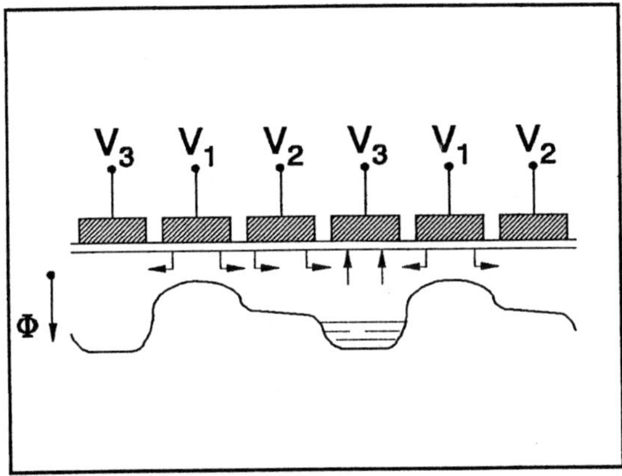

FIGURE 1.13. Schematic illustration of the effect of the surface states : the trapping of information charges and their subsequent release.

totally different, the energy of the surface states determines the time constant of the emptying action, τ_{empty}, which ranges from 10^{-11} sec to 10^{-3} sec.

In actual fact, minority carriers from a CCD charge packet can be easily trapped by surface states. The more charge carriers available, the faster this trapping process takes place. The opposite process, namely releasing these charge carriers again, is in most cases a much slower process :

$$\tau_{empty} >> \tau_{fill} .$$

[1.46]

There may be a lapse equal to several hundreds of μsec between the moment of capture and the moment of release. By the moment of release, however, the charge packet of which electrons are trapped in the surface states will have already been transferred further on, perhaps over a distance corresponding to several CCD cells. The electrons will be freed and added to a charge packet which has no relation to the one the electrons came from. Charges can consequently be mixed up as between two charge packets. This also has the effect of lessening the transfer efficiency.

To study this process of transport in relation to the surface states, the charge-transfer inefficiency ϵ is defined as follows : it is the ratio between the amount of charge left behind after a CCD transfer on one hand, and the amount of charge originally contained in the charge packet which had to be transferred, on the other hand.

The value of the charge-transfer inefficiency, depending on the presence of the surface states, can be empirically written as (Tompsett 73):

$$\varepsilon = \frac{q.k.T}{C_{ox}} . \frac{N_{ss}}{\Phi_S - \Phi_{Sdd}} . \ln(m+1) \qquad\qquad [1.47]$$

with :
- N_{ss} : the number of surface states (/cm^2.eV);
- m : the number of gates in a single CCD cell, or the CCD is defined as being an "m-phase" charge-coupled device.

A typical (state-of-the-art) value for the number of surface states is :
- N_{ss} = 5*10^8/cm^2.eV for silicon with a <100> crystal orientation;
- N_{ss} = 10^9/cm^2.eV for <111> oriented silicon.

From the above relation and from these values, it will not be surprising that all CCDs are processed on <100> material. The reason for a higher value of N_{ss} in the case of the <111> silicon can be found in the fact that this material has many more Si atoms at the surface compared to the situation with <100> orientation.

To keep the transport inefficiency low, the number of surface states has also to be kept low. Much effort is put into CCD processing and technology for the purpose of reducing the number of surface states. But although the final number attained nowadays is in fact quite low, it is impossible to manufacture CCDs without surface states. (Not only the transfer efficiency but also the dark current are negatively influenced by the presence of surface states. This effect will be described in section 3.2.)

One way to minimize the effect of transport inefficiency can be found in the so-called "Fat Zero" technique. This method makes use of a bias charge in the CCD and aims to avoid :
- the transport (by diffusion) of the "last" charge carriers in a packet;
- a difference between τ_{fill} and τ_{empty}.

With the "Fat Zero" method, the relative zero level of the content of the charge packet is no longer equal to zero electrons in the charge packet but, for instance, to a charge packet containing 20 % of $Q_{n,sat}$. With a fat zero of this value, two effects can be established :
- the surface states will be almost continuously filled and, if they release an electron, always enough electrons are available (in all packets, even with no information charge) to refill them immediately;
- in the case of a fixed but finite transport time, the number of charges which are left behind in the transport process are almost the same for all transports, despite the amount of information charge.

In addition to much better transport efficiency, the introduction of the fat zero has another advantage : negative information can also be handled. As already mentioned, the zero level for the information contained in the charge packet is shifted to 20 % of $Q_{n,sat}$. If the content of the charge packet is less then 20 % of $Q_{n,sat}$, the information corresponds to a negative value.

The disadvantage of the fat zero is its limitation to the charge-handling capability, because each packet contains a certain number of dummy carriers in order to define zero level.

It should be clear from this discussion that high transport efficiencies are only possible :
- if the number of surface states is low;
- if interaction with these surface states is minimized (or preferably absent);
- if the transport of the charges is fast enough (or the fringing fields are high).

These important characteristics (low interaction with surface states and high fringing fields) can be obtained if the charge transport can take place in the bulk of the silicon instead of transporting the charges along the Si-SiO$_2$ interface. The devices which are able to do so are described in the section which follows.

WORTH MEMORIZING

CCD transport with high transfer efficiency can only be achieved if the time available to complete the transport is long enough. If this is not the case, the design, the fabrication technology, and the application have to be optimized toward :
- high fringing fields, to speed up the transfer speed;
- a low number of surface states, to minimize their interaction with the charge carriers.

1.5. Buried channel CCD

The first publications on the buried-channel CCD (BCCD) date from 1972 (Walden 72, Esser 72). Only a few years after the invention of surface-channel charge-coupled devices, researchers were already reporting on new types of charge-transfer devices which were characterized by a construction permitting the transport of charges packets, not now at the Si-SiO$_2$ interface, but through the bulk of the silicon substrate. If the transport channel is designed such that the minority carriers cannot interact with the surface states, there will be no transfer loss due to these traps. And by moving the transport channel to deeper in the bulk, transport is also guided by larger fringing fields. This is shown schematically in Figure 1.14, where the fringing fields are drawn at different levels in the silicon. The charge-coupled device has a three-phase structure with the gates biased respectively at 10 V, 5 V, and 0 V. The waveforms represent the following situations :
- A-A' : shows the fringing fields very close to the Si-SiO$_2$ interface. The waveform can be approximated by straight lines. The minimum value of the fringing

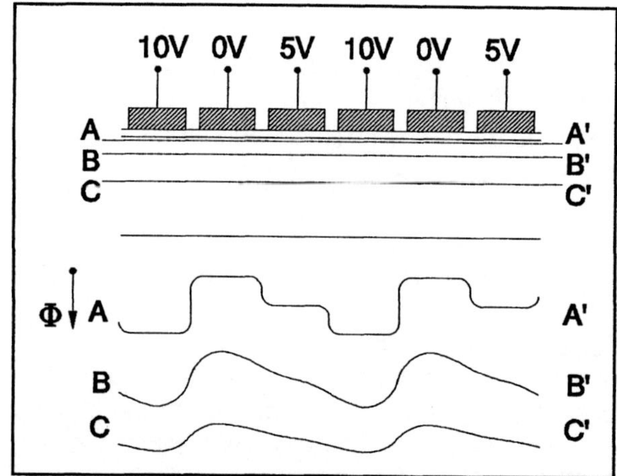

FIGURE 1.14. Illustration of the fringing fields at the Si-SiO$_2$ interface (A-A'), at a shallow depth (B-B'), and deeper (C-C') in the silicon bulk.

field underneath the gate biased at 5 V is almost zero;
- B-B' : at a certain (small) depth in the silicon the influence of the neighboring gates on the electric field underneath the middle gate (at 5 V) is at its highest level. The minimum fringing field is at its maximum value;
- C-C' : burying deeper in the silicon bulk will lead to a decreasing influence of the fields generated by the neighboring gates. At these depths the various gates are almost the same distance apart, and the minimum value of the fringing field will decrease.

From Figure 1.14 it can be seen that the minimum fringing field under a gate will be maximum at a certain depth. It is, of course, best to design the buried-channel CCD in such a way that the transport channel is located close to this depth so as to derive maximum benefit from the fringing fields, thereby enhancing the speed of the CCD transport.

1.5.1. FROM SCCD TO BCCD

To keep the minority-charge packet separate from the Si-SiO$_2$ interface, a channel with a potential minimum in the bulk has to be generated. This can be done by an extra n-type top doping of the p-type silicon substrate. If the doping concentration is chosen such that the n-type layer is fully depleted during operation of the CCD, a situation of the electrostatic potential as shown in Figure 1.15c can be created. The ionized positive ions of the n-type top layer transform the positive gate voltage to an even higher channel potential in the silicon. As illustrated, the potential well in the n-type Si now has its maximum not at the interface but deeper in the bulk.

The corresponding situation for a surface-channel CCD, with empty potential wells or $Q_n = 0$, is shown in Figure 1.15a. Observe that the surface potential Φ_S in the SCCD is lower than the gate voltage V_G, while, in the case of the BCCD, the channel potential Φ_{CH} is higher than the gate voltage V_G.

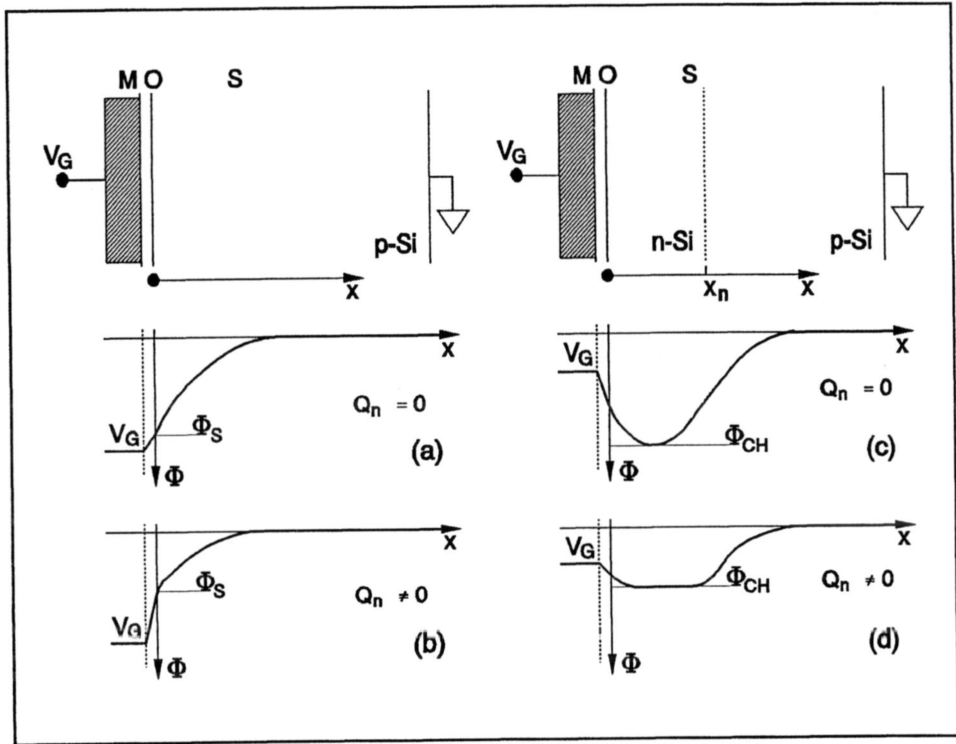

FIGURE 1.15. Illustration of the surface and channel potential for respectively a SCCD ((a) and (b)) and a BCCD ((c) and (d)), with empty ((a) and (c)) and filled wells ((b) and (d)).

The situation in which a charge packet is stored in the BCCD is schematically illustrated in Figure 1.15d ($Q_n > 0$). Additional minority-charge carriers lower the channel potential, and the absolute value of the channel potential Φ_{CH} approaches the value of the gate voltage V_G. Filling the potential well of the BCCD brings the electrons closer to the Si-SiO$_2$ interface. This can be seen by comparing Figures 1.15c and 1.15d. At a certain level of well-filling, the electrons will start to interfere with the surface states again. Charges will no longer be stored completely in the buried mode, but now partly also in the surface mode, recreating the disadvantages of the surface-channel CCD (Esser 73). The interaction with the surface states will be small or negligible if the "voltage distance" between the Si-SiO$_2$ interface and the charge packet is greater than kT/q (= 25.8 mV at room temperature).

This boundary condition renders arbitrary the definition of a full well in a buried-channel device.

To complete the comparison with the surface-channel CCD in Figure 1.15, the situation for a partly filled potential well in the surface-channel CCD is included in Figure 1.15b.

A more analytical result of the channel potential versus the effective gate voltage with the charge content as a parameter is shown in Figure 1.16 (Beynon 80). The relation between the channel potential and the effective gate voltage is almost a straight line, which is about the same as for surface-channel CCDs. The correlation between both parameters for a surface-channel CCD is also given for comparison purposes. The SCCD is simulated with the same values for the oxide thickness t_{ox} and doping concentration N_A as the BCCD has been, excluding the n-type top doping and with an empty well. Numbers of these parameters are listed in Figure 1.16, with :
- x_n : the thickness of the n-type top layer;
- N_D : the doping concentration of the buried-channel implant.
Observe that in all cases the channel potential Φ_{CH} of the BCCD is greater than the surface potential Φ_S of the SCCD at the corresponding gate voltage. For a buried-channel CCD with empty wells, the following relation applies:

$$\Phi_{CH} > V_G \qquad\qquad [1.48]$$

while for a surface channel, with empty wells :

$$\Phi_S \approx V_G . \qquad\qquad [1.49]$$

1.5.2. FRINGING FIELD AND TRANSFER TIME

Not surprisingly, parameters such as fringing fields and transfer time are key characteristics in the study of the buried-channel CCD. Several papers concerning analysis of these points have been published (Collet 74, Hanneman 75, de Meijer 81, Bakker 91). The main question to which all the authors try to give an answer is : "To what depth in the silicon have the charges to be transported in order to benefit as much as possible from the fringing fields ? " To give some insight into the various aspects of this discussion the situation illustrated in Figure 1.17 is analyzed : the gate electrodes are isolated from the substrate by a silicon-oxide insulation layer with a dielectric constant ϵ_{ox} and thickness t_{ox}. The substrate dielectric constant is represented by ϵ_{si}. The length of the gates is equal to L, as also is the pitch of the gates : P. The gaps between the various gates are so small that their influence is considered negligible. Note the difference in the origin of the depth coordinate; in all previous analysis x = 0 was located at the SiO_2-Si interface, while

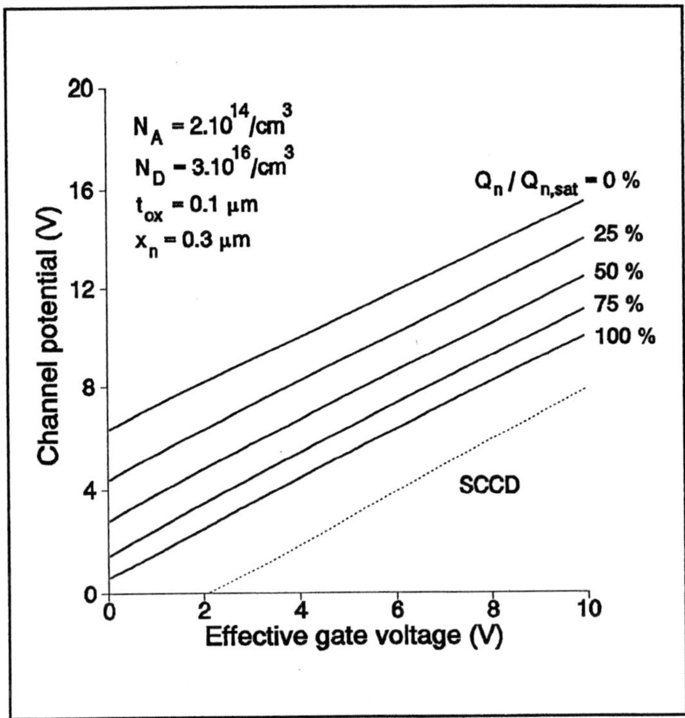

FIGURE 1.16. Channel potential of a BCCD as a function of the effective gate voltage, the parameter being the relative amount of charge in the potential well.

in this study it is sited at the gate-SiO_2 interface. (For this reason the depth coordinate is denoted as x'.)

To find the optimal transfer depth of the structure sketched in figure 1.17, the depth at which the fringing field reaches a maximum, namely under the middle of the negative-going transfer gate, is taken as an indication of the optimal transfer depth. This definition has the advantage that the fringing field at this position does not depend on the actual voltage of the transfer gate, but only on the fringing fields induced by the potential differences of the neighboring gates. Moreover, for a three-phase CCD with gates of equal length, the fringing field at this position gives an upper limit for the transfer time tt. The transfer time tt is defined as the time it takes the last electron to cross the transfer gate, at the half-way potential for the negative-going transfer gate and for a fixed transport depth of the electron.

With all these definitions in mind, the results of the analytical study are shown in Figures 1.18a and 1.18b (Bakker 91). In both figures the transfer time as a function of transfer depth is shown, Figure 1.18a for a three-phase CCD and Figure 1.18b

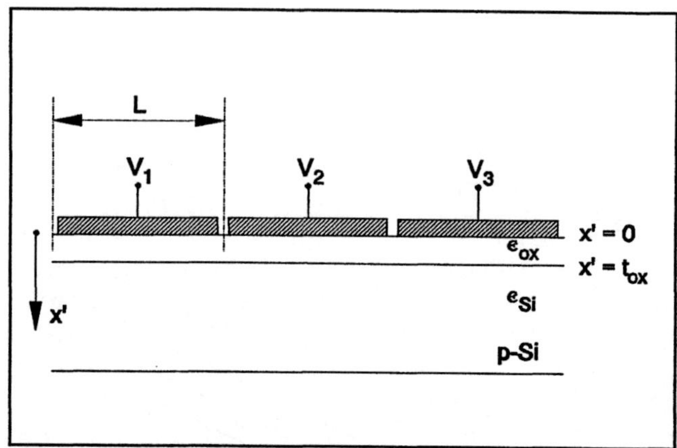

FIGURE 1.17. Two-dimensional device set-up used in the study of fringing fields and transfer times.

for a four-phase CCD. In both illustrations devices with a 0.1 μm gate insulator are used, and the gate length L serves as parameter. (Other constants used in the study are : the electron mobility : μ_n = 500 cm²/V.sec, difference in gate potential ΔV : 1 V.)

From Figures 1.18a and 1.18b it can be seen that :
- for L = 2 μm the optimal transport depth is calculated to be 0.44 μm for the three-phase CCD and 0.43 μm for the four-phase CCD;
- for L = 4 μm, neither has a deep minimum : once a minimum depth of 0.6 μm is reached, the exact value no longer matters;
- for L = 10 μm, a transport depth of even 2 μm into the silicon is not the optimum value;
- for depths less than the optimum value, the transfer time increases as the inverse of the depth;
- for depths greater than the optimal values, the transfer time grows exponentially by $\pi.x'/L$.

Expressed in general terms, the optimum charge-transport depth x'_{opt} for a three-phase or four-phase device is given respectively by x'_{opt} = 0.271L and x'_{opt} = 0.267L, and the minimum transfer time corresponding to the transport at this depth is approximated with the relation (Bakker 91) :

$$tt_{opt} = \frac{3.L^2}{\mu_n.\Delta V} \, . \tag{1.50}$$

To complete the discussion about fringing fields and optimum transport depth and transfer times, it should be noted that the value of the transport time given by [1.50]

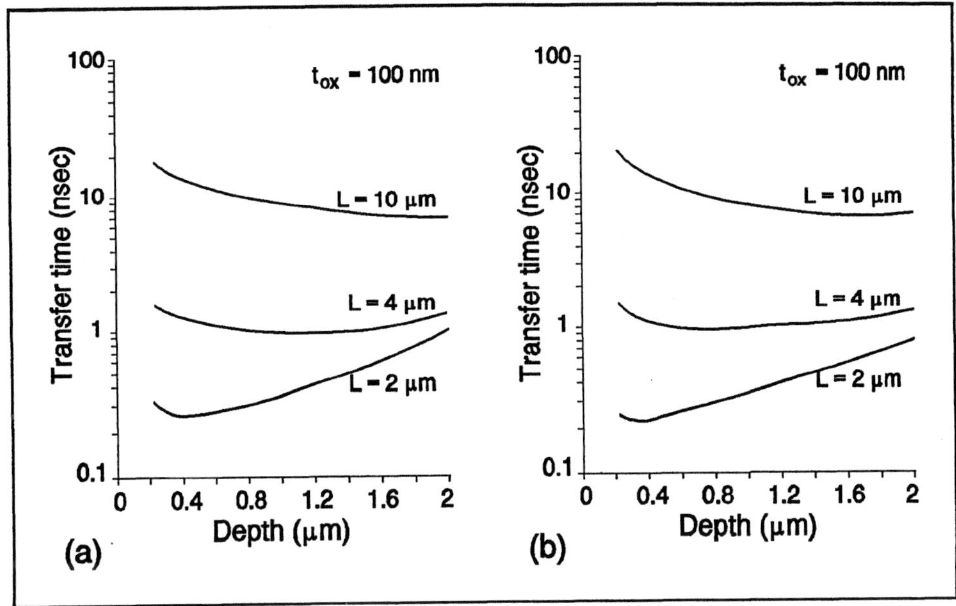

FIGURE 1.18. Transfer time as a function of depth for a three-phase (a) and a four-phase (b) CCD. The gate length serves as a parameter.

is an indication of the best-case situation. Especially for the situation in which large amounts of charge are to be transported in the CCD, an important part of ΔV will be devoted to the storage of the charge and ΔV is not completely available for the charge transfer. Consequently, the voltage swing available to generate the fringing fields will be reduced by a proportional amount and the transfer time will be increased.

1.5.3. CHARGE-HANDLING CAPABILITY

Compared to surface-channel charge-coupled devices, the buried-channel CCDs make optimal use of the fringing field effect by transferring the charge packet through the bulk. A second important advantage of this mode of operation is the separation between the charges to be transported and the surface states. In other words, the transport of a BCCD is faster and more complete compared to the SCCD. But precisely due to the fact that buried-channel charge-coupled devices store their charge in the bulk of the silicon, their charge-handling capability is smaller. In some applications where the cell size of the CCD is fixed and/or the clock swing available on the gates is limited, this may be a severe limitation of the BCCD. Figure 1.19 is included to explain this effect. It shows cross sections of CCD cells from an SCCD (Figure 1.19a) and BCCD (Figure 1.19b) register (Beynon 80). (The thick lines in the figures represent the center of gravity of the stored charges.)

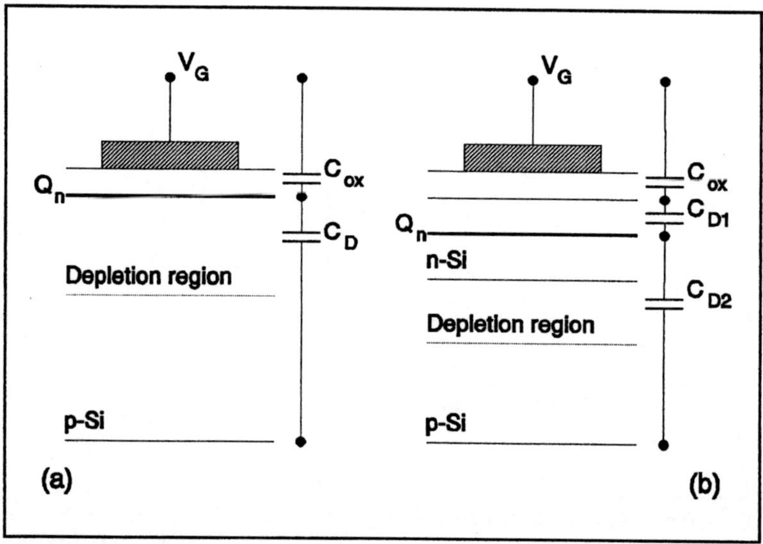

FIGURE 1.19. Indication of the charge storage site in a CCD cell from a surface-channel CCD (a) and a buried-channel CCD (b) register.

In the case of a surface-channel CCD the signal charge resides at the node between the oxide capacitance C_{ox} and the depletion capacitance C_D, exactly at the SiO_2-Si interface. The charge-handling capability is given by [1.27], or the amount of charge equals the product of the effective gate voltage across the oxide capacitance :

$$Q_{n,sat} = C_{ox} \cdot (V_G - V_T) = \frac{\epsilon_{ox}}{t_{ox}} \cdot (V_G - V_T) \, . \qquad [1.51]$$

Whereas the capacitance above the charge packet in the SCCD is only C_{ox}, in a BCCD this capacitance above the charge packet is a series combination of the same oxide capacitance C_{ox} and the depletion capacitance C_{D1}, as shown in Figure 1.19b. In analogy to [1.51], the charge-handling capability for the BCCD is given by :

$$Q_{n,sat} = \left(\frac{C_{ox} \cdot C_{D1}}{C_{ox} + C_{D1}} \right) \cdot (V_G - V_T) \, . \qquad [1.52]$$

If the exact value for C_{D1} is known, $Q_{n,sat}$ can easily be calculated. In reality, however, the depletion capacitance C_{D1} depends very much on the doping profile used to define the buried channel and consequently on the location of the charge packet in the bulk of the silicon. For a first estimation the doping profile of the buried-channel implant can be considered to be constant and the depth at which the charge packet is located can be approximated by $x_n/2$. A more realistic approach to the doping profile is a Gaussian dope profile. In this situation the charge packet comes slightly closer to the interface than in the situation of an uniformly doped buried channel. The depth in the silicon at which the charges are transported is about equal to $x_n/3$ (Esser 81). In this situation the depletion capacitance can be written as :

$$C_{D1} = \frac{3 \cdot \varepsilon_{Si}}{x_n} \cdot$$
[1.53]

Combining [1.52] and [1.53], the charge-handling capability of a more or less standard buried-channel device can be written as :

$$Q_{n,sat} = \frac{3 \cdot \varepsilon_{ox} \cdot \varepsilon_{Si}}{\varepsilon_{ox} \cdot x_n + 3 \cdot \varepsilon_{si} \cdot t_{ox}} \cdot (V_G - V_T) \cdot$$
[1.54]

The ratio in charge-handling capability between a surface-channel CCD and a buried-channel CCD (with the same device parameters) can be obtained by dividing [1.51] by [1.54], which yields :

$$1 + \frac{\varepsilon_{ox} \cdot x_n}{3 \cdot \varepsilon_{Si} \cdot t_{ox}} ,$$
[1.55]

and with $t_{ox} = 0.1$ μm and $x_n = 2$ μm the ratio is about 3.
(A more accurate analysis can be found in the literature (Esser 93)).

As already mentioned, the smaller charge-handling capability of the buried-channel devices can mean a limitation on its application field. To circumvent this drawback, the doping profile of the buried channel can be optimized toward charge handling. Where the charge packet comes closer to the interface, the handling capability is increased by lowering C_{D1} in relation [1.52]. To obtain a buried-channel device of this kind, even a double top-layer implantation is possible. This technique is applied in P^2CCDs (Esser 73) :
  - a relatively deep, but lower-doped implant defines the buried channel in which the charges are transported at a relatively large depth in the bulk but with a low handling capability;
  - a very shallow, but higher-doped implant which acts as a barrier toward the interface, shields the charge packets completely from the interface, and takes care of the larger charge-handling capacity.
Although the P^2CCD has a higher charge-handling capability, combination of the two implantations does not make the production or fabrication technology any easier.

To summarize the characteristics of the buried-channel CCDs :
  - the buried channel has a higher transfer efficiency owing to the absence of the surface-state interaction;

- higher operation frequencies are possible thanks to the higher fringing fields,
- buried-channel CCDs have intrinsically a lower noise level : interaction with the interface states is absent and the number of bulk states is much lower (see 3.4.2.);
- the technology used to produce the devices is slightly more complex than that for surface-channel CCDs;
- the charge-handling capability of the buried channels is reduced in relation to the surface-channel charge-coupled devices.

Notwithstanding the disadvantages, all charge-coupled devices used in today's image-sensing applications are of the buried-channel type.

WORTH MEMORIZING

Buried-channel charge-coupled devices have been developed to circumvent the charge-transfer shortcomings of surface-channel CCDs. The charge transport of the charge carriers takes place in the bulk of the silicon. In this way, the device profits from :

- the maximum value of the fringing fields to speed up the charge transport to a maximum;

- the absence of any interaction of the charge carriers with the interface states, thus minimizing the transfer inefficiency.

On the other hand, the price which has to be paid for these advantages is the charge-handling capability. Compared to surface-channel charge-coupled devices, the buried-channel alternative can handle only about one-third of a charge packet. P^2CCDs combine the advantages of both the SCCD and the BCCD.

1.6. One-dimensional potential analysis

During the further study of the charge-coupled devices in general, but also in their application as solid-state imagers, the description of the working principle of one or another effect will often be based on the potential analysis of the device. In this section a one-dimensional potential analysis will be given which is valid for any type of CCD, whether it is a surface-channel or a buried-channel device. For this simple study, the device will be considered as being not limited in its dimension along the charge-transport region, and will be fully depleted in the direction under study : the x direction perpendicular to the charge-transport channel. This situation is shown schematically in one of the top diagrams comprising Figure 1.15 (in the analysis given in Figure 1.15 the gate is limited and not extended to infinity, in this study the opposite is being considered).

The potential $\Phi(x)$ for any value of x can be calculated for an arbitrary dope level throughout the electric field E(x) and using the Poisson equation :

$$\frac{dE(x)}{dx} = \frac{Q(x)}{\varepsilon_{Si}} .$$ [1.56]

Because the device is considered as being fully depleted, the charge $Q(x)$ is completely determined by the fixed bulk charge introduced through the doping profile with concentration $N(x)$:

$$Q(x) = q.N(x)$$ [1.57]

and

$$\frac{d^2\Phi(x)}{dx^2} = -\frac{dE(x)}{dx} = -\frac{q.N(x)}{\varepsilon_{Si}} .$$ [1.58]

The above expression [1.58] gives the relation between the electrostatic potential $\Phi(x)$ and the doping profile $N(x)$ perpendicular to the transport channel in the silicon. Note that any doping profile can be considered using the parameter $N(x)$: a constant profile for a surface-channel CCD, a buried-channel CCD with an implanted top layer, and even a P^2CCD with different implantations on top of each other.

In the gate dielectric the doping level is of course equal to zero : $N_{(x<0)} = 0$, consequently E_{ox} is constant and $\Phi_{(x<0)}$ is a linear function of depth. This is shown in Figures 1.15a and 1.15c. For the electric field across the oxide E_{ox} , the following value is found :

$$E_{ox} = \frac{V_G - \Phi_{(x=0)}}{t_{ox}} = \frac{V_G - \Phi_S}{t_{ox}}$$ [1.59]

with V_G the applied gate voltage and Φ_S the surface potential ($x=0$).

In the silicon bulk and at a depth equal to the depletion depth x_d, the following two boundary conditions are valid :

$$E_{(x=x_d)} = 0$$ [1.60]

and :

$$\varepsilon_{Si}.E_{(x=0)} = \varepsilon_{ox}.E_{ox} ,$$ [1.61]

indicating that the electric displacement in the oxide layer is equal to that in the silicon at $x=0$. Integrating [1.58] with the above boundary conditions yields (Esser 81) :

$$\Phi(x) = V_G + \frac{q.t_{ox}}{\epsilon_{ox}} \int_0^{x_d} N(x).dx + \frac{q.\gamma}{\epsilon_{Si}} \int_0^x N(x).dx + \frac{q.x}{\epsilon_{Si}} \int_x^{x_d} N(x).dx \quad . \qquad [1.62]$$

In this expression the flat-band voltage V_{FB} is set to zero. If that is not the case, the value of V_G has to be corrected for it.

The newly introduced parameter γ represents the center of gravity at which the fixed bulk charge between 0 and x might be assumed to be concentrated. Its value is given by :

$$\gamma = \frac{\int_0^x x.N(x).dx}{\int_0^x N(x).dx} \quad . \qquad [1.63]$$

From [1.62] it can be seen that the potential $\Phi(x)$ at any given value of x is equal to :

- the externally supplied gate potential V_G, plus;
- the potential drop across the oxide as a result of the fixed bulk charge in the depleted area between $x=0$ and $x=x_d$ (second term on the right-hand side of [1.62]), plus;
- the potential drop caused by the fixed bulk charge between 0 and x, and located at its center of gravity with a capacitance ϵ_{Si}/γ to the surface, plus;
- the potential drop caused by the fixed bulk charge between x and x_d, and located at x with a capacitance ϵ_{Si}/x to the surface.

The sum of the first two terms on the left-hand side of equation [1.62] gives the expression for the surface potential Φ_S.

Relation [1.62] can be used for all types of charge-coupled devices if there is no mobile charge in the area $0 < x < x_d$. If there is a mobile charge at the surface, then a term has to added to equation [1.62], namely :

$$- \frac{t_{ox}}{\epsilon_{ox}} . Q_n \quad , \qquad [1.64]$$

representing the charge Q_n at the oxide capacitance C_{ox}.

It will be clear that $N(x)$ has a positive value in the donor situation (n-type silicon) and a negative value in the acceptor situation (p-type silicon).

In real situations the MOS gate will, of course, not extend to infinity and in an actual charge-coupled device the potential at a considerable distance from the gate will be directly influenced by the potentials of the neighboring gates as well. Thus in reality a two-dimensional analysis may be necessary for a proper understanding of some CCD effects. Moreover, in a situation where the gates also become very small in width (the direction perpendicular to the charge-transport direction), even a two-dimensional analysis might not be suitable and a three-dimensional study will be inevitable. For these complicated studies, no analytical expression can be written down and numerical computer simulations have to lend a hand.

WORTH MEMORIZING

Simple analytical expressions for a one-dimensional potential analysis can be derived. For more realistic structures two-dimensional and even three-dimensional studies are required in order to fully understand and design charge-coupled devices. In these situations potential analyses are done by means of numerical computer simulations.

1.7. Conclusions

The basic concept of a charge-coupled device, which is a simple series connection of MOS capacitors, is described. The individual capacitors are physically located very close to each other. To operate the CCD, charge packets are transported from one capacitor to its neighbor by means of digital pulses on the top plates of the MOS structure. This transport of isolated charge packets can be accomplished almost perfectly, without any noticeable deterioration of the charge content.

Charge storage is studied starting from the fundamental physics of a single MOS capacitor. Expansion from this single capacitor to a group of four is used to introduce the transport mechanism. Various charge-transfer mechanisms are investigated, all of which apply to real CCDs : thermal diffusion, drift by self-induced electric fields, and drift by fringing fields.

Special attention is paid to incomplete charge transfer. To circumvent this problem an almost ideal charge-transfer device is introduced : the buried-channel CCD. The most interesting features of the buried-channel charge-coupled device are reported : its fringing fields, its transfer time, and its charge-handling capability. These aspects are not all in favor of the buried-channel CCD as compared to the surface-channel CCD.

This chapter ends with the derivation of an analytical expression for the electrostatic-potential analysis for all types of CCD.

Top : a linear image sensor with 6000 pixels on a single row. The imager can be operated at a speed of 20 MHz (courtesy of Dalsa). Middle left : the application of solid-state image sensor in broadcast cameras for studio use (courtesy of BTS). Middle right : frame-transfer image sensor with 1000 pixels/line and 600 interlaced lines. Dynamic Pixel Management adds the option of a switchable aspect ratio to the imager (courtesy of Philips). Bottom left : top view of a frame-interline-transfer CCD with 2M pixels and dedicated to HDTV imaging. The total chip size is 284 mm² (courtesy of Sony). Bottom right : cross section of an interline-transfer image cell : in the middle of the photograph a double layer of poly-silicon covered with a tungsten light shield can be recognized. Two photodiodes are located on the left and right sides (courtesy of NEC).

Chapter 2

INTO, THROUGH AND OUT OF A CHARGE-COUPLED DEVICE

A charge-coupled device has the capability to transport charge packets by means of digital pulses on the CCD gates. The content of the charge packets represents an analog signal. The basics of charge-coupled devices, as far as charge transport from one gate to the next and charge storage underneath a single gate, are concerned, have been described in the previous chapter. In this chapter, the input of charge packets into a charge-coupled device, their transfer through the CCD by means of the digital clock signals on the gates, and the output of the charge packets are studied. The various sections of the charge-coupled device are illustrated

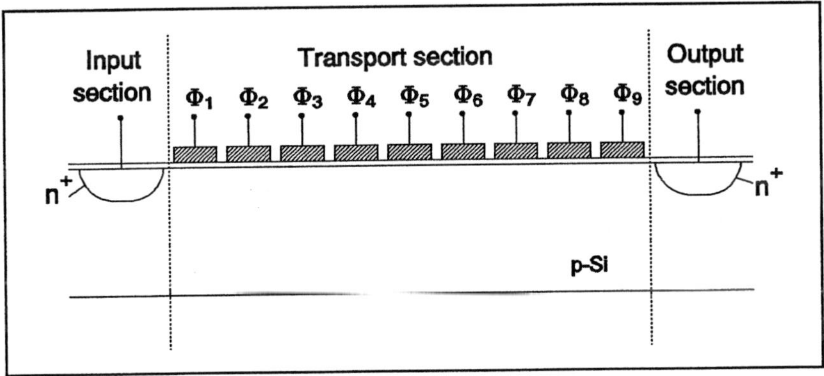

FIGURE 2.1. Cross section of a CCD, with its input section, transport section and output section.

in Figure 2.1 : the input section, represented by an n^+-p diode, the transport section, which is mainly a connection of closely spaced MOS capacitors, and the output section, represented again by an n^+-p diode. Input and output of charges is realized by means of a converting stage : at the input an external signal, for instance an electric voltage, is converted into a charge packet, while at the output the charge packet is converted back to a measurable quantity, for instance an electric current or a voltage.

Although not in a logical sequence, several transporting mechanisms are compared with each other in a first section of this chapter. In reality the transport section of the charge-coupled device is the biggest part as to silicon area. The charge packets are forced through the CCD by the transport system, but they also have to be kept sideways (perpendicular to the transport direction) in the CCD channel. How this is done forms part of the second section. After the transport operation has been explained, the input structure of the CCD is reviewed. In this application, the feeding in of charge packets representing an externally applied voltage is considered. The electrical input structures are important only in applications involving CCD delay lines and CCD memories. After the input structures, the output stages are examined. Conversion of the charge packets into a measurable quantity is important for all CCD applications.

2.1. Transport systems

Various transport systems are described proceeding from a classical four-phase system, via a three-phase, a two-phase, and a one-and-a-half-phase to a single-phase clocking scheme. All of these are applicable to solid-state image sensors. Most popular, as far as imaging is concerned, are the four-phase and the two-phase systems. The former is mostly found in the two-dimensional part of the imager, and the latter is widely used in one-dimensional output registers which have to be driven at high speeds.

To cover all the different transport schemes in this study, a ripple-clock mechanism is also included. Although not so often used, ripple clocking has some typical advantages over the others.

2.1.1. FOUR-PHASE SYSTEM

In a four-phase-transport system, the gates of the charge-coupled device are connected to four clock-pulse generators : Φ_1, Φ_2, Φ_3 and Φ_4 as shown in Figure 2.2. In the starting situation, three of the four signals, Φ_2, Φ_3 and Φ_4, are biased to a high voltage (e.g. 10 V), while the fourth, Φ_1, is kept to a low voltage (e.g. 0 V). This sequence is illustrated at time t_1 in Figure 2.2. By time t_2, clock Φ_2 has changed state, namely from 10 V to 0 V. This process makes the potential wells smaller (from three gates to two gates) and the barriers between two potential wells broader (from a single gate to two gates). If the number of information charges is not too high, they are kept inside the potential wells although the wells have decreased in volume.
In a next step at t_3, the contrary process takes place : phase Φ_1 goes high and the potential wells extend again to a width of three gates. The charges redistribute again across the total width of the potential wells. Comparing the situations at time points t_1 and t_3, shows how the potential wells have moved over a distance equal to the length of one CCD gate.
Reduction of the buckets is repeated at t_4, while broadening them again is repeated at t_5. The combination of these two actions is repeated over and over again and charge packets are transferred along the Si-SiO$_2$ interface (SCCD) or through the

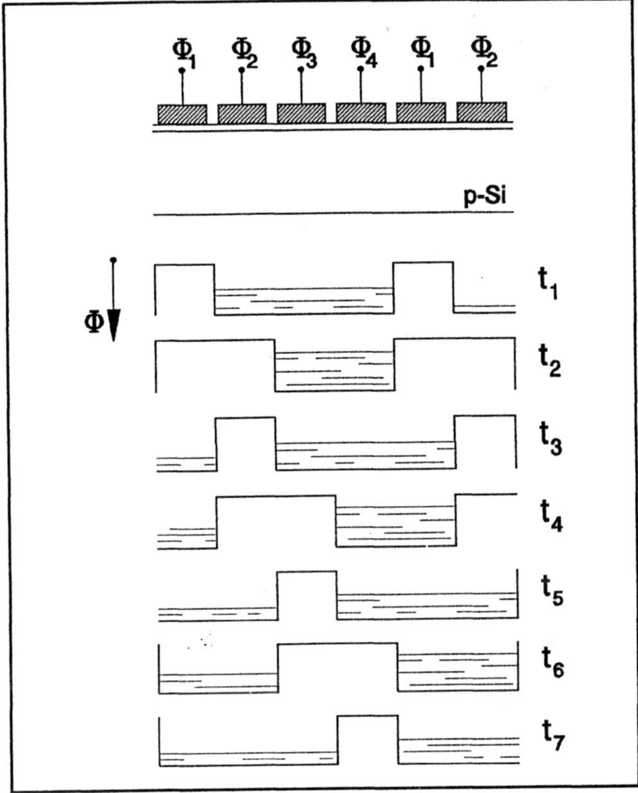

FIGURE 2.2. Cross section of a CCD transport section driven by a four-phase-clocking system.

bulk (BCCD) of the charge-coupled device.

The timing diagram of the four-phase clock-pulse generator is shown in Figure 2.3. The indicated time instants correspond to the different situations illustrated in Figure 2.2. Note that the clocks run in a sequence with a duty cycle of 62.5 % (= 5/8). With correct mutual timing and overlap, the potential wells in the CCD always have a width equal to at most three times the CCD-gate length and at least twice the CCD-gate length. Bearing in mind that a CCD cell is defined by four CCD gates, it is easy to define the worst-case situation for the charge-handling capability. The charge storage is limited to a potential well with a width equal to 50 % of the CCD cell.

An alternative timing diagram with antiparallel pulses is shown in Figure 2.4. All clocks run with a duty cycle equal to 50 %. The generation of the various pulses is simplified in comparison to the previous clocking scheme : the pulses of Φ_3 and

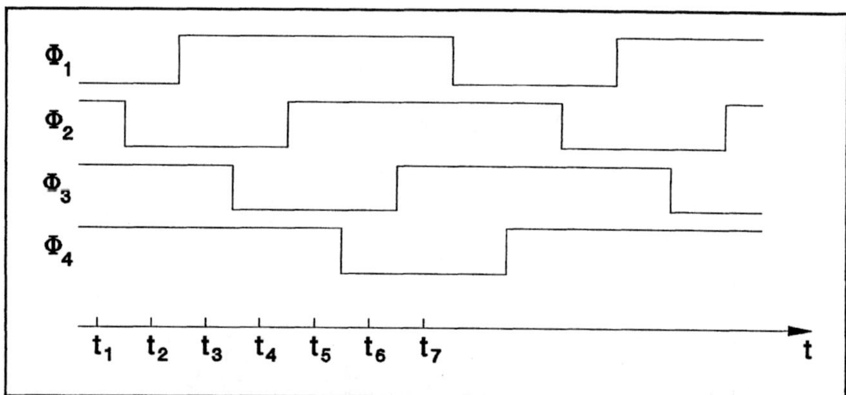

FIGURE 2.3. Timing diagram of the transport system shown in Figure 2.2.

Φ_4 are the inverses of Φ_1 and Φ_2, respectively. This straightforward generator design is the main advantage of this system, but if the delays of the different clocks are not equal, the situation can occur that one of the clocks is earlier negative than its counterpart is positive. This worst-case situation greatly reduces the charge-handling capability : instead of 50 %, only 25 % of the CCD cell can be used to store the information.

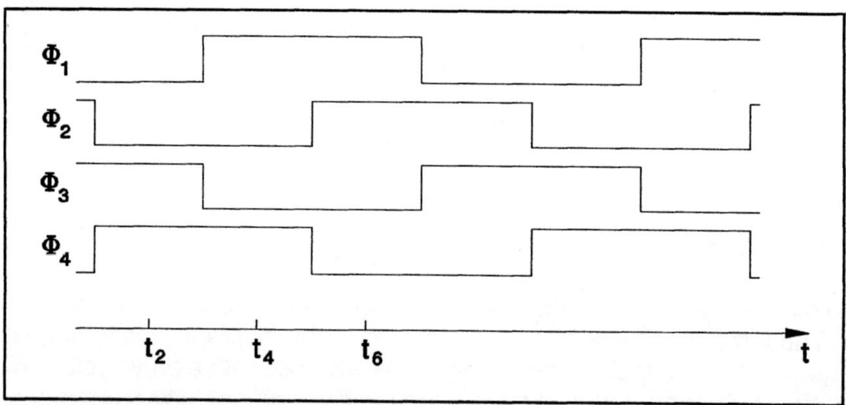

FIGURE 2.4. A four-phase clock system with antiparallel pulses.

Precautions can be taken to avoid this drawback : the clock driver can be designed to slow down the fall time of the clock a little, and in such a way that a charge-handling capability of 50 % is guaranteed. For the sake of completeness, the corresponding time points already shown in Figure 2.2 are indicated again in Figure 2.4.

2.1.2. THREE-PHASE SYSTEM

A three-phase system is very similar to a four-phase system. Its shaping and reshaping of the CCD potential wells are illustrated in Figure 2.5. The redundancy

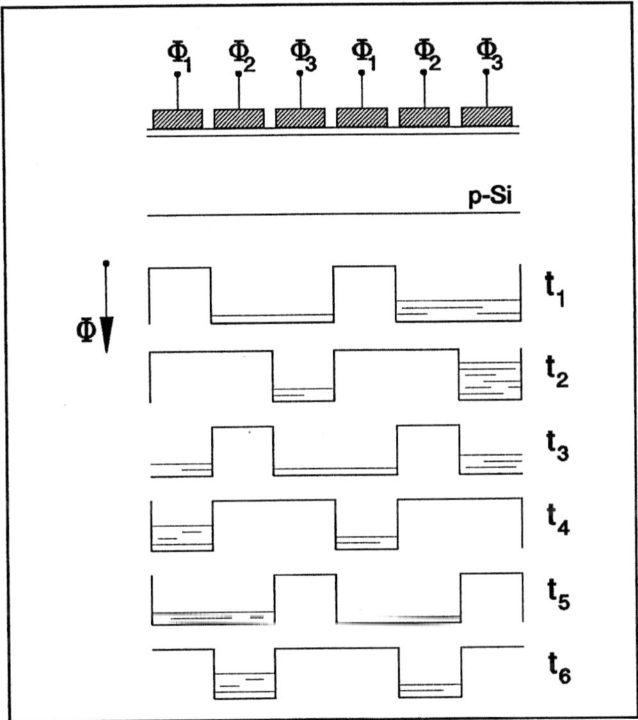

FIGURE 2.5. Cross section of a CCD transport section driven by a three-phase-clocking system.

which apparently existed in a four-phase system (at least 2 gates out of 4 defined the potential well) is removed : one gate takes care of the charge storage (phase Φ_3 at time point t_2), one gate is used for the well separation (phase Φ_1 at time point t_2) and the third forces the charge transport through the CCD (phase Φ_2 at time point t_2). In this simple construction it seems that the minimum number of gates required to implement CCD transport is three.
The timing diagrams of the three clocks are shown in Figure 2.6. All pulses have a duty cycle of 50 % and their mutual phase shift is 120°.

In the three-phase CCD the CCD cell is limited to just three gates but, as Figure 2.5 clearly shows, the charge-handling capability is also restricted. The worst case is when only one gate is available to store the information charge, and the charge-

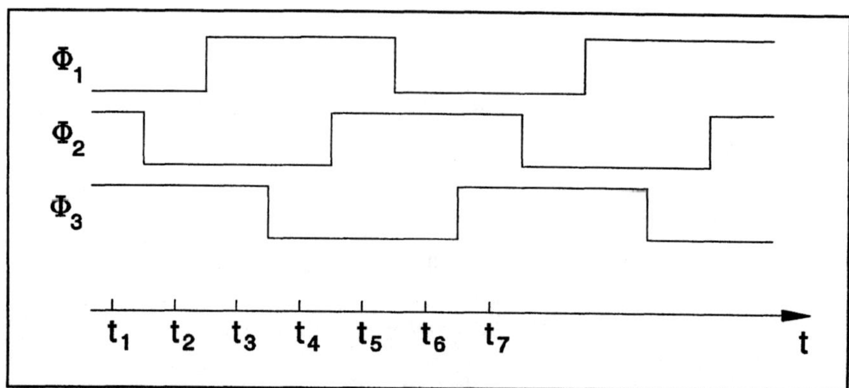

FIGURE 2.6. Timing diagram of the transport system shown in Figure 2.5.

storage area is only 33 % of a CCD cell. This rather small charge-storage capability can be a drawback if different transfer systems are compared to each other but, as will be described later, other clocking schemes can be even worse. An advantage of the three-phase system over the four-phase CCD is the fact that smaller CCD cells can be used for equal length of the CCD gates. So if compactness is of prime importance, a three-phase CCD is favorable.

2.1.3. TWO-PHASE SYSTEM

It might be obvious that three gates per CCD cell are really a minimum to effect the CCD transport. Nevertheless, transport systems with even two clocks are possible, but in such cases part of the loss due to the missing third (and fourth) clock can be compensated in the design of the CCD itself. For instance, Φ_2 may be used for the charge storage, and Φ_1 for the separation of the different charge packets while the direction of transport is built in by means of a preference direction achieved with an oxide thickness variation. This mode is illustrated in Figure 2.7 after a first charge transfer from t_1 toward t_2. As described in section 1.2, the thinner oxide thickness underneath the right-hand part of the gates creates a deeper potential well than to the left-hand part of the gates. Charges are stored preferably at the right-hand side of the CCD gates where the gate dielectric is thinnest.

When charges are stored only underneath the Φ_1 gates, with empty wells underneath the Φ_2 gates, a situation as shown at time point t_1 is created. Note that both clocks are at a low level. Nevertheless, the charge packets remain isolated. After pulsing Φ_2 to a high level, the potential well underneath this gate is pushed to a deeper level. If a continuously decreasing potential curve exists from the left side Φ_1 toward the right side Φ_2, the information charges are transported over a distance equal to the length of a single CCD gate. This condition is valid at time point t_2.

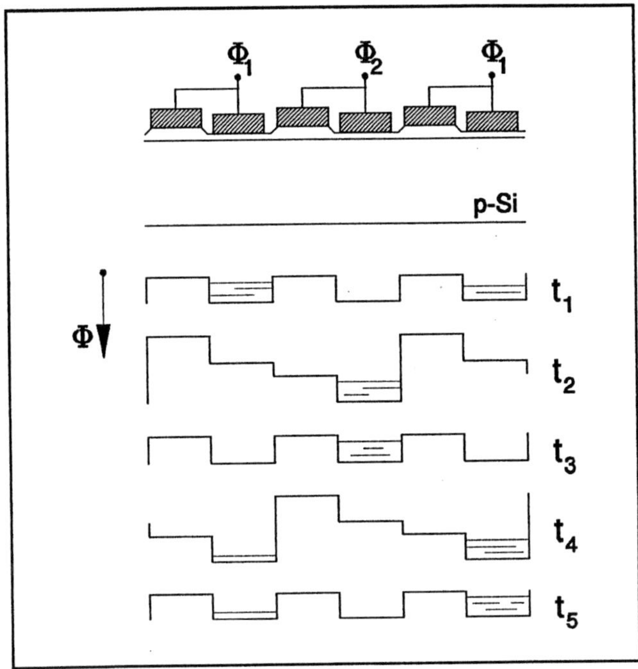

FIGURE 2.7. Cross section of a CCD transport with a two-phase clocking system. Its preference transport direction is defined by means of thickness variations of the gate dielectric.

At a later point both clocks are low again and the charge packets remain stored underneath Φ_2. After pulsing Φ_1 high, the charges can move on again to the next CCD gate.

Note that the starting situation in Figure 2.7 is made ideal by potential wells underneath Φ_2 being empty. If this is not the case, charge packets will be simply added up by the operation at t_2 and as early as time point t_3 the ideal condition of all wells underneath one phase being empty is created.

The timing diagram for the two-phase transport system described above is included in Figure 2.8. Clock generation is very simple with clocks which have a duty cycle of about 25 % and no critical overlaps. The only restriction is the non-overlapping condition.

With this two-phase construction, a CCD cell has a width of only two gates, and the charge-handling capability is limited to 25 %. In real CCD designs, however, the widths of the thicker oxide parts are made as small as possible and are mostly less than the widths of the thinner oxide parts. This layout detail increases the charge-handling capability above 25 %, but hardly ever higher than, say, 40 %.

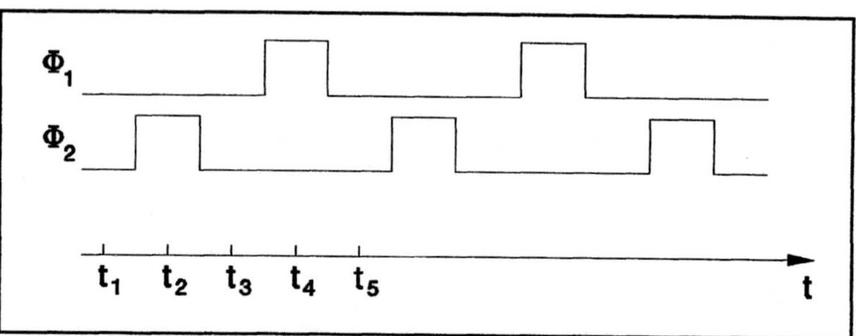

FIGURE 2.8. Timing diagram of the transport system shown in Figure 2.7.

An alternative two-phase CCD is illustrated in Figure 2.9. The preferential transport direction is no longer created by means of different oxide thicknesses, but by an extra n-type-doped region, located under parts of the CCD gates. These n-Si regions act in the opposite way to the thicker gate oxides : they make the potential well locally deeper. Charges are stored preferably at these deeper levels of the potential

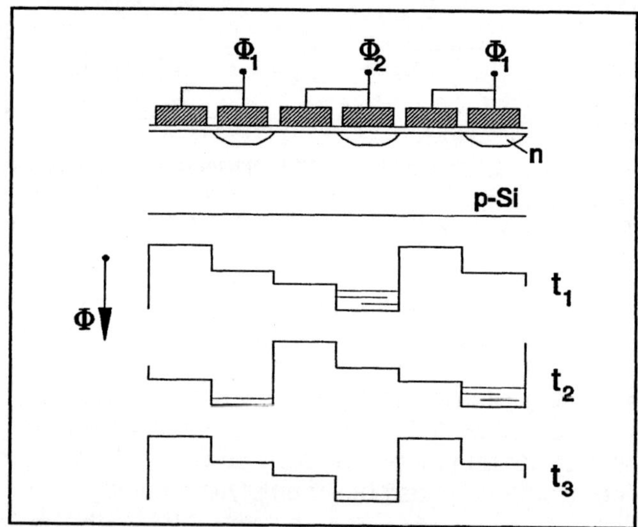

FIGURE 2.9. Cross section of a CCD transport section driven by a two-phase clocking system. Its preference direction of transport is defined by means of extra ion implantations.

wells. (Compare this to the buried-channel process.) The technology for adjusting the depth of the potential wells or shaping the potential wells under the CCD gates is simpler and more effectively reproducible if done by means of ion implantation than it is by oxide-thickness variation. Actually, this tuning of the potential wells is exactly the same technique as the one used to adjust threshold voltages of the MOS transistors in conventional MOS technology.

The timing diagram fitting into the clocking scheme shown in Figure 2.9 is drawn in Figure 2.10. The two phases are pulsed by two antiparallel clocks : Φ_2 is the waveform of Φ_1 inverted. Overlaps are not critical : the charge packets can only move from a negative-biased gate towards a positive-biased one. If both gates are at a low value or if both gates are connected to a high voltage, the charge packets stay at their location.

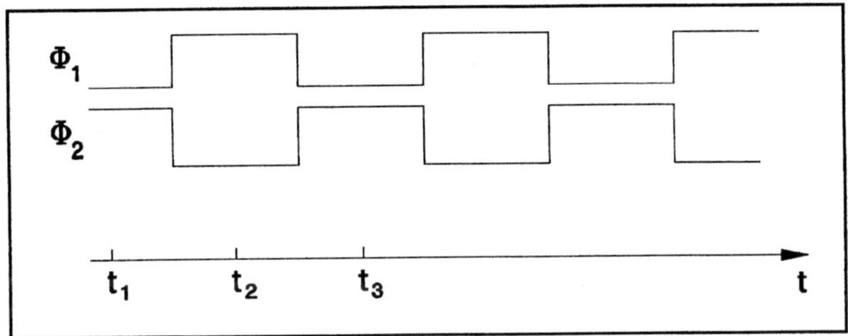

FIGURE 2.10. Timing diagram of the transport system shown in Figure 2.9.

As far as the two-phase transport systems are concerned, both structures described (gate-oxide thickness difference and preferred transport direction by the implant) and the two clocking schemes (true two-phase and antiparallel pulses) can be mixed. The antiparallel pulses are not restricted to the structure using potential barriers built in by ion implantation, but are applicable as well to the other structure with different gate-oxide thicknesses.

Compared to the four- and three-phase transport systems, the two-phase needs a more complicated CCD structure and has a smaller charge-handling capability. On the other hand, the timing of the pulses is not very exacting and the pulse generator can be made very simple. One restriction on the two-phase transport system but not on the other two is its unidirectional transport. In the figures 2.7 and 2.9, the charge packets can only travel from left to right. With a four-phase and a three-phase system it is quite easy to change the transfer direction of the charge packets. In the case of the four-phase clocking system (see figure 2.2), interchanging Φ_1 and Φ_3 or Φ_2 and Φ_4 reverses the direction. The same result can be obtained with the three-phase clocking scheme (see figure 2.5) by changing the mutual phase shift of 120° to -120°.

2.1.4. ONE-AND-A-HALF-PHASE SYSTEM

An extra extension to the two-phase system is the one-and-a-half-phase transport scheme. The CCD employing this transfer method is shown in Figure 2.11. Compared to the full two-phase system (with internal potential barriers created by means of an extra ion-implantation), the device concept is identical, the only difference being the driving method. The second clocking gate (Φ_2) is replaced by a DC signal (Φ_{DC}). This DC signal is biased somewhere between the positive and negative levels of the clocking-phase Φ_1 : its level should be higher than the low level of the Φ_1 to allow transport from Φ_1 towards Φ_{DC} (t_1), but it should also be lower than the high level of Φ_1 to allow transport from Φ_{DC} towards Φ_1 (t_2).

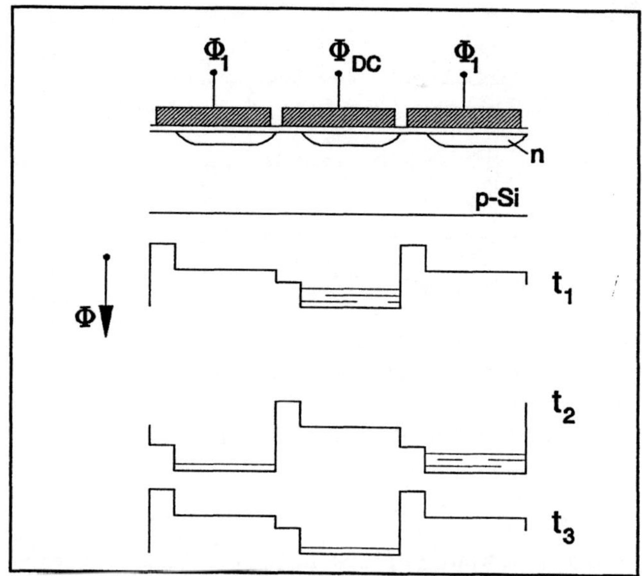

FIGURE 2.11. Cross section of a CCD transport section driven by a one-and-a-half-phase clocking system.

The corresponding timing diagram is presented in Figure 2.12. Although timing has become very simple, this one-and-a-half-phase system has a severe drawback compared to the full two-phase system : its lower charge-handling capability. The mechanism of biasing Φ_{DC} about halfway between the two clocking values of Φ_1 limits the charge storage to about 50 % compared to the two-phase clocking system (with the same clocking amplitude). To increase the charge-storage capacity the amplitude of Φ_1 has to be increased, but this move also raises the internal electric fields and the power dissipation.

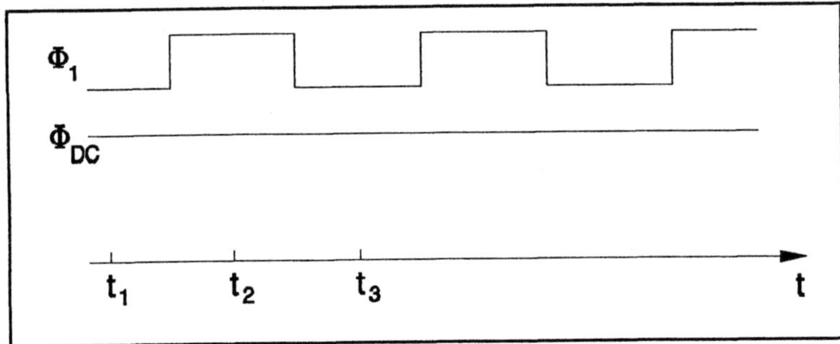

FIGURE 2.12. Timing diagram of the transport system shown in Figure 2.11.

2.1.5. VIRTUAL-PHASE SYSTEM

The virtual-phase transport system is no more than a one-phase clocking system. It looks very similar to the previous one-and-a-half phase but, instead of biasing the non-clocking phase with an external DC voltage, the non-clocking phase of the virtual-phase system is internally DC-biased. This is illustrated in Figure 2.13. The p-Si channel implant in the virtual phase is internally tied to the p-Si substrate and fixed to the substrate voltage. In this case, the external gate (Φ_{DC} of the one-and-a-half phase) can be left out (Hynecek 81). This makes the CCD gate configuration very attractive because, only a single layer of gate material needs to be used. There is no need for overlapping or closely-spaced gates and, even from the point of view of image sensors, covering the silicon surface only partially with gate material is very attractive as far as the photosensitivity of the device is concerned (see section 7.3.2).

Charges are stored underneath Φ_1 or in the virtual well. The gate potentials corresponding to the various time points indicated are shown in Figure 2.14 : only a single clock is left. The same drawback as already mentioned when discussing the one-and-a-half phase is valid for the virtual-phase system : limitation of the charge-storage capacity. Increasing the amplitude of the Φ_1 can solve the problem, but can also lead to a quite large electric field in the silicon substrate.

2.1.6. RIPPLE CLOCK

The ripple clocking method, as illustrated in Figure 2.15, is characterized by its very compact structure. Each packet is stored underneath a single gate and the separation between two packets is also effected by a single gate. The complete transporting system is built around (2.I + 1) gates, where I represents the number of charge packets in a unit cell : I gates are used for storage and I gates are used to define the barriers. One extra CCD gate is needed for transportation. The mechanism is clearly shown in Figure 2.15. The example has 7 clocks : $\Phi_1 \ldots \Phi_7$ and can store 3 packets in a unit cell. As indicated, at time t_1 two packets stay

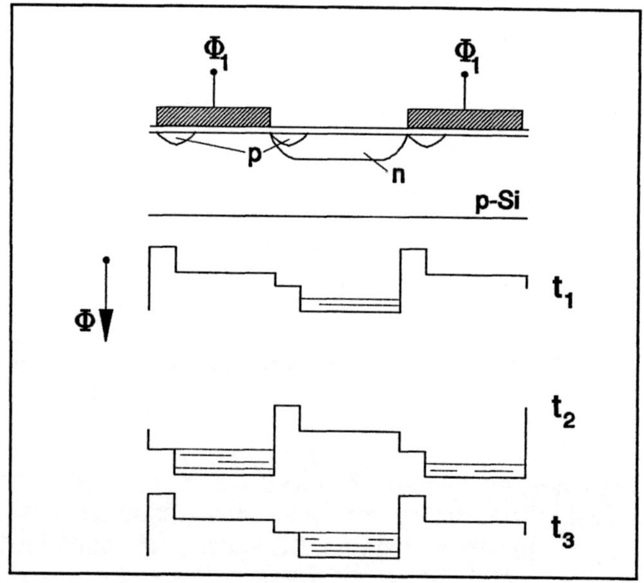

FIGURE 2.13. Cross section of a CCD-transport section driven by a virtual-phase clocking system.

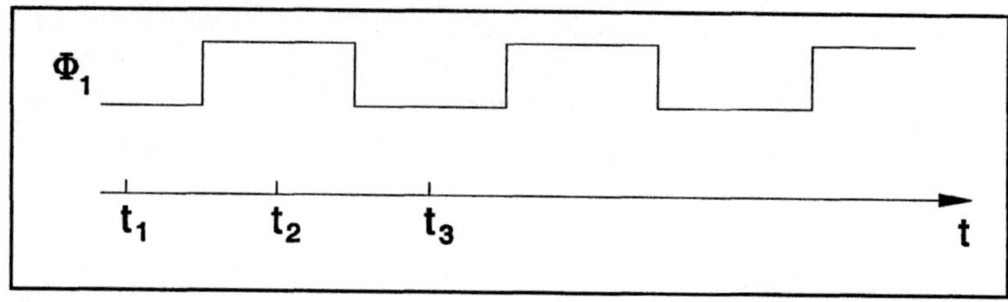

FIGURE 2.14. Timing diagram of the transport system shown in Figure 2.13.

at their location and the third is moving from Φ_7 toward Φ_1. At the next time point the transport is completed and the following one is prepared : the barrier underneath Φ_6 is extended toward Φ_7, time point t_3 is equivalent to time point t_1, etc. The complete timing diagram of this example is shown in Figure 2.16.

The biggest advantage of the ripple-clocking scheme is its compactness. The charge-storage capacity is slightly less than 50 % of the total silicon area used, and one CCD cell is about two gates.

But besides this advantage, the method has several disadvantages : the clocking itself is quite cumbersome and slow. Several clocks need to run through a complete cycle to transfer the charge packets over a distance of only one CCD gate. The number of clocks can also put a restriction on the package design (in which the

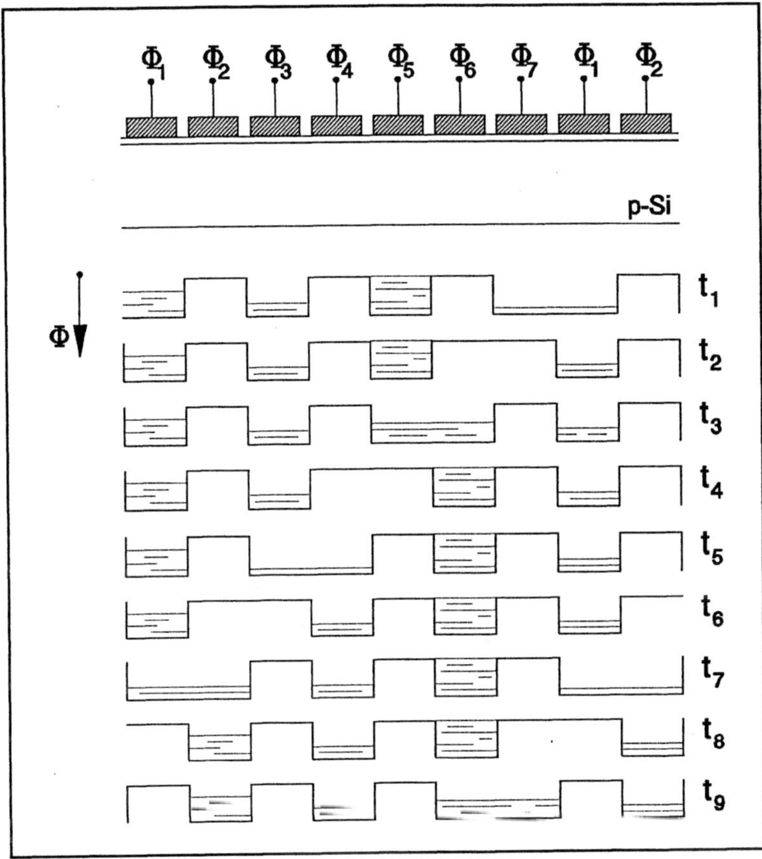

FIGURE 2.15. *Cross section of a CCD transport section driven by a ripple clocking method.*

CCD chip has to be mounted) because each clock needs a separate pin. The ripple-clocking method is used only in special applications.

WORTH MEMORIZING

Charge packets are transported through a charge-coupled device under the influence of digital pulses applied to the CCD gates. These CCD gates are arranged into groups or phases to minimize the number of drivers needed. According to the configuration, charge-coupled devices can be driven in a four-phase, a three-phase, a two-phase, a one-and-a-half-phase, a virtual-phase or a ripple-clock mode. Each of these systems has its pros and cons as far as compactness,

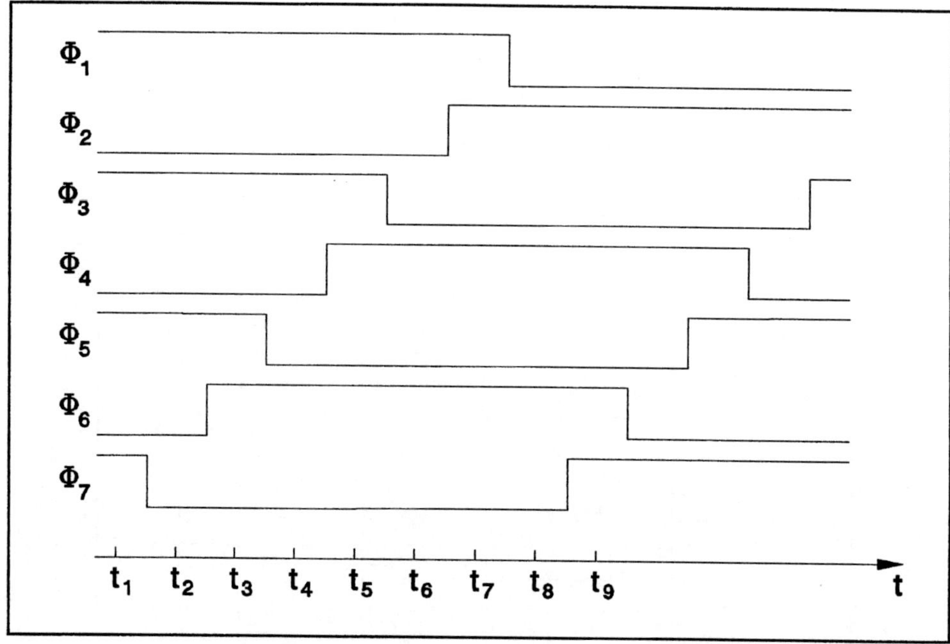

FIGURE 2.16. Timing diagram of the transport system shown in Figure 2.15.

charge-handling capability and clock waveform restrictions are concerned.

2.2. Channel definition

In the previous section the continuous movement of the minority-charge packets through the CCD channels has been studied. But if this movement is compared to driving a car on a road, a very important boundary condition for driving the car from A to B is to keep the car on the road. The same is true of a CCD. As illustrated by Figure 2.17, if the charge packets have to be transported from the input of the device to its output, it is not enough to choose an appropriate transporting mechanism; it is also necessary to keep the charge packets inside the CCD channels. This lateral definition of the CCD channels is described in this section.

The theory defining CCD channels is based on the concept already explained in chapter 1 by means of Figure 1.7, where the influence of the background doping, the thickness of the gate oxide and the voltage on neighboring gates is shown. Minority charges can be easily kept inside the CCD channel by manipulating the structure using the aforementioned parameters. This is shown in Figure 2.18, where

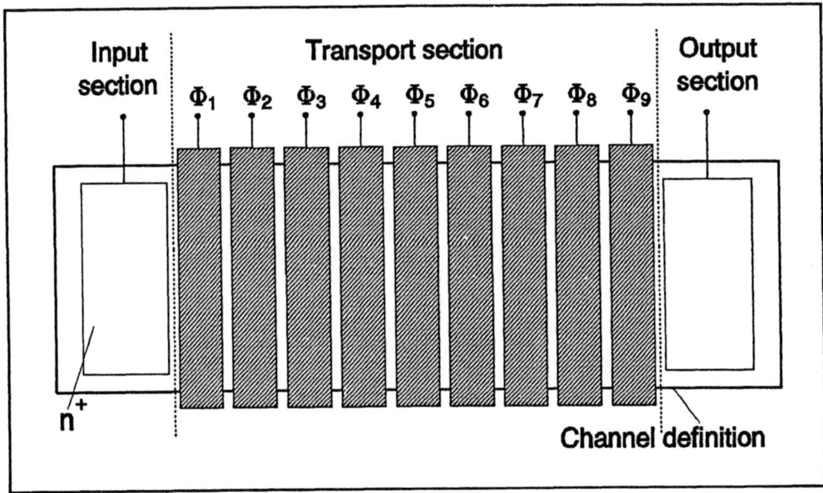

FIGURE 2.17. Top view of a CCD delay line (the cross section of which is shown in Figure 2.1) with its input section, transport section and output section.

three possible cross sections of the CCD channel are shown. The cross sections are taken perpendicular to the charge transport direction. The issue is how to keep the charge packets inside the channel, or how to isolate the left and right sides of the channel from the outside world.

In Figure 2.18a the channel is defined by means of channel-stopper implantations : on the p-type substrate p^+-regions are defined at both sides of the structure, and the CCD channel is enclosed between these regions. The highly doped p^+-regions locally increase the threshold voltage so that with the gate voltages applied a channel is created only between the two implantations. Similar techniques are also used in standard MOS technology to isolate different MOS transistors from each other. (An alternative to the implantation of barriers to separate the CCD channels exists : the opposite implantation to define deeper CCD channels in a uniformly doped substrate.)

Differences in the oxide thickness can also influence the threshold of the MOS structure. This effect (called LOCOS = LOCal Oxidation of Silicon) is illustrated in Figure 2.18b and is used to isolate the CCD channel from the outside world.

Figure 2.18c demonstrates the working principle of the field-shield isolation. Extra gates, DC-biased at a low voltage, are incorporated to separate the CCD channel from the external silicon area. Making use of field shields has a typical advantage : if the connections to the field shields are led to the outside, the user can influence the potential on the field shields. If two CCD channels are placed close to and isolated from each other by means of such a field shield, depending on the

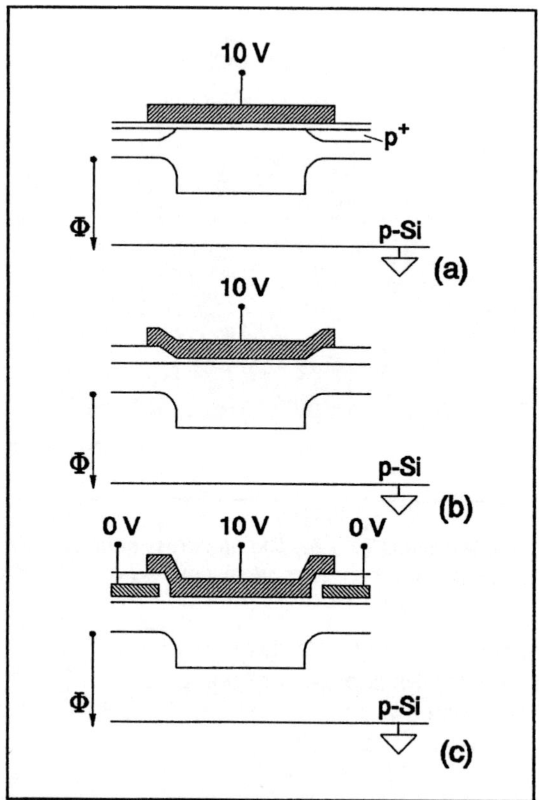

FIGURE 2.18. CCD channel definition by means of (a) channel-stopper implant, (b) oxide isolation, and (c) DC-biased field shields.

application, it might be attractive to remove the isolation between the two channels, enabling the two CCD channels to communicate with each other. This option can be applied when charge packets have to be added, for instance. (Summing charge packets in a CCD is called charge binning.)

Combinations of these techniques are possible also. Particularly the use of LOCOS and channel-stopper implantations is quite popular.

WORTH MEMORIZING

Channel definition of the charge-coupled devices can be done very effectively with :
- channel-stopper implantations;
- variations in gate-dielectric thickness;

- DC-biased field shields;
or even a combination of these techniques.

2.3. Input structures.

It might be surprising to find a description of (electrical-) input structures in an overall study of solid-state imaging with CCDs. In the design of charge-coupled image sensors, however, electrical-input structures can be a very viable tool for the test engineer to evaluate a certain number of characteristics of his devices only by an electrical signal forced into the CCD. For this reason an overview of electrical input structures can have a certain interest. In all other applications, e.g. delay lines, memories or filters, all signals are fed into the CCD by means of electrical input structures.

All CCD-input configurations make use of a diode to supply minority carriers for MOS capacitors working in deep depletion. This diode acts as a source for these minority carriers in the same way as the source of an MOS transistor does. The parts of the MOS transistor of interest in this analysis, namely the source and the gate of the device, are shown in Figure 2.19.

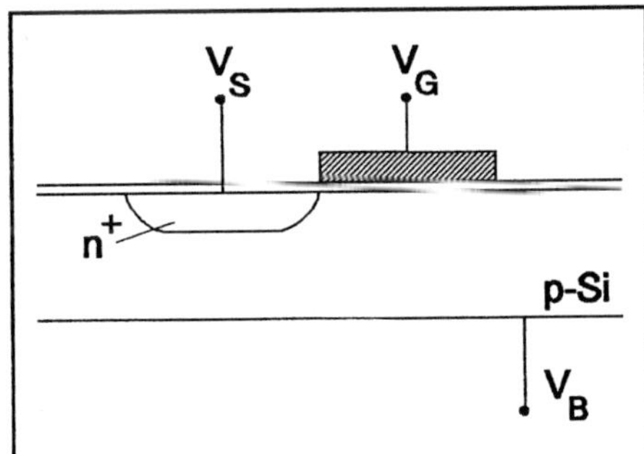

FIGURE 2.19. The basic input configuration of a CCD : the input diode (which acts as the source of electrons) and the input gate (which controls the amount of injected charges).

In this example, the source (at 5 V) will supply free electrons to the potential well underneath the gate (at 10 V), causing a charge packet of minority carriers to be built up. As learned from section 1.1.3, the voltage available to store these charges is equal to $|V_G - V_T|$.
The maximum amount of charge $Q_{n,sat}$ which will be supplied to and can be stored on the capacitor can be calculated with [1.27], in which the threshold voltage V_T for a biased silicon bulk is given by [Carr 72] :

$$V_T = V_{TO} + V_{SB} + K \cdot \sqrt{V_{SB} + |2.\Phi_F|} - \sqrt{|2.\Phi_F|} \qquad [2.1]$$

with :
- V_{TO} : the threshold for zero bias between the source and the bulk;
- V_{SB} : the bias between source and bulk of an MOS transistor;
- K   : the body factor of an MOS transistor given by (Veendrick 90) :

$$K = \frac{1}{C_{ox}} \cdot \sqrt{2.q.\epsilon_{Si}.N_A} \; . \qquad [2.2]$$

Inserting [2.1] into the relation [1.27] yields :

$$Q_{n,sat} = C_{ox} \cdot \left| V_G - V_{TO} - V_{SB} - K \cdot \sqrt{V_{SB} + |2.\Phi_F|} - \sqrt{|2.\Phi_F|} \right| \; . \qquad [2.3]$$

With the parameters chosen as : t_{ox} = 100 nm, N_A = $8*10^{14}$ /cm^3, K = 0.5 V$^{1/2}$, and V_{SB} = 5 V, the number of electrons stored underneath the gate can be calculated as $Q_{n,sat}/q$ and thus equal to $9.4*10^{11}$ electrons/cm^2. For a gate area of 10 μm x 10 μm, the number of electrons to be stored is $9.4*10^5$ electrons.

Of more importance in this discussion are the various parameters influencing the amount of stored charge $Q_{n,sat}$. From the technological point of view, the amount of charge depends on the oxide thickness, the gate dielectric material, the threshold voltage of the MOS structure, the body factor, and the Fermi potential. Once the CCD is fabricated these parameters are fixed. However, the user of the CCD can determine the number of charges under the gate using parameters such as the source potential V_S, the back bias V_B, and the gate potential V_G. In applications where an electrical signal has to be fed into the CCD, the input signal can be connected to one of these three terminals. In practical situations only V_S and V_G are of interest. The most common structures for the two alternatives are the "diode cut-off" technique for the input through V_S and the "fill and spill" method for the input through V_G.

2.3.1. DIODE CUT-OFF

With the diode cut-off technique the AC input signal is fed into the CCD by means of an input diode which is mutually "connected" to and "disconnected" from the transfer channel by a gated MOS capacitor. The basic construction is depicted in Figure 2.20, where, from left to right, the input diode (connected to the AC input signal V_{in}, the input gate (V_G), and three clocks of the CCD (Φ_1, Φ_2, Φ_3) are shown. In the lower part of the illustration the surface potential is drawn at different time points. At t_1 the empty well underneath Φ_1 is fully isolated from the input structure and from the rest of the CCD channel by means of the blocked input gate V_G and the next clocking gate Φ_2. At time t_2 the barrier underneath the input gate is removed and the diode, acting as a source of electrons, fills the entire potential well extending across V_G and Φ_1 with minority carriers. The level up to which the well is filled equals the diode potential or V_{in}. The signal V_{in} is the electrical signal which has to be passed through the CCD when the device is used as a delay line.

At t_3 the voltage at V_G is lowered again and the filled potential well under Φ_1 is again

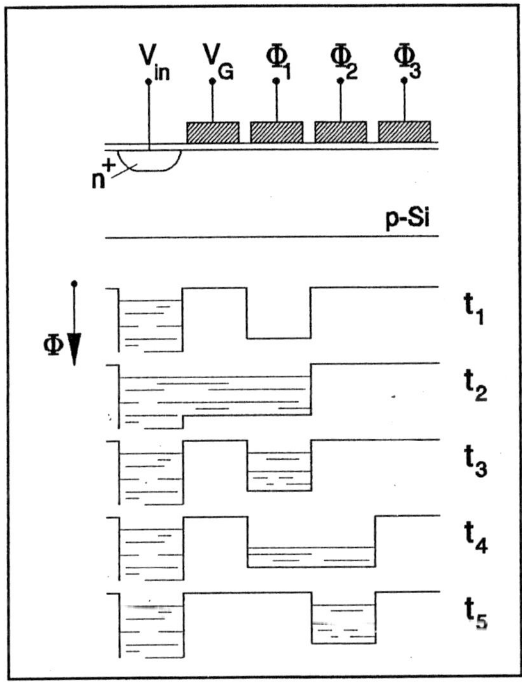

FIGURE 2.20. *The diode cut-off technique used to feed an electrical signal into a CCD.*

separated from the input diode and the input signal. The charge packet is fully isolated underneath the first clocking gate of the CCD. At further time steps t_4 and t_5 the classical three-phase transport takes place to shift the created charge packet into and through the CCD. After the next interval when the situation is the same as that shown for the time t_2 the next sample is taken from the input signal and fed into the CCD and so on.

The timing diagram of the different clocks and of the input signal are shown in Figure 2.21. The various time points indicated correspond to the situation depicted in Figure 2.20. Note the relatively short pulse on V_G compared to the length of the pulses on the transporting gates. The time during which the V_G pulse is high has to be shorter than the time for which the Φ_1 pulse is high and lies completely "inside" the Φ_1 pulse. Despite its simplicity, the diode cut-off technique is not frequently

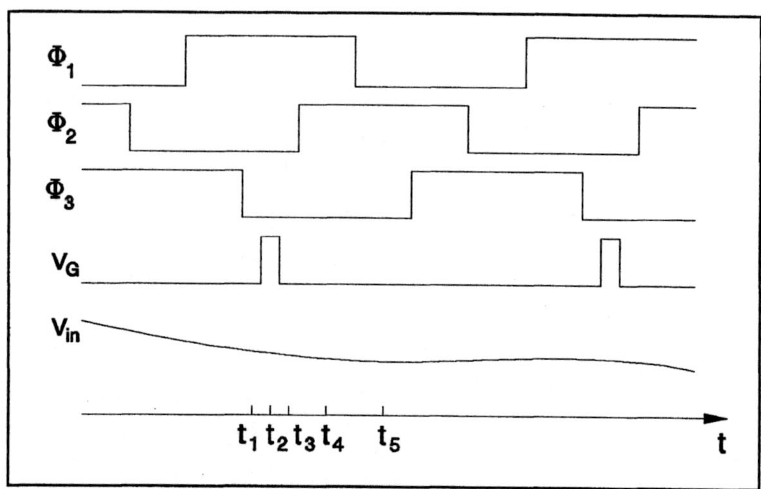

FIGURE 2.21. Timing diagram corresponding to the diode cut-off method illustrated in Figure 2.20.

used. Restrictions to this method are the noise and the linearity. Noise is specifically generated at the moment when V_G is pulsed low again after the filling of the potential wells (see t_3). The charge carriers under V_G when this gate is at a high voltage level have to be drained back to the diode at the moment the gate is pulsed low. This draining action is not ideal. Most of the electrons are indeed drained to the diode but a certain proportion is also pushed to the right side of V_G, to the potential well underneath Φ_1 and added to the charge packet representing the input signal. This noise phenomenon is called partitioning noise and is almost unavoidable with the diode cut-off technique.

A second limitation of the diode cut-off method is its limited linearity. If the diode voltage V_{SB} in relation [2.3] is replaced by the input voltage V_{in} :

$$Q_{n,sat} = C_{ox} \cdot \left| V_G - V_{T0} - V_{in} - K \cdot \sqrt{V_{in} + |2.\Phi_F|} - \sqrt{|2.\Phi_F|} \right| \qquad [2.4]$$

the nonlinearity between the amount of charge $Q_{n,sat}$ and the input signal V_{in} is directly demonstrated.

(Observe the notation of $Q_{n,sat}$ in [2.3] and [2.4]. This $Q_{n,sat}$ denotes the maximum charge packet underneath V_G but does not necessarily stand for a full charge packet in the CCD. This is only the case if V_G equals the high level of Φ_1, and if the area of the input gate and the clocking gate are also equal.)

2.3.2. FILL-AND-SPILL

The fill-and-spill technique makes use of two input gates and one input diode to get rid of the noise and non-linearity typically associated with the diode cut-off method. The basic working principle is illustrated by Figure 2.22, which shows from left to right, the input diode V_D, the DC-biased input gate (V_{DC}), the AC input gate (V_{in}), and a few clocking gates of the CCD (Φ_2, Φ_3) are shown. The starting situation at time point t_1 shows a relatively small, but empty well under the AC input gate. This well is isolated from the CCD channel by a blocking Φ_2 gate and from the input diode, the source of electrons, by the DC-biased input gate. At t_2 the diode is rendered active by lowering its voltage. The available potential well is completely filled with electrons (t_2), but quite rapidly all excess electrons are spilled back into the diode which is pulsed high again (t_3).

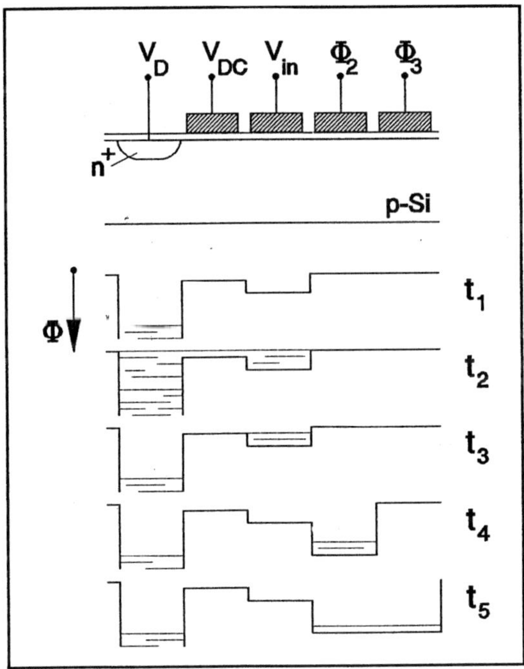

FIGURE 2.22: The fill-and-spill technique used to feed an electrical signal into a CCD

The potential well underneath V_{in} is completely filled with electrons and its charge content is given by :

$$Q_{n,sat} = C_{ox}.(V_{in}-V_{DC}) \ . \qquad [2.5]$$

Formula [2.5] demonstrates the full linear relationship between $Q_{n,sat}$ and the input signal V_{in}. At times t_4 and t_5 the classical three-phase CCD transport feeds the charge packets into the CCD channel and transports it further.

The timing diagram corresponding to Figure 2.22 is illustrated in Figure 2.23. (For the sake of completeness the clock Φ_1 is included in the timing diagram while it is omitted on Figure 2.22 : Φ_1 is the right-hand neighbor of Φ_3.) It looks quite similar to the one shown in Figure 2.21, with comparable boundary conditions for the pulse

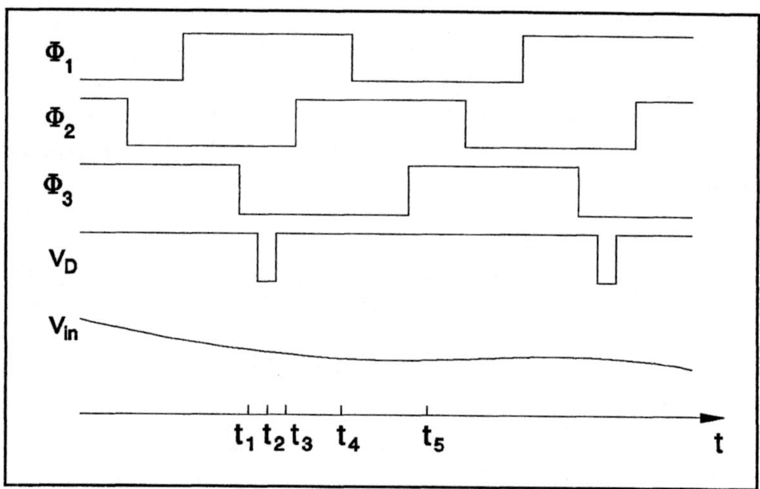

FIGURE 2.23. *Timing diagram corresponding to the fill-and-spill method illustrated in Figure 2.22.*

on V_D : the negative pulse on this gate should lie completely "inside" the low level of the pulse on Φ_2. The DC level at which V_{DC} is biased has to be chosen such that the difference between the low level of the CCD clocks (especially Φ_2) and the DC level of V_{DC} is small (e.g. 0.5 V). (This V_{DC} gate can also be clocked, but a DC bias helps to avoid clock-feedthrough noise added to the input signal).

A serious drawback of this method can be deduced from the situation sketched at time t_5 in Figure 2.22 : at the moment when Φ_2 is pulsed low to push the charge packet toward Φ_3, the charges have to be transported in the preferred direction and must not be shifted back again underneath V_{in}. To fulfil this condition, the surface-potential level of the partially filled well underneath Φ_2 and Φ_3 has to be higher in value (or lower on the drawing) than the potential level under V_{in}, otherwise charges can flow backwards. This boundary condition can be translated in a practical situation as a maximum V_{in} level somewhere halfway between the V_{DC} setting and the high level of the CCD clocks. In addition to this limitation, this boundary condition also restricts the charge-handling capability of the fill-and-spill method. On the other hand, however, any limitation of this kind can easily be circumvented by a local increase in the charge-handling capability of the input structure. In practice this can be done by :

- special shaping of the input structure : the input gates and the diode are designed twice as wide as the rest of the CCD channel. A restricted charge-handling capability in a relatively wide CCD input structure can be adapted to the nominal charge-handling capacity of the CCD channel itself;
- choosing a surface-channel CCD input structure in combination with a buried-channel transporting channel. The higher transport inefficiency in the surface channel can be neglected because the surface mode is restricted to only a few gates.

Although the diode cut-off method is simpler, the most-widely used input technique is the fill-and-spill method. The main reason for this preference is the superior linearity of the latter (Sankara 91).

WORTH MEMORIZING

The conversion of an electrical input signal into a charge packet to be transported in a CCD can be achieved by combining a simple diode and one or two extra MOS capacitors. The most elementary construction (diode cut-off) suffers from noise and non-linearity problems; the more sophisticated has a limited charge-handling capacity. The latter drawback can be got round by increasing the gate area of the input MOS capacitors.

2.4. Output structures

CCD input structures are only of interest for applications outside the solid-state imaging domain, but CCD output structures are of importance for every application. The charge carriers contained in a potential well have to be measured and a relevant output signal fed to the outside world. In almost all situations, the measuring consists in converting the charge content into a voltage or current. In the context of this study, only the conversion into a voltage will be described.

Two different output structures are studied : the floating-diffusion output with reset and the floating-gate output without reset. In both cases the converting medium (floating diffusion or floating gate) is buffered to the outside world by means of source followers.

2.4.1. FLOATING DIFFUSION WITH RESET

The most widely used output structure is the floating-diffusion output with reset. The charge packet to be sensed or converted is dumped on a capacitor defined by means of a floating diffusion. The construction of this output configuration is illustrated in Figure 2.24. The last clocking CCD gates (Φ_2, Φ_3), the DC-biased output gate, and the n^+-floating diffusion are shown from left to right. The last of these is connected to the positive supply voltage V_{DD} via a reset transistor acting as a switch and controlled by the reset pulse Φ_R. The voltage on the floating-diffusion output is sensed by a single-stage source follower. The output node V_{out} is fed to the outside world.

The operating principle of this output stage is explained by the potential diagrams included in Figure 2.24, starting at time t_1 with a charge packet stored underneath Φ_3 and a high value at Φ_R. In this way the floating diffusion is connected to V_{DD} and reset to its reference value. The voltage at the output node V_{out} is almost the same as the voltage on the floating-diffusion output itself. At time t_2 the reset switch closes again and the diffusion becomes definitely floating. Owing to some clock feedthrough from the reset clock (see below), the voltage on the n^+ region is slightly decreased. This effect also occurs at time t_2 and the small voltage difference is measured by the source follower. At time instant t_3 the clock Φ_3 is low and the charge packet is dumped across the DC-biases output gate and onto the floating diffusion. The extra electrons which are stored on the floating-diffusion capacitance will decrease the potential of this n^+ region still further. The same applies to the V_{out} voltage. After the sensing action described it is time to shift the next charge packet to the output node. This is done by the classical three-phase clocking system illustrated in Figure 2.24.

The corresponding timing diagrams of the various CCD clocks and the reset clock are featured in Figure 2.25.
Note the staircase of the V_{out} signal : resetting to the reference voltage, the clock feedthrough of the reset clock, and finally the representation of the CCD output signal. The way in which the AC signal is extracted from this time-discrete output signal is part of the subject matter of chapter 6.

The output swing at the floating-diffusion node and at the output pin, namely ΔV^*_{out} and ΔV_{out} respectively for a given charge content Q_n, can be easily calculated. The electrical representation of the floating-diffusion node is given in Figure 2.26 : the capacitance C_{FD} (which represents the capacitance of the floating-diffusion node) connected to the power supply V_{DD} via the reset transistor. The latter is driven by the reset pulse Φ_R. The corresponding definitions and diagrams of the individual voltage variations are illustrated in Figure 2.27. The values of the capacitance between the gate of the reset transistor and the floating diffusion are symbolized by C_{FD} and C_p, respectively. The variation ΔV^*_{out} as a function of Q_n is given by :

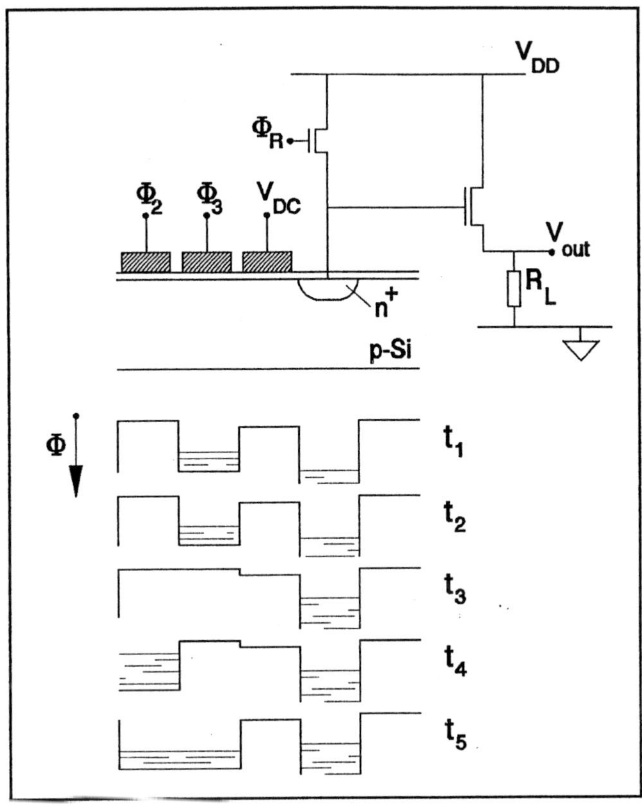

FIGURE 2.24. *The floating-diffusion output with reset used to measure the content of the charge packet transported through a CCD.*

$$\Delta V_{out}^{*} = \frac{Q_n}{C_{FD} + C_p} \, .$$

$$[2.6]$$

Taking into account the amplification of the source follower A_{SF}, the voltage changes at the output node can be easily deduced to be :

$$\Delta V_{out} = \frac{Q_n}{C_{FD} + C_p} . A_{SF} \, .$$

$$[2.7]$$

($A_{SF} < 1$ is valid for all source-follower configurations.)

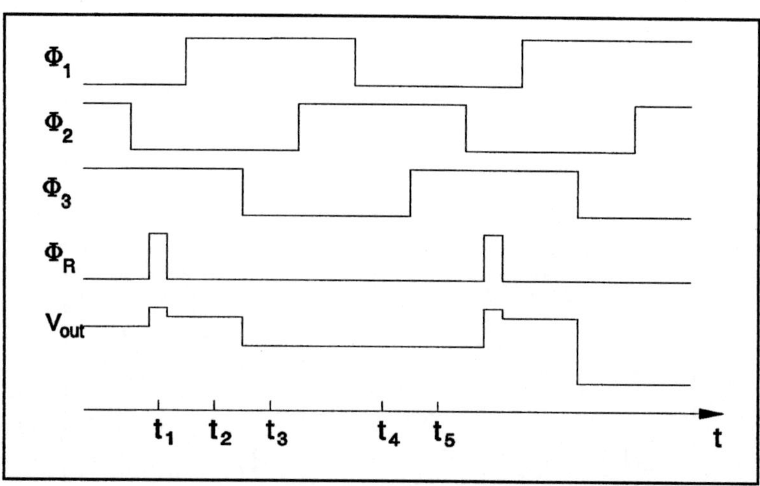

FIGURE 2.25. *Timing diagram corresponding to the floating diffusion read-out method illustrated in Figure 2.24.*

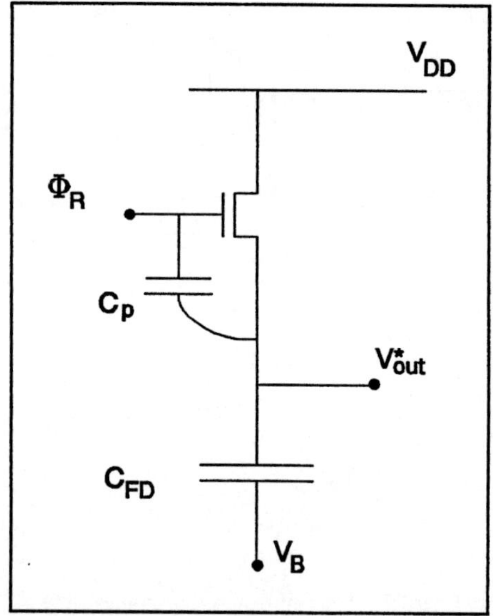

FIGURE 2.26. *Electrical representation of the floating-diffusion output node.*

The value of the reset-clock feedthrough ΔV_R can be written as a function of the ratio of the parasitic capacitance and the total capacitance :

$$\Delta V_R = \frac{C_p}{C_{FD}+C_p}.\Delta\Phi_R.A_{SF}.$$ [2.8]

From the two relations it is easy to appreciate that the parasitic capacitance should be made as small as possible in all cases. However, the same is true for the capacitance value of the floating diffusion. A smaller value for C_{FD} makes the conversion factor (expressed in μV/electron) higher.

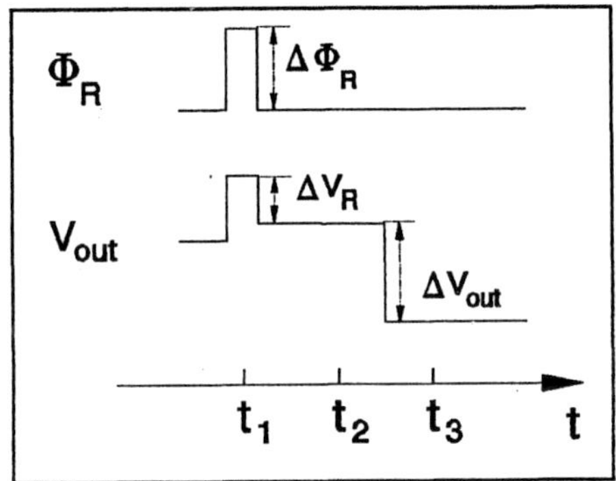

FIGURE 2.27. Definitions of the different voltage variations corresponding to the circuit depicted in Figure 2.26.

In practical designs the output stage is constructed in such a way that the floating diffusion acts simultaneously as the source of the reset transistor and vice versa. Only a single n^+ region is included to minimize C_p and C_{FD} (examples of these types of constructions will be given in section 8.4). In present-day devices, very low values can be obtained for the total capacitance, e.g. 10 fF ... 15 fF, corresponding to conversion factors of 15 μV/electron ... 10 μV/electron at the floating-diffusion node. Taking into account a value of 0.9 for A_{SF}, conversion factors at the output pin of the package of the sensor are in the range of 13 μV/electron ... 9 μV/electron.

This section is restricted to the basics of the floating-diffusion output. However, this output stage is of such great importance to solid-state imaging applications that it will be treated more extensively in Chapter 8. At that point of the discussion special attention will be paid to the noise and bandwidth of the output amplifier.

2.4.2. FLOATING GATE WITHOUT RESET

As already mentioned in the description of the floating-diffusion output stage, that type of output configuration is very widely used. It is very simple to operate, has

a high conversion factor, and can be made very linear. One drawback, however, is its destructive manner of readout. After the sensing of the voltage on the floating-diffusion node, the latter one is reset time after time, and the electrons applied to the floating-diffusion node are drained off to the reference power supply.
This effect of destructive readout can be avoided by using a floating-gate amplifier. The basic configuration of such an output stage is shown schematically in Figure 2.28.

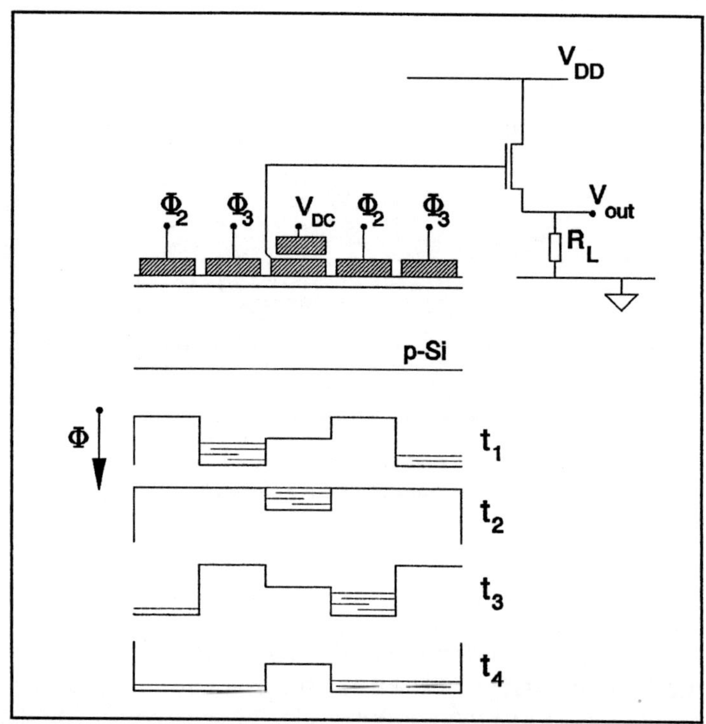

FIGURE 2.28. *The floating gate without reset method used to measure the content of the charge packet transported through a CCD.*

In the classical configuration of, for instance, a three-phase CCD clocking scheme, a Φ, gate is replaced by a floating sensing gate. The voltage on this gate is measured by a source follower. The floating gate needs an appropriate bias in order to operate the source follower at its optimum working point, but also to give the floating gate the opportunity to act as a barrier gate and a storage gate. For this purpose, a second gate which is DC-biased and is located on top of the floating gate, sets the floating gate capacitively to its correct DC value.

The overall working principle is also illustrated in Figure 2.28. At t_1 the charge packet concerned is stored underneath Φ_3. One period later, at t_2, Φ_3 goes to a low voltage and forces the charge packet to move under the floating gate. The packet of charge carriers induces a voltage change on the floating gate, which is sensed by the source follower and fed to the output node. When Φ_2 is ready to receive the charge packet, the well underneath the floating gate is emptied and the original value of the electrostatic potential on the floating gate is restored again. Further CCD transport transfers the charge packet through the CCD as shown by the illustrations at the other time points.

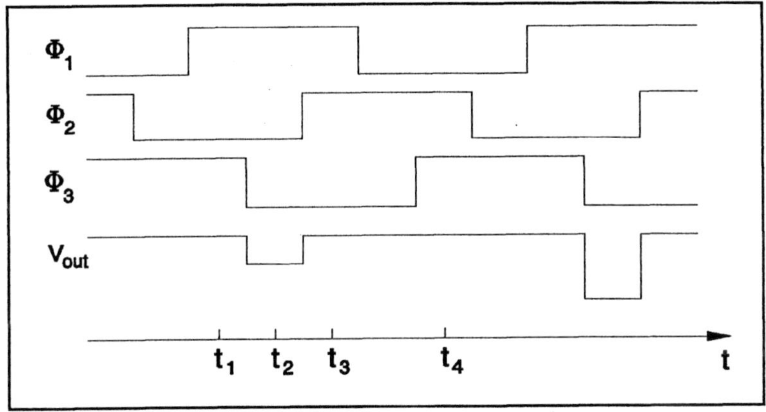

FIGURE 2.29. Timing diagram corresponding with the floating-gate readout method illustrated in Figure 2.28.

Figure 2.29 represents the timing diagram which corresponds to Figure 2.28. The various clocks are quite straightforward and self-explanatory.

A closer look at the floating-gate configuration and its electrical equivalent is presented in Figure 2.30 (Beynon 80). In this diagram, the symbols C_D, C_{ox1}, C_{ox2} and C_G represent respectively the depletion capacitance underneath the floating gate, the floating-gate capacitance toward the silicon substrate, the capacitance between the floating gate and the biasing gate, and finally the input capacitance of the source-follower stage (including the wiring capacitance). The transport of a charge packet with a content of charge Q_n from Φ_3 toward the potential well underneath the floating gate causes the following change in surface potential $\Delta\Phi_s$ at this location (the charge Q_n being forced on the series combination of C_{ox1}, C_{ox2} and C_G which is arranged in parallel to C_D) :

$$\Delta\Phi_s = \frac{Q_n}{C_D + \dfrac{C_{ox1}\cdot(C_{ox2}+C_G)}{C_{ox1}+C_{ox2}+C_G}} \cdot \qquad [2.9]$$

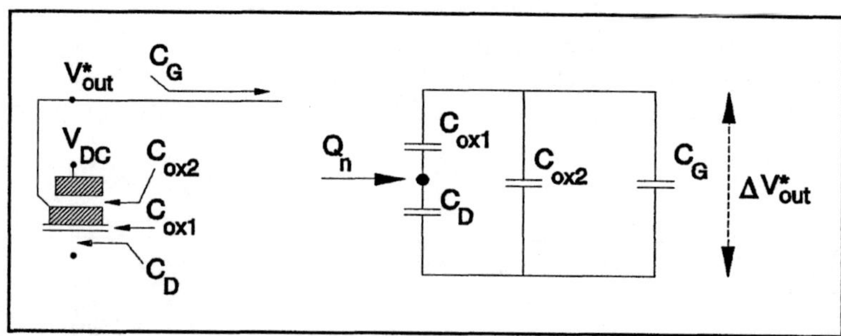

FIGURE 2.30. A "close-up" of the floating-gate configuration and its electrical equivalent.

This variation in surface potential causes a variation in potential on the floating gate. The output node responds to changes of this kind with a similar shift in voltage. This interaction is given by the following relation :

$$\Delta V_{out} = \Delta V_{out}^* \cdot A_{SF} = \frac{C_{ox1}}{C_{ox1}+C_{ox2}+C_G} \cdot \Delta\Phi_S \cdot A_{SF} \cdot \qquad [2.10]$$

(All symbols have the same meanings as before.)
Combining both equations [2.9] and [2.10] yields the relation between Q_n and ΔV_{out} :

$$\Delta V_{out} = \frac{Q_n}{C_{ox2}+C_G} \cdot A_{SF} \cdot \qquad [2.11]$$

Note that ΔV_{out} is independent of the capacitances C_D and C_{ox1}. Only the intergate capacitance and the parasitic of the source follower are of importance.

The most important feature of the floating-gate output stage is the nondestructive manner in which it measures the charge packets. After the sensing operation, the charge packets are clocked further through the CCD and can be sensed once again at another stage, or can be manipulated by one means or another. Not surprisingly, therefore, the floating-gate amplifier is a very attractive alternative in applications where nondestructive sensing is important.

WORTH MEMORIZING

The conversion of a charge packet into an electrical output signal (e.g. voltage) can take place via a floating-diffusion stage with reset or a floating-gate construction without reset. The former is the most widely used output node while the latter is applied only in situations where nondestructive readout is essential.

2.5. Conclusions

In all practical applications of a charge-coupled device, CCD shift registers form the biggest part of the device as far as silicon area is concerned. Nevertheless, the input and output structures of the CCDs are of at least equal importance. In this chapter the input of the charge packets into a charge-coupled device, its transfer through the CCD by means of digital clocks on the gates, and the output of the charge packet have been discussed.

The first section of the chapter describes the various transport mechanisms. Four-phase, three-phase, two-phase, one-and-a-half-phase, virtual-phase, and ripple-clock systems are explained. Their basic transport characteristics are compared with each other, revealing aspects such as charge-handling capability and clock-waveform definition. A three-phase transport system seems most attractive for straightforward clocking, and a four-phase system more attractive for charge-handling properties, while depending on other boundary conditions, a CCD designer might opt for yet another alternative.

In the survey of channel definitions, channel confinement by stopper implantations, dielectric thickness variations, and field-shield isolations have been investigated.

As already mentioned earlier, electrical input structures for a CCD are not important for CCD-imaging applications, but a more detailed description of a couple of different input techniques, for forcing charges into delay lines or into a wide variety of CCD shift registers, might be of interest. The output structures are extremely important. Two different alternatives have been studied with the imaging applications in mind, namely the most widely used output-node configuration, i.e. the floating-diffusion amplifier with reset and the floating-gate configuration without reset. The latter is an attractive alternative when nondestructive read-out of the CCD is needed. In both cases formulas are provided in which the variation of the output voltage is given as a function of the amount of charge fed to the output stage and as a function of different device parameters.

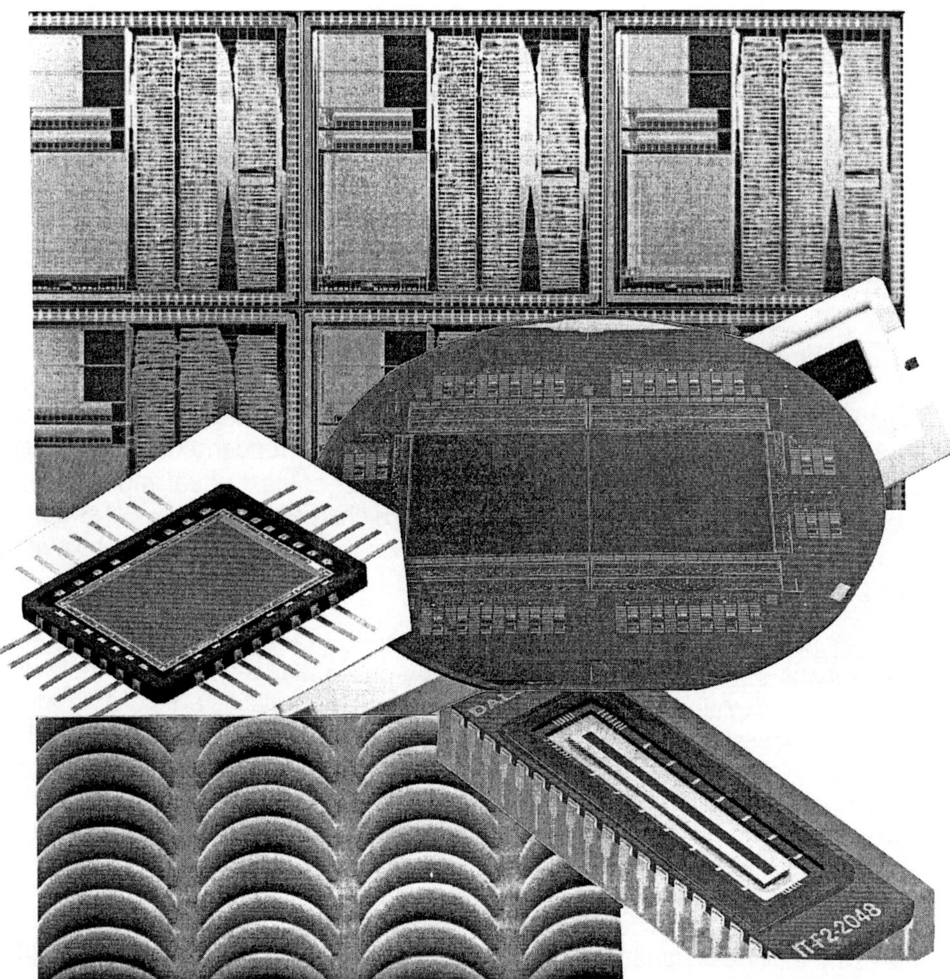

Top : ASIC Vision, a single chip CMOS imager (256 x 256 pixels) designed for a fingerprint acquisition and matching system (courtesy of VVL). Middle left : an IR imager based on the concept of Charge-Sweep Devices. The imager has 512 x 512 pixels of the PtSi Schottky-barrier type (courtesy of Mitsubishi). Middle right : a four inch wafer containing two 2k x 2k full-frame imagers. The devices are buttable and fabricated on back-side thinned substrates (courtesy of Thomson CSF). Bottom left : submicron spaced lens array for a high photosensitive interline-transfer CCD (courtesy of Matsushita Electric Company). Bottom right : a 2048 x 96 TDI image sensor. The device is provided with a bidirectional TDI option with 8 parallel outputs to increase its overall speed (courtesy of Dalsa).

A REAL CCD DELAY LINE

As far as the theory of charge-coupled devices is described in the previous chapters, it relates to perfect and ideal charge-coupled devices : transport was assumed to be perfect without any losses, and generation of minority carriers by generation centers at the interface or in the bulk was assumed to be zero. But in practice nothing is perfect, and that also holds for the charge-coupled devices. They have to cope with nonideal characteristics. These can be found in transport inefficiency, in dark-current generation, dark-current nonuniformities, various noise sources and, even more fundamental, in the sampling theorem.

For an ideal delay line, as schematically illustrated in Figure 3.1, the output signal equals the input signal, but delayed over a known length of time. This delay depends on the number of stages in the CCD and on the clocking speed of the device.

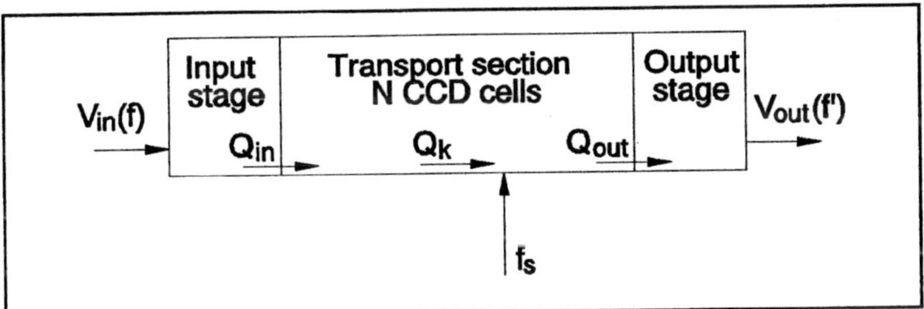

FIGURE 3.1. Illustration of the basic parts of a CCD delay line : the input, transport and the output section. The CCD is clocked with a frequency f_s.

With the following conventions :
- the number of stages in the CCD equals N;
- the clocking frequency is f_s, with a period of T_s;
- the input signal is Q_{in};
- the output signal is represented by Q_{out};
- n is a random integer;

the relation between the output signal Q_{out} in a single charge packet and the input signal Q_{in} in a single charge packet can be written as :

$$Q_{out}(nT_s) = Q_{in}(nT_s - NT_s) \; . \qquad\qquad [3.1]$$

This relation can be stated verbally as follows : the output signal Q_{out} at a random point of time (nT_s) equals the input signal Q_{in} at time nT_s minus the delay time. The last of these can be written as the number of CCD cells, N, multiplied by the clocking period T_s, or NT_s.

In reality, however, the expression above is never valid and :

$$Q_{out}(nT_s) \neq Q_{in}(nT_s - NT_s) \; . \qquad\qquad [3.2]$$

As already mentioned, the reason why the output voltage is different from the delayed input voltage is due to :
- the nonideal transport efficiency : charge packets lose part of their information because the time allowed to transport the charge packets is limited; and some of the electrons are withdrawn from the actual charge packet and added to the following one because of their interaction with surface states;
- dark current : thermally generated minority carriers at the $Si-SiO_2$ interface and in the silicon bulk are added to the charge packets and deform the information which is stored in and transported through the CCD cell;
- dark-current nonuniformities : dark-current generation is not the same for all CCD cells. As will be shown, this effect adds a kind of fixed pattern to the information signal, especially in cases where the information packets are not only transported in the CCD but also stored for several clock periods;
- the presence of noise deteriorates the analog signals at several locations : shot noise of the dark current and trapping noise in the CCD channel, kTC noise at the input and the output stages and noise generated in the output amplifier;
- the sampling action of the input stage which imposes certain extra requirements on the electrical input signal. If the input waveform does not comply with these requirements, the signals passing through the delay line will also be affected.

In this chapter on a real CCD delay line, all these aspects will be described separately and their influence on the signal waveform will be discussed. Although at this stage only a delay line is used as the subject of study, all CCD limitations and imperfections described here are also applicable to imagers.

3.1. Effect of transport inefficiency

The reasons for imperfect transport have been discussed in Chapter 1. They are the finite time allowed for transportation and the charge trapping by surface states. This section offers an answer to the question : "What effect does the transport inefficiency have on the signal waveform if a perfect sine wave is passed through a real CCD delay line with a transport inefficiency of ϵ for each gate-to-gate transfer ?"

Theoretical studies of charge-transfer devices can be tackled quite easy in the z-domain because the z-domain theory was developed to deal with delay and delay stages. Adaption of this mathematical aid leads to the transformed expression [3.1] :

$$Q_{out}(z) = Q_{in}(z) \cdot z^{-N} \qquad [3.3]$$

where z^{-1} denotes the delay occurring in a single CCD cell, and z^{-N} in N cells. In the z-domain the transfer function from the input waveform to the output waveform $H(z)$, is then equal to :

$$H(z) = \frac{Q_{out}(z)}{Q_{in}(z)} = z^{-N} \cdot \qquad [3.4]$$

Transformed back to the frequency domain, the transfer function $H(\omega)$ can be written as :

$$H(\omega) = e^{-j\omega N T_s} = e^{-2\pi j \frac{f}{f_s} N} , \qquad [3.5]$$

taking into account that for the analog input signal :

$$\omega = 2\pi f \qquad [3.6]$$

(the relation which links the frequency of the input signal to its angular equivalent ω) and for the clocking signal :

$$T_s = \frac{1}{f_s} \cdot \qquad [3.7]$$

These various expressions indicate simply that the output signal is the same as the input signal, but is delayed a certain amount of time : NT_s. Or that the amplitude of the output signal is equal to the amplitude of the input signal, but that a phase shift occurs between the two. This phase shift $\Delta\Psi$ equals :

$$\Delta \Psi = \omega N T_s = 2\pi \frac{f}{f_s} N . \qquad\qquad [3.8]$$

However, moving over to a real CCD delay line, the amplitude and phase-shift relation between the output and input signal deviate from the ideal case. If the transfer inefficiency for each CCD transport is equal to ϵ (the transfer inefficiency is defined in section 1.3.3), and the number of gates (or phases) per CCD cell equals m, the transfer inefficiency for a complete CCD cell ϵ_m is given by :

$$\epsilon_m = m\epsilon . \qquad\qquad [3.9]$$

With this definition, a charge packet in the CCD delay line at location $(k+1)$ at time nT_s is equal to that at location k at the previous time point $(nT_s - T_s)$ plus the charges left from the incomplete transport from the previous charge packet at location $(k+1)$. The charge content of the packet located at site k at the previous time point $(nT_s - T_s)$ and transferred to the location $(k+1)$ can be written as :

$$(1-\epsilon_m) . Q_k(nT_s - T_s) , \qquad\qquad [3.10]$$

while the charge left from incomplete transport from location $(k+1)$ to location $(k+2)$ equals :

$$\epsilon_m . Q_{k+1}(nT_s - T_s) , \qquad\qquad [3.11]$$

or, combining both expressions, the actual charge content at location $(k+1)$ is equal to :

$$Q_{k+1}(nT_s) = \epsilon_m . Q_{k+1}(nT_s - T_s) + (1-\epsilon_m) . Q_k(nT_s - T_s) . \qquad\qquad [3.12]$$

If the same path is followed as in the case of the ideal delay line, the z-transform results in :

$$Q_{k+1}(z) = \epsilon_m . Q_{k+1}(z).z^{-1} + (1-\epsilon_m) . Q_k(z).z^{-1} . \qquad\qquad [3.13]$$

To extract the transfer function from stage k to stage $(k+1)$, this expression can be rewritten as :

$$Q_{k+1}(z) = \frac{(1-\epsilon_m).z^{-1}}{1-\epsilon_m.z^{-1}} . Q_k(z) . \qquad\qquad [3.14]$$

If there are N delay stages, the relation between the output signal and the input signal can be simply derived by taking N times the previous transfer function :

$$Q_{out}(z) = \left[\frac{(1-\epsilon_m).z^{-1}}{1-\epsilon_m.z^{-1}}\right]^N . Q_{in}(z) .$$ [3.15]

Or in the case of a non-ideal delay line, with a transport inefficiency equal to ϵ_m per CCD cell, the transfer function can be written :

$$H_{\epsilon_m}(z) = \left[\frac{1-\epsilon_m}{1-\epsilon_m.z^{-1}}\right]^N . z^{-N} = z_{\epsilon_m}^{-N} .$$ [3.16]

For very small values of the transfer inefficiency ϵ_m, the expression for $z_{\epsilon m}$ can be simplified to :

$$z_{\epsilon_m} = \frac{1-\epsilon_m.z^{-1}}{1-\epsilon_m} = [1+\epsilon_m.(1-z^{-1})].z = e^{\epsilon_m(1-z^{-1})}.z$$ [3.17]

With z and $z_{\epsilon m}$ transformed into

$$z = e^{2\pi j.f.T_s} , \qquad z_{\epsilon m} = e^{2\pi j.f'.T_s} ,$$

the previous relation can be rewritten as :

$$e^{2\pi jf'T_s} = e^{\epsilon_m\left(1-e^{-2\pi jfT_s}\right)}.e^{2\pi jfT_s} ,$$ [3.18]

f' being the frequency of the output signal after the input signal with frequency f has passed through the real CCD delay line.
Or, after reshuffling with [3.7] :

$$f' = f+\frac{\epsilon_m f_s}{2\pi}\left(\sin(2\pi\frac{f}{f_s})-j.[1-\cos(2\pi\frac{f}{f_s})]\right) .$$ [3.19]

Since the transfer function in the frequency domain $H_{\epsilon m}(\omega)$ is defined as being equal to :

$$H_{\epsilon_m}(\omega) = e^{-2\pi jf'T_s} ,$$ [3.20]

the transfer function, expressed with a real and an imaginary part, is found to be :

$$H_{\epsilon_m}(\omega) = e^{-\epsilon_m N\left[1-\cos(2\pi\frac{f}{f_s})\right]} \cdot e^{-2\pi jN\frac{f}{f_s}\left[1+\frac{\epsilon_m f_s}{2\pi f}\cdot\sin(2\pi\frac{f}{f_s})\right]} . \qquad [3.21]$$

From this expression it can be deduced that the presence of a nonideal transfer, expressed by ϵ_m, has a double effect on the output signal :

1. An amplitude deformation, which is expressed by the real part of the transfer function given by [3.21]. The ratio of the amplitudes of the output signal Q_{out} and the input signal Q_{in} as a function of the frequency f of the input signal, normalized to the clocking (or sampling) frequency f_s is by definition the modulation transfer function or MTF. The expression for the MTF is also given by the real part of [3.21] :

$$MTF = e^{-\epsilon_m N\left[1-\cos(2\pi.\frac{f}{f_s})\right]} \qquad [3.22]$$

The characteristic MTF versus the normalized frequency of the input signal is shown in Figure 3.2. Note that for f = 0 and f = f_s the above expression

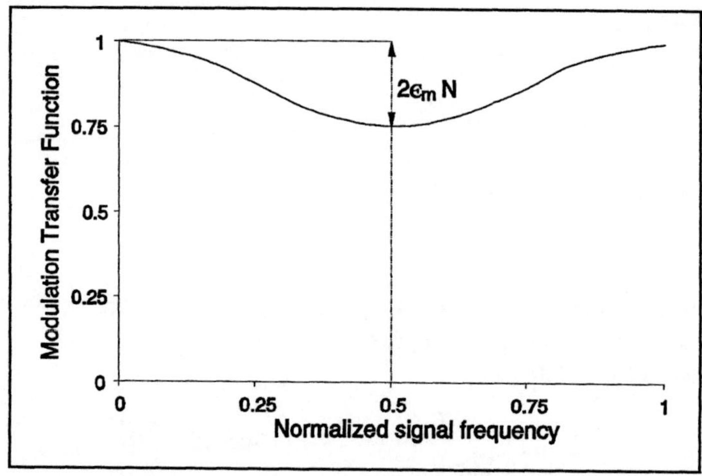

FIGURE 3.2. Modulation Transfer Function as function of the electrical signal frequency f, normalized to the sampling frequency f_s.

for the MTF [3.22] reduces to MTF = 1. This effect can be understood quite easily : for a DC input (f = 0) every charge-transfer loss of a packet is compensated by the loss from the previous packet, and the net result of output amplitude matches the amplitude of the input signal. Also for

$f = f_s$ the signal which is forced through the CCD is a DC signal and consequently MTF = 1 again. (The phenomenon of generating a DC signal from an input signal with a frequency f will be explained in more detail during the discussion of the sampling theory in section 3.5.) For $f = f_s/2$ a value relating to the MTF can be calculated, namely :

$$MTF = 1 - 2\epsilon_m N . \qquad [3.23]$$

This value has a quite important practical use : if a delay line is clocked with a frequency f_s and the input signal frequency f obeys the relation $f = f_s/2$, measurement of the ratio of the output signal amplitude to the input signal amplitude can give the exact value of the transfer inefficiency of the device via the above relation. This way of measuring ϵ or ϵ_m is relative simple.

2. A phase shift different from the phase shift in the ideal case and given by the imaginary part of the transfer function [3.21] :

$$\Psi(MTF) = -2\pi N\frac{f}{f_s} - N\epsilon_m.\sin(2\pi\frac{f}{f_s}) . \qquad [3.24]$$

This phase shift contains two components of which the first one corresponds to the phase shift introduced by an ideal delay line (see also expression [3.8]), and the second takes into account the nonideal characteristics of the device.
In Figure 3.3 the extra phase shift, compared to the one in the ideal case is plotted as a function of the normalized frequency of the input signal. In this case there are also a few situations which yield interesting results. For $f = 0$, $f = f_s/2$, and $f = f_s$ the extra phase shift equals zero. For the intermediate values of f ($f = f_s/4$ and $f = 3.f_s/4$) the additional phase shift is maximum, but in one case negative, and in the other positive, but in both cases with the same absolute value of $\epsilon_m.N$.

In charge-coupled devices used nowadays, the optimized production technology and the introduction of buried-channel CCDs have decreased the transfer inefficiency to a level where it no longer degrades the output signal so much as to make it unacceptable or even unmeasurable in practical devices. Linear array imagers with more than 8,000 CCD cells are fabricated today whose transfer inefficiency is so small as to be imperceptible in all their applications.

WORTH MEMORIZING

Incomplete charge transport by a charge-coupled device degrades the signal passed through the device. For a given charge-transfer

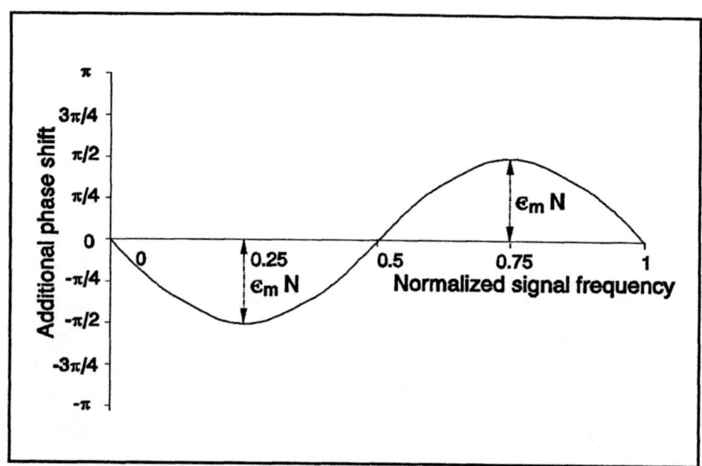

FIGURE 3.3. *Additional phase shift, compared to the ideal case, as a function of the electrical signal frequency f, normalized to the sampling frequency f_s.*

inefficiency an analytical expression can be formulated for the change in amplitude and phase shift of the electrical output signal compared to the input signal. Conversely, if the amplitude decrease is known (e.g. by measurement), the value of the transfer inefficiency can be easily calculated.

3.2. Effect of dark current

A nonzero value of transfer inefficiency is manifested by an amplitude decrease and the addition of an extra phase shift to the output signal compared with the ideal-case situation. The presence of dark current will not deteriorate the AC behavior of the output signal of the CCD but will add a DC component to it. This effect reduces its charge-handling capability for an input signal.

As already described in the first chapter when discussing the basic working principles of the charge-coupled devices, the MOS capacitors of a CCD operate in the nonequilibrium condition of deep depletion. This operating condition means that the potential wells of the CCD cells will collect any available minority carriers so as to build up the charge packets. These minority carriers can be introduced into the CCD channel by means of a special input structure but can also be made available to the CCD cells by thermal generation of minority carriers in the charge-coupled device itself. The source of these charge carriers is referred as the "dark current", J_{dark}, expressed in A/cm^2.

This dark current can be generated at different locations in the CCD but has in all cases to do with irregularities in the fundamental crystal structure of the silicon. For instance, metal impurities (gold, copper, iron, nickel, cobalt, ...) and crystal defects (silicon interstitials, oxygen precipitates, stacking faults, dislocations, ...) are known to be thermal generation sites of charge carriers in silicon. Generation sites of dark current can be located (Slotboom 91) :

  - at the SiO_2-Si interface. This is where the largest irregularity in the Si-crystal structure occurs and, not surprisingly the interface is considered to be the principal source of dark-current generation. CCD's production technology is critically optimized in respect of the properties of the Si-SiO_2 interface. More details of these technological measures will be given in section 8.2, Most of them are concerned with avoiding free or dangling Si or O bonds;
  - in the bulk of the silicon and still inside the depletion layer defined by the CCD potential wells. Any minority carrier generated by a crystal defect or impurity located inside the depletion layer will be collected in the potential well of the CCD;
  - in the bulk of the silicon but outside the depletion layer defined by the CCD potential wells. This case is quite similar to the previous one, but after the minority carriers are generated, they have first to diffuse to one of the potential wells of the CCD before they can be collected. If their diffusion length is short enough they will be recombined in the silicon bulk and cannot reach the potential wells. In this situation, two conflicting arguments play an important role : keeping the generation rate of minority carriers as low as possible and keeping their diffusion length as short as possible. Semiconductor processing of CCDs takes care of decreasing the generation rate as much as possible and also the number of the dark-current generation sites in the top part of the silicon bulk.

If the dark current of a given CCD is equal to J_{dark} (A/cm^2) and, if its clocking period T_s ($= 1/f_s$) and the area of its unit cell A_{cell} are known, the dark current Q^1_{dark} added to the charge packets in a single CCD cell is given by :

$$Q^1_{dark} = A_{cell} \cdot J_{dark} \cdot T_s \; . \tag{3.25}$$

With summation, the dark current Q^n_{dark} added to the charge packet located at a CCD cell with number n or, n cells after the input section is given (assuming that J_{dark} and A_{cell} are constant for every CCD cell) by :

$$Q^n_{dark} = A_{cell} \cdot J_{dark} \cdot n \cdot T_s , \tag{3.26}$$

and the dark current added to every charge packet after it has been transferred through all N CCD stages and passed to the output of the CCD, is :

$$Q_{dark}^{N} = A_{cell} \cdot J_{dark} \cdot N \cdot T_s \, . \qquad\qquad [3.27]$$

This relation is valid for all charge packets and gives the amount of charge added to the information charge. The effect of the dark-current's presence on the charge-handling capability will be illustrated by means of an example in the next section.

WORTH MEMORIZING

Dark current is defined as the addition of a (small) DC component to the signal fed into and through the CCD. The presence of dark current will not adversely affect the AC content of the delayed signal as far as amplitude or phase shift is concerned, but the available charge-handling capability of the device may be reduced by the presence of the dark current.

3.3. Effect of dark-current nonuniformities

As sketched above during the discussion of dark current, the behavior of a real CCD delay line is slightly more complicated. This is because the dark current generation is hardly uniform from cell to cell. The generation centers of the dark current are statistically distributed through the silicon. This means that not all cells have the same number of generation centers. On the other hand the generation rate of each center can also vary from type to type. All these variations make the dark current no longer uniform as was supposed above in the previous section. But on the other hand, if a CCD delay line is used with a continuously clocking system at a constant frequency, all charge packets pass at the same speed through every CCD cell. There will be no difference in dark current collection from charge packet to charge packet. They all have "seen" all generation centers during the same time.

If the dark current, nonuniform by nature, at CCD cell n is denoted by $J_{dark}(n)$, the total quantity of dark-current carriers collected in a charge packet passed through the CCD delay line can be written as the summation of the contribution of all CCD cells individually :

$$Q_{dark} = A_{cell} \cdot \sum_{n=1}^{N} J_{dark}(n) \cdot T_s \, . \qquad\qquad [3.28]$$

(To indicate that the dark current is no longer uniformly distributed, the total quantity of carriers is indicated by Q_{dark} instead of Q_{dark}^{N}.)

As already stated and formulated in the above expression, even if dark-current generation is not uniform, the output signal for all packets is the same if clocking of the delay line is continuous. However, things change drastically if clocking is not continuous. In that case the total clocking scheme may be, for instance, a combination of transport (pulses on the clock lines), integration (DC bias on the clock lines) and transport again (pulses on the clock lines). The dark-current signal added to the charge packets for such a clocking system is a combination of :

- a uniform dark current, the same for all charge packets, which is collected during the charge-transport sequences. All charge packets pass through all cells with the same speed during the two transport phases and consequently this part of the dark current is the same for all of them;
- a nonuniform dark current, which is collected during the integration phase. The amount of charge collected depends on the local presence and distribution of dark-current generation sites. These vary from CCD cell to CCD cell and this situation can also be recognized in the nonuniform distribution of dark-current generated charge carriers.

Uniform dark-current generation adds a DC component to the signal passed through the delay line. So does a nonuniform dark-current generation, but only when the clocking of the CCD is effected continuously. If an integration period is also included in the clocking program, not only a DC component will be added to the signal but also some (so-called) fixed-pattern noise (FPN). This fixed-pattern noise is very hard to remove. This is only possible if the distribution of this nonuniform spurious signal is known. One measure which can be taken is a dummy readout of the delay line without an electrical input, storing the obtained information (which is equal to the dark current) from this readout into a memory and subtracting it from every readout signal from the delay line during its actual use.

The following example will illustrate the influence of the dark current on the dynamic range of the CCD : a BCCD delay line with ($N =$) 1000 cells is clocked with a frequency of ($f_s =$) 1 MHz, its cell area is ($A_{cell} =$) 10 x 10 μm^2, its dark current ($J_{dark} =$) 0.5 nA/cm² at room temperature (RT = 300 K), and its charge-handling capability ($Q_{n,sat}/q =$) 200,000 electrons. The number of dark-current generated electrons can easily be calculated from :

$$\frac{Q_{dark}}{q} = \frac{1}{q} \cdot A_{cell} \cdot J_{dark} \cdot \frac{N}{f_s} . \qquad [3.29]$$

Entering in all parameters, one finds for Q_{dark}/q = 300 electrons or 0.15 % of the total charge-handling capability. This value may seem pretty small, but things turn out worse if an integration period of 20 msec is added to the delay. In that case T_{int} = 20 msec, the ratio of dark-current generated electrons and total charge-handling capability becomes 2.25 %.

With the above-mentioned characteristics of the delay line the dark current may even equal the charge-handling capability after a maximum integration time $T_{int,M}$ of 670 msec. Or, after an integration time of only 670 msec, the potential wells of the CCD are completely filled with dark-current generated electrons.

A very important factor in the process of determining the dark current is the temperature of the environment in which the CCD is operated. In general the dark current doubles with every 8°C.

For instance, if the device is the same as that just described, is considered at 60°C, its dark current $J_{dark,60}$ is related to the dark current at room temperature $J_{dark,RT}$ by :

$$J_{dark,60} \simeq J_{dark,RT} * 32 . \qquad\qquad [3.30]$$

Clocking the CCD without integration gives a dark-current signal equal to 4.8 % of the total charge-handling capability, or a maximum integration time, $T_{int,M}$ equal to only 21 msec.

On the other hand if the CCD delay line is cooled to -40°C, its dark current $J_{dark,-40}$ is given by :

$$J_{dark,-40} \simeq \frac{J_{dark,RT}}{256} . \qquad\qquad [3.31]$$

Operating the delay line at this temperature results in a very low dark current which is only 0.0006 % of the total charge-handling capability, and to fill the charge packets completely with dark-current-generated electrons, an integration time $T_{int,M}$ of about 3 min is needed.

The two examples discussed here might seem to be only theoretical, but both situations are very common. What about a video camera which is kept in the back of a car, parked in the sun ? In this situation the temperature of the CCD used as an imager can easily rise above 50°C. On the other hand, in applications where the CCD has to cope with very small signals (e.g. imaging for astronomical applications), the devices can be cooled down to -40°C by means of, for instance, a Peltier element, so as to lower the dark current and to permit very large integration times.

WORTH MEMORIZING

Dark current adds only a DC component to the signal passed through the CCD. However, if generation of dark current is not uniform, a fixed-pattern noise component will be superimposed on the signal if the CCD is clocked in a noncontinuous mode.

This fixed-pattern noise part of the signal can only be removed by complicated signal processing.
Dark-current generation is highly temperature-dependent, doubling with every temperature increase of 8°C. This effect makes control of the fixed-pattern noise even more complex.

3.4. Effect of noise

Basically, the charge-coupled device is an electronic, analog component. And, like all such circuits, a CCD is not free from noise. The presence of noise sources degrades the analog signals passed through the CCD at several locations : dark-current shot noise and trapping noise in the CCD channel, kTC noise at the input and output stages, and noise generated in the output amplifier. This list of noise sources holds only for the delay-line applications. In other applications, there may be additional noise sources. In the case of imager use, photon shot noise is another very important source.

In this section only these general noise parameters typical of a delay line will be briefly described (Carnes 72a, Barbe 75). A more complete study of the noise phenomena is added to the discussion in Chapter 7. Then possible measures to suppress and eliminate noise are also described. However shot noise, trapping noise, kTC noise, and amplifier noise are investigated as part of this section.

3.4.1. SHOT NOISE

Shot noise is a noise source associated with the presence of dark current. The generation of dark current is a random process. In the first place generation sites are randomly distributed, but the dark-current generation mechanism is also purely random as a function of time. The number of electrons generated and collected in a charge packet traveling through the CCD can be described with the Poisson probability distribution. The variance of this phenomenon is equal to the mean for the Poisson distribution. Or the number of electrons representing the shot noise n_{shot} on the dark current is given by the square root of the number of dark-current generated electrons :

$$n_{shot} = \sqrt{\frac{Q_{dark}}{q}} \, . \qquad\qquad [3.32]$$

For example, in a BCCD with a dark-current content of 400 electrons in a charge packet (at room temperature), the shot noise of the dark current is equal to 20 electrons.

3.4.2. TRAPPING NOISE

Trapping noise results from the fact that not every surface state or other trapping site is active in the same way as the others when minority carriers are trapped and re-emitted. Traps are more or less active, depending on their characteristics. As in the case of dark-current generation, these trapping and emission processes are similarly random.

For a surface-channel CCD, the number of traps in the surface states is quite high and their distribution is very extensive. A very wide range of all kinds of traps exists. For these reasons the trapping noise in a surface-channel CCD $n_{tr,SCCD}$ is relatively constant :

$$n_{tr,SCCD} \simeq constant \, . \tag{3.33}$$

The situation for a buried-channel CCD is completely different. The bulk traps of these devices are highly localized, and the number of traps and their characteristics depend on the frequency at which the particular CCD is operated and on the amount of charge transported through the CCD, so that the number of electrons $n_{tr,BCCD}$, representing the trapping noise in a BCCD can be written as :

$$n_{tr,BCCD} = g(f_s, Q_n) \, . \tag{3.34}$$

Due to the fact that the emission time-constants of the traps vary widely, the number of charges released in a well-defined packet is greatly dependent on the time available for the packet to stay in the environment of the trap. If the packet is passed quickly (at a high frequency), slow traps will not release very many electrons to the packet. On the other hand, if the packet stays a longer time in the "capture environment" of the trap (operating the CCD a low frequencies), the number of traps which will release charges is much higher.

Trapping noise in a BCCD is due to the bulk traps in the device. The number of traps which can be active to release charges is determined partly by the volume of the charge content of a packet. The more electrons in it, the bigger the volume of the charge packet, and traps which are located inside the physical volume of the charge packet will only capture electrons and on balance there will be no release. Thus the activity of the traps, and consequently also the trapping noise, is influenced by the volume of the charge packet and by the amount of charge in it.

In addition to the fact that the trapping mechanism and the generation of trapping noise are completely different from those of a surface CCD and a bulk CCD, the trapping noise of the buried-channel device is always much smaller than the trapping noise of a surface-channel device, a 10-factor difference being no exception :

$$n_{tr,BCCD} \ll n_{tr,SCCD} \, . \qquad\qquad [3.35]$$

Alongside their now familiar transport properties, this characteristic of buried-channel devices is one of the biggest advantages over the surface alternative. In practical devices used in today's applications, the number of bulk states in a buried-channel device is so low that the trapping noise is generally negligible.

3.4.3. kTC NOISE

The kTC noise is important at all locations in a CCD at which capacitors are charged and discharged through a resistor. This is done at the input structure and at the output node of a CCD. At the input stage, the MOS capacitors are charged with electrons supplied by the input diode via the resistive CCD channel. At the output stage, the floating diffusion capacitor is charged with electrons supplied by the reset transistor via the channel of the same transistor. Due to the voltage fluctuations across the capacitors when they are charged through the resistor, kTC noise is introduced. The value of this noise component n_{kTC} in conjunction with a capacitor C is given by :

$$n_{kTC} = \frac{\sqrt{kTC}}{q} \, , \qquad\qquad [3.36]$$

or, at room temperature :

$$n_{kTC,RT} = 400 . \sqrt{C(pF)} \, . \qquad\qquad [3.37]$$

For a typical value of the floating-diffusion capacitance of 40 fF, the kTC component of the noise equals 80 electrons. Compared to the shot noise of the dark current in the previous example, this is a relatively high value. It will not be surprising to learn that in practical applications quite a lot of effort is put into reduction of this noise component. (In applications where only the kTC noise of the output is important, this noise component is often referred to as reset noise.)

3.4.4. AMPLIFIER NOISE

This noise component is associated entirely with the output amplifier and, of course, depends strongly on the type of output stage used. In the ideal case where a noise-free output amplifier would be applicable, the amplifier noise component would be zero. In reality, like other domains, that of output amplifiers is not ideal. In typical MOS source-follower configurations the amplifier noise comprises mainly two components :
    - 1/f noise : at lower frequencies MOS transistors display a noise spectrum which is typically described as a 1/f function. This behavior is explained

by the interaction between the charge carriers and the traps (surface and bulk) present in the transistor channel;
- thermal noise : at higher frequencies the 1/f noise diminishes and the remaining noise component is due to thermal noise generated in the resistive channel of the MOS transistor.

Shot noise and trapping noise are typically technology-related, but kTC noise (for the output stage) and amplifier noise are also technology-dependent, but much more related to the design of the device and to signal processing after the CCD. In practical applications, a great deal of attention is paid to appropriate design of the CCD device itself but also to the most suitable processing electronics. These items are studied in much more detail in the section on noise in chapter 8, which deals with Advanced Imaging.

WORTH MEMORIZING

During the passage of a charge packet into, through and out of a CCD, various noise sources will have an influence on the electrical signal. The most important are :
- shot noise on the dark current due to the random process which the generation of dark current is;
- trapping noise due to the inconsistent nature of the interaction between the surface states and the charge packet;
- kTC noise at the input and output nodes due to the charging and discharging of capacitors through a resistor;
- amplifier noise consisting of 1/f noise and thermal noise in the MOS transistors.
The first two sources are strongly technology-related, while the last two are more linked to the design of the device and the signal processing applied after the CCD output amplifier.

3.5. Sampling of an electrical signal

Although not explicitly discussed, sampling has already been introduced in the description of the CCD's input structure in the previous chapter. The input configurations are designed in such a way that they can measure or sample the electrical input signal. The input stage measures the value of the electrical signal and feeds a representative amount of charge into the CCD. This measuring process takes place at certain well-defined points in time. Timing is driven by the clocking of the CCD. Together, all the discrete samples fed into a delay line represent the complete electrical signal passed through the device. A continuously varying electrical signal is digitally segmented into separate, time-discrete but analog samples and fed into the CCD.

If the sampling process itself is well controlled and done in accordance with the sampling theorem, no information in the electrical input signal is lost due to the sampling itself. The original waveform can be reconstructed completely. A simple

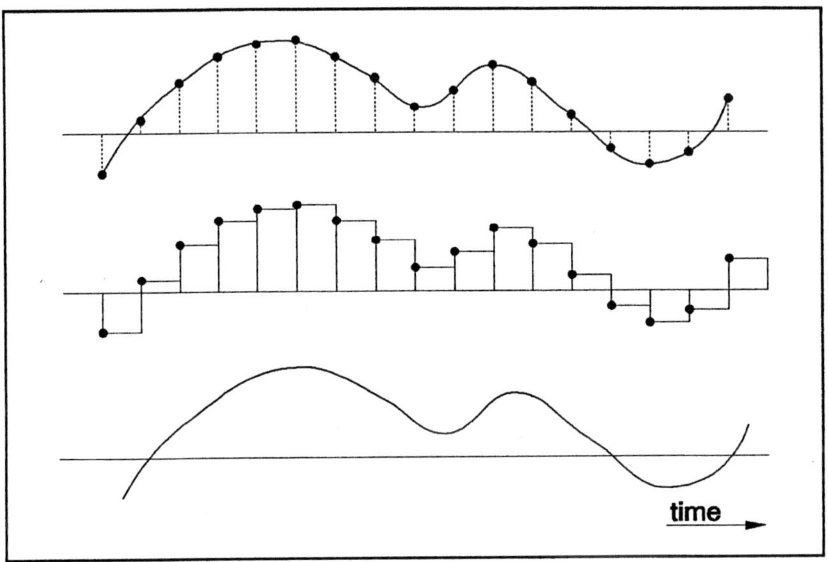

FIGURE 3.4. *Illustration of the sampling action : the analog value of the input signal is measured at discrete time points, and reconstructed from these data by means of filtering.*

example of the sampling action in general is illustrated in Figure 3.4 (Pohlmann 87). The original signal, plotted in the top illustration as a function of time, has a continuous waveform. It is sampled at time points separated from each other by a period T_s or with a frequency f_s given by :

$$f_s = 1/T_s .$$

[3.38]

The result of this sampling operation is given in the middle illustration in Figure 3.4. The dots represent the value of the original waveform at the corresponding points in time. Between the dots the signal is kept at its value when last sampled. The original continuously varying signal is thus transformed into a staircase function, in which the steps of the staircase are located at points defined by the original waveform. Note that the locations of the staircase steps (in the vertical direction) are of a fully analog nature. They can assume any arbitrary value between defined limits.

If the said staircase representation of the signal is passed through a low-pass filter with a well-defined bandpass, the waveform shown in the bottom illustration in Figure 3.4 is generated. If filtering is done effectively, the output waveform will be exactly the same as the original input signal. The overall sampling process will not have any noticeable effect on the electrical waveform of the signal.

As already mentioned several times, appropriate sampling is needed, followed by more or less ideal filtering to keep the output signal the same as the input signal. These conditions are not always fulfilled. This can be seen in Figure 3.5 where different sinusoidal signals are sampled at identical points in time (Pohlmann 87). In all situations the original signal is indicated by a solid line, and the output signal reconstructed from the sampled one, is represented by a dotted line.

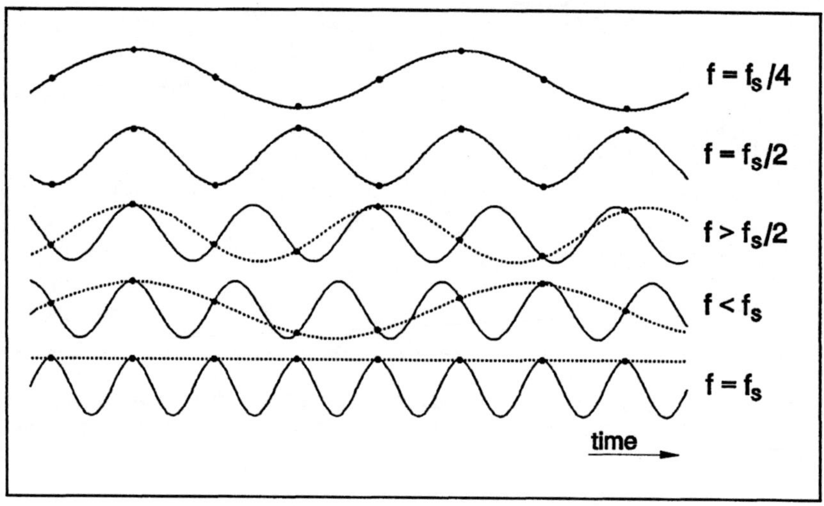

FIGURE 3.5. Illustration of the output signal (dotted lines) if the frequency of the input signal f is increasing toward the sampling frequency f_s (dots).

In the graph at the top the frequency of the sampled signal f is much lower than the sampling frequency f_s, in the example $f = f_s/4$. The dots represent the sampling points, and the original signal can be exactly reconstructed from these dots.

In the second graph the frequency of the analog waveform is $f_s/2$. In this situation the sampling dots are localized at the extreme values of the sine. As in the first situation, an exact copy of the original signal can also be reconstructed here.

In the third graph, the frequency of the input signal is further increased and a situation is reached in which $f > f_s/2$. The output waveform, which is represented also by dots is still a sine wave, but its frequency is completely different from the frequency of the input signal. The high frequency of the input signal cannot be reconstructed and, instead of it, an output signal with a lower frequency is generated.

The same applies to in the fourth illustration, where the input frequency is again slightly increased. As a result of this the output signal drawn through the sampling dots is even lower in frequency than in the previous situation.

The fifth and last example in Figure 3.5 shows the effect of sampling with a frequency f_s of an input signal with the same frequency $f = f_s$. If this relation is maintained, the output signal after the sampling is a DC signal. The amplitude of this DC signal is no longer primarily related to the input signal, but primarily to the phase shift between the sampling signal and the input signal. In the case shown in the figure, the amplitude of the DC output signal is maximum.

From the above discussion and from the various illustrations in Figure 3.5 the fundamental sampling theory, also known as the Nyquist theorem, can be formulated, namely : "The sampled input signal can be fully reconstructed from the sampling points if the frequency of the input signal f is at most as high as half of the sampling frequency f_s", or :

$$V_{in} \equiv V_{out} \quad if \quad f \preceq f_s/2 . \qquad [3.39]$$

The Nyquist theorem can be proved by a simple example, illustrated in Figure 3.6 (Pohlmann 87). The input signal v_{in} is sampled with a frequency f_s, e.g. 1 MHz.

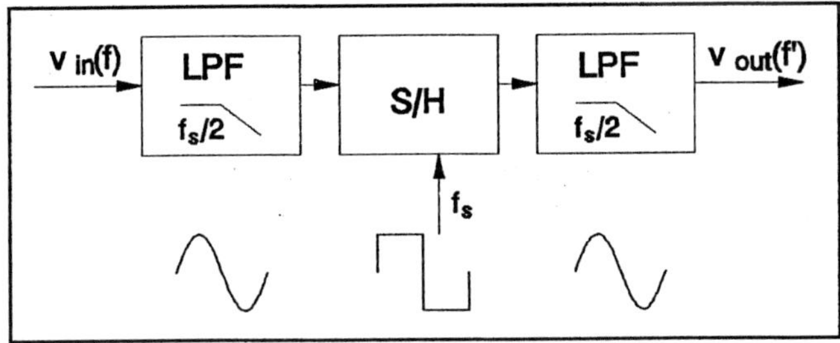

FIGURE 3.6. The fundamental sampling theory or the Nyquist theorem : the input frequency f could at most be equal to $f_s/2$.

Before the input signal is fed to the sampler, it is bandpassed by a filter with a cut-off frequency of $f_s/2 = 500$ kHz. This operation ensures that the highest frequency present in the input signal is $f_s/2$.

For the theoretical limit a 500 kHz sine wave is fed to the sampler and sampled with a clock frequency of 1 MHz. In this situation the signal at the output of the sampler is a square wave, also with a repetition frequency of 500 kHz. After bandpassing through a 500 kHz cut-off filter, only the first harmonic from the square wave is retained, namely a sine wave with a frequency of 500 kHz. Thus, even with the theoretical limit of $f = f_s/2$, the input signal can be fully reconstructed after sampling with a frequency f_s by putting the correct bandpass filters before and after the sampler.

The complete sampling theorem is summarized in Figure 3.7 (Pohlmann 87), where the sampling process is sketched in the time and frequency domain. In the first row of figures the original input signal is shown at the left, and its frequency spectrum on the right. Note the band-limited spectrum : all frequencies available in the input signal are lower than $f_s/2$. The sampling signal is illustrated in the second row of Figure 3.7. The time and frequency spectrum are shown again, respectively on

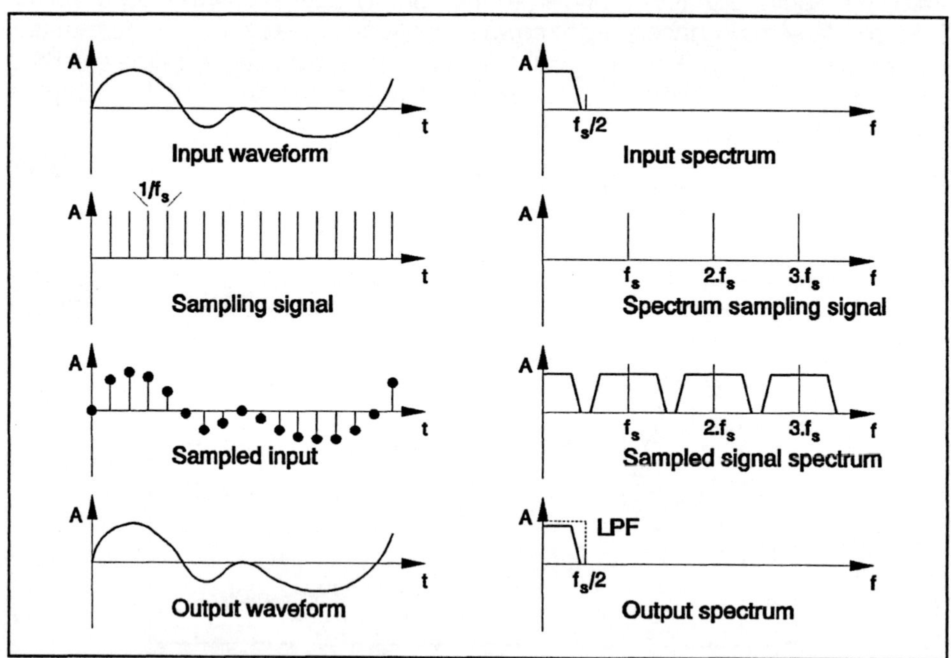

FIGURE 3.7. Summarization of the complete sampling theorem for a band-limited input signal; signals in the time and frequency domains are illustrated (all are given in A.U.).

the left and right. The sampling pulses are delta signals with a period of $1/f_s$ or repetition rate f_s and every multiple of f_s. The effects of sampling are shown in the third row. In the time domain the sampled points are shown as the result of the sampling, which produces in the frequency domain as the basic frequency spectrum repeated and mirrored around each multiple of the sampling frequency f_s. The final result of filtering the sampled signal is illustrated at the last row of diagrams in Figure 3.7. The output signal is an exact copy of the input signal, while its frequency spectrum after filtering is the same as the original one. The frequency response of the low-pass filter (LPF) is represented by the dashed line in the last diagram in Figure 3.7.

The example described in Figure 3.7 holds for a signal which satisfies the Nyquist theorem. In the opposite situation, when the frequency of the input signal is not limited to $f_s/2$, aliasing components are added during the sampling process and this is illustrated in Figure 3.8, which contains exactly the same information as Figure

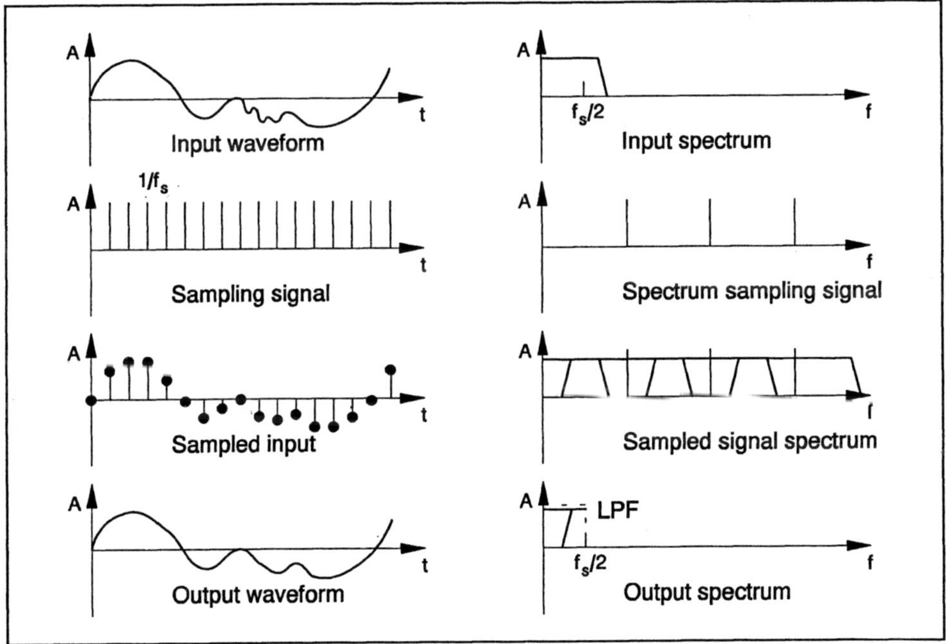

FIGURE 3.8. *Summarization of the sampling theorem when the input signal is not band-limited; signals in the time and frequency domains are illustrated (all are given in A.U.).*

3.7 has already shown, but, as can be seen from the first frequency spectrum, the input signal has components outside the band limitation at $f = f_s$. If the time-domain pictures in Figures 3.7 and 3.8 are compared, this frequency extension will be found in the negative-going part of the input waveform. The sampling signal is identical in both situations, but the results of the sampling are completely different. As can be seen in the third row of Figure 3.8, the dots no longer represent exactly the same waveform as the original one. Some high-frequency information is lost. This can also recognized in the frequency spectrum of the sampled signal. The original spectrum is repeated and mirrored around the multiples of f_s and these spectra overlap each other : one frequency band originating from, for instance, mirroring around f_s, coincides with the original one. Bandpassing the sampled waveform to reconstruct the output signal shows that the reproduced signal lacks some high-frequency information which is replaced by a low-frequency signal. This effect can also be seen in the frequency spectrum : the low-pass filter has cut off part of the original frequencies (above f_s) and added to it something extra with a lower frequency.

This effect of aliasing can only be avoided by appropriate filtering of the input signal. In all situations where the correct filtering is not applied, aliasing components can degrade the output signal and, once this has happened, the aliasing signal can no longer be separated from the original signal because the frequency band of the aliasing components coincides completely with the lower frequency band of the (unfiltered) input signal.

WORTH MEMORIZING

Sampling of an electrical signal can be seen as a continuous operation of measuring an analog signal at discrete time intervals. If the time intervals are chosen small enough, the original signal can be fully reconstructed after sampling by means of appropriate low-pass filtering. If the time intervals are too large, aliasing occurs and the sampling action introduces a deformation of the original input signal.
Sampling theory is completely described by the Nyquist theorem : sampling does not deteriorate an original signal as long as the sampling frequency is at least twice the highest frequency available in the original signal.

3.6. Conclusions

Nobody is perfect and neither is a charge-coupled device. The imperfections of a surface-channel charge-coupled device are reasonable well remedied in the buried-channel alternative, but even this type of CCD is driven to its limits in some of its applications. Although its transfer inefficiency can be very small, even a very small transfer inefficiency multiplied by the number of stages in a very large CCD shift register might result in a relatively high overall inefficiency. The effect of this incomplete charge transfer on the amplitude and phase characteristics of the signal passed through a charge-transfer device are described. Compared to the original input signal, even a perfect charge-coupled device will introduce a certain phase shift between the input and the output signal. This should not be surprising because the actual purpose of a delay line, for instance, is to introduce a phase shift between the input and the output signal.

On the other hand if the charge transfer is not perfect, not only will the theoretical phase shift be altered but the amplitude will also be changed due to the transfer through the device. Both interactions between the transfer inefficiency on one side and the amplitude degradation and phase shift on the other side can be described by the introduction of the modulation transfer function (MTF). An analytical expression can be derived for both parameters in relation to the transfer inefficiency, which is also done in this chapter by making use of the charge transport in the z-domain.

Charge-coupled devices are operated in the deep-depletion mode, which is fundamentally a nonequilibrium stage. The device itself will tend toward an equilibrium situation and this can be achieved by the thermal generation of minority carriers in the depletion layer but only in the bulk of the CCD. Charges generated in the depletion layer will all be added to the charge packet passed through the CCD; charges generated deeper in the bulk of the device have first to diffuse to the potential wells of the CCD. Independently of the mechanism, however, thermally-generated charges or dark current will be added to the information charge present in the charge packets. As explained in this chapter, the effect of the dark current on the characteristics of the CCD depends on the uniformity of the dark current and on the clocking of the device. A uniform dark current does not influence the amplitude or phase characteristics of the signal fed through the CCD, but it can reduce the charge-handling capability of the device. On the other hand a nonuniform dark current in combination with noncontinuous clocking of the charge-coupled device can completely destroy the amplitude information contained in the charge packets passing through the CCD channel. The resultant effect is the addition of a fixed-pattern noise. To minimize the effect of the dark-current nonuniformities, effort is devoted in the device fabrication technology reducing the number of dark-current generation sites. Alternately, the device can also be cooled during operation to lower the thermal-generation rate of the charge carriers.

Other noise parameters which can have an adverse effect on CCDs are the shot noise, the trapping noise, the kTC noise (or reset noise), and the amplifier noise. The first two of these are related to the random processes in dark-current generation and the trapping characteristics of the interface states. To reduce the effect of shot and trapping noise, dark-current generation has to be minimized and the number of interface states reduced.

The last two noise components, kTC or reset and amplifier noise, are especially associated with the processes taking place at the output node of the charge-coupled device. The charging and discharging (= resetting) of the floating diffusion via the resistive channel of the MOS transistor is responsible for the kTC noise. The MOS transistor of the output amplifier itself generates the 1/f and thermal noise components. Appropriate design of the output stage as a whole, including the floating-diffusion configuration, is of extreme importance for reducing the effect of the various noise sources at the output amplifier.

The last section of this chapter deals with the sampling theorem. The CCD is an analog device driven by digital clock pulses. Information is presented in analog form, but in view of the digital driving pulses, the information is determined or measured at discrete time points. This measuring process is referred to as sampling. Sampling theory or the Nyquist theorem fully describes the sampling process : sampling does not degrade an original analog input signal as long as the sampling frequency (or clocking frequency of the CCD shift register) is at least twice the highest frequency available in the original analog signal. Violation of the Nyquist theorem introduces aliasing components in the reconstructed signal after sampling.

A complete family of CCD imagers, top left : a back-side illuminated frame-transfer imager with 1260 x 1152 pixels (courtesy of Thomson CSF); top right : a full-frame CCD having 3k x 2k pixels, each of 9 µm x 9 µm (courtesy of Kodak); middle : a low-cost frame-transfer sensor having 270,000 pixels on an image section with a 3.2 mm diagonal (courtesy of Sanyo); bottom left : an HDTV image sensor with an interline-transfer architecture, having 2M pixels and a diagonal of the image section of 11 mm (courtesy of NEC); bottom right : a frame-transfer CCD intended for broadcast application with 500,000 pixels (courtesy of Philips).

Chapter 4

SOLID-STATE IMAGING AT A GLANCE

In this chapter on the basics of solid-state imaging some elementary aspects of the theory of photon sensing and imager configurations will be explained. After a discussion of the fundamentals of charge-coupled imaging devices in general, the first aspects of solid-state imaging will be examined.

There are two main sections in this chapter. The first, dealing with photon sensing, can be further subdivided into two parts : the photon-conversion mechanism and the subsequent electron collection. Only a brief introductory overview of both mechanisms is given, providing just an adequate basis for the second section of this chapter. This discussion of imager configurations will consider the linear devices first, followed by several possibilities for sensor configurations in the array-imaging domain. This study will conclude with an overview of the advantages and disadvantages of the various architectures.

Starting the description of solid-state imaging with the imager architectures will enable the reader to study subsequent chapters almost independently of each other. The other way around, starting with in-depth discussion of solid-state imaging aspects would nullify this freedom to continue freely, in random order, through the various remaining chapters.

4.1. Photon sensing

Solid-state imaging is based on the physical principle of converting quanta of light energy (photons) into a measurable quantity (voltage, electric current). The link between the photons at the input side of an imager and the voltages at the output side of the device is the electrons. Of great importance in this chain is the generation, capture, and transport of these electrons. In this section attention will be paid to all elements involved in the basic operation of solid-state imaging.

Energetic photons impinging on and penetrating into a semiconducting substrate can transfer part of their energy to the substrate. This energy transfer can take place by the generation of electron-hole pairs : if the energy content of the photons is high enough, electrons will be released from the valence band and swept into the conduction band, in the case of a p-type semiconductor, leaving behind a hole in the valence band. This action reduces the energy of the incoming photon by an amount equal to the energy difference between the conduction band and the valence band : the bandgap (= 1.1 eV for silicon). In the opposite case, the energy of the photon has to be higher than the bandgap to generate an electron-hole pair.

If this principle is applied to solid-state imaging, photons can create electron-hole pairs in silicon if their energy is higher than 1.1 eV, or if their wavelength is shorter than about 1000 nm. After the generation of the charge carriers in the bulk of the semiconductor, the negatively-charged electrons have to be separated from the positively-charged holes. The easiest way to do this is to apply an electric field by which the electrons are captured and the holes drained (or vice versa, depending on the type of substrate). On the other hand, if separation is successful, it is almost impossible to transport a single electron or hole to an output stage and to convert it to a measurable quantity : its energy content is much too small. This problem can be circumvented by an integrating operation : instead of transferring each single electron or hole, the charge carriers are locally integrated into a charge packet over a certain amount of time. In this way the number of charge carriers in such a packet can be made large enough to represent an amount of energy which is easily detectable by the output of the device.

The separation of the electrons from the holes and the integration of charge carriers, both operations in the bulk of the semiconductor can be done by means of a small capacitor. Two alternatives are shown in Figure 4.1 : a metallurgical n^+p junction (reverse-biased photodiode) and an externally induced np junction (MOS capacitor). Both illustrations are based on a p-type-silicon substrate. The separation of the electrons and the holes is effected by the electric field across the np junctions : electrons are caught and holes are drained to the p-type substrate. The integration of the electrons takes place on the reverse-biased and electrically isolated junction capacitances (Collet 86). After the collection of electrons, the voltage across the capacitance will decrease, and so consequently will its depletion region width, depending on the quantity of electrons integrated or collected.

The next link in the imaging chain is the transport of the charge packets from the integrating sites toward the output of the device. Two alternatives are again possible and both are shown in Figure 4.2 : an MOS switch with a sensing line and a CCD shift register. In both cases the imaging cell or pixel is a photodiode built on a p-type-silicon substrate. The choice between the MOS switch combined with the sense line and the CCD shift register is somewhat application-dependent because both have their advantages and disadvantages. For the MOS switch with the sensing line, the fabrication technology is fairly simple, but the MOS switch connects the small capacitance of the pixel to the relatively large capacitance of the sensing line. That is why the signal-to-noise performance of this construction is rather poor. On the other hand, the technology for the CCD shift register is somewhat more complicated, and the transfer of the charge packet from the pixel toward the output diffusion needs appropriate clocking. But the charge packet taken from the small pixel capacitance is transferred to the small capacitance of the output diffusion.

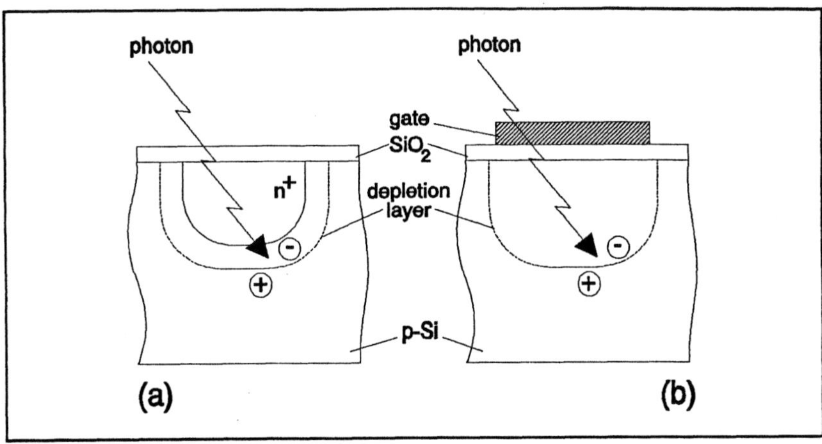

FIGURE 4.1. A metallurgical n⁺p junction (a) and a voltage-induced np junction (b), each acting as a photon-converting site.

This incorporates the potential for a relatively high signal-to-noise performance of the imager construction with a CCD shift register as the charge-transport medium.

Figure 4.2 shows only the combination of a metallurgical n^+p junction with the alternative readout schemes. An MOS capacitor can, of course, also be combined with a CCD shift register.

The conversion from a charge packet to a voltage at the output pin of the imager is done in a classical way : sensing of the voltage changes on a floating n^+-region by means of a source-follower amplifier. This is also illustrated in Figure 4.2. The technique of converting the charge packets into a series of voltages has already been covered in the general description of the CCD output stages (see section 2.4).

WORTH MEMORIZING

Solid-state imaging is based on the following basic steps :
   1. converting the incoming photons into electron-hole pairs;
   2. separating the electrons from the holes;
   3. integrating the charges into a charge packet;
   4. transferring the charge packet to an output amplifier;
   5. converting the charge packet into a measurable quantity.

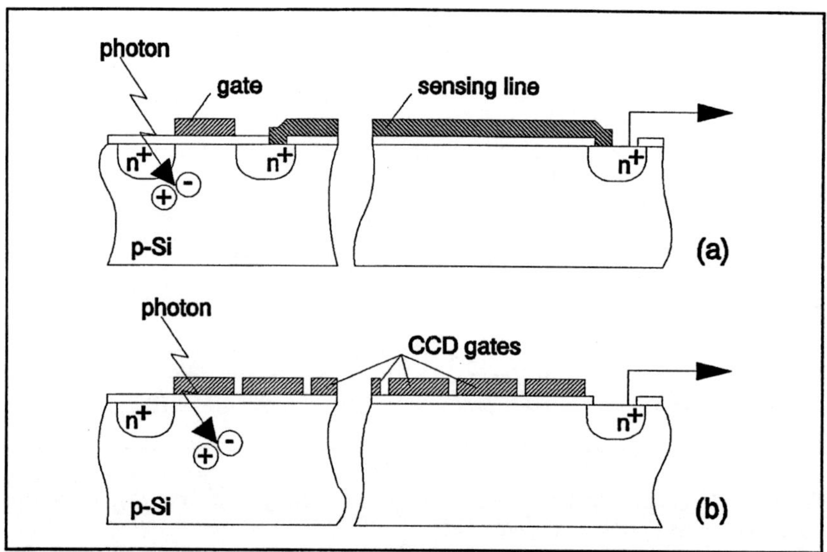

*FIGURE 4.2. Readout structure for connecting the photodiode to the outside world :
an MOS switch (a) and a CCD shift register (b) are shown.*

4.2. Imager configurations

In the foregoing section only the operation of a single image cell or pixel is explained. In practical applications, images can be build up in a one-dimensional way (e.g. facsimile) but most of them in a two-dimensional configuration (e.g. home video). In this paragraph imager architectures will be discussed ranging from simple linear, bilinear, over quadrilinear to two-dimensional arrays : frame-transfer, interline-transfer, frame-interline-transfer, MOS-XY imagers and charge-injection devices.

4.2.1. LINEAR IMAGERS

The simplest organization of pixels in a solid-state imager is a single line of photosensitive elements, photodiodes, or MOS capacitors, as shown in Figure 4.3a. Located next to the row of pixels is a CCD shift register required for readout of the charge packets; the length of a CCD unit cell is equal to the pitch of the pixels. The isolation between the pixels and the CCD is done by means of a transfer gate (e.g. Aoki 85). After the integration of charge carriers in the photo-sensitive elements, the transfer gate is raised to a high voltage, the pixels are emptied at the same time in a parallel manner toward the CCD shift register as indicated by the arrows in Figure 4.3a. Next the transfer gate is set to a low voltage. A new integration period can start and in the mean time the charge packets can be transferred through the CCD shift register toward the output of the device in a serial way. The CCD shift register must be shielded from the incoming light to avoid disturbance in the number of carriers in the charge packets. The overall transport of the charge packets out of the photoconversion sites toward the output structure of the CCD is a parallel-serial transfer : parallel into the CCD, serial toward the output.

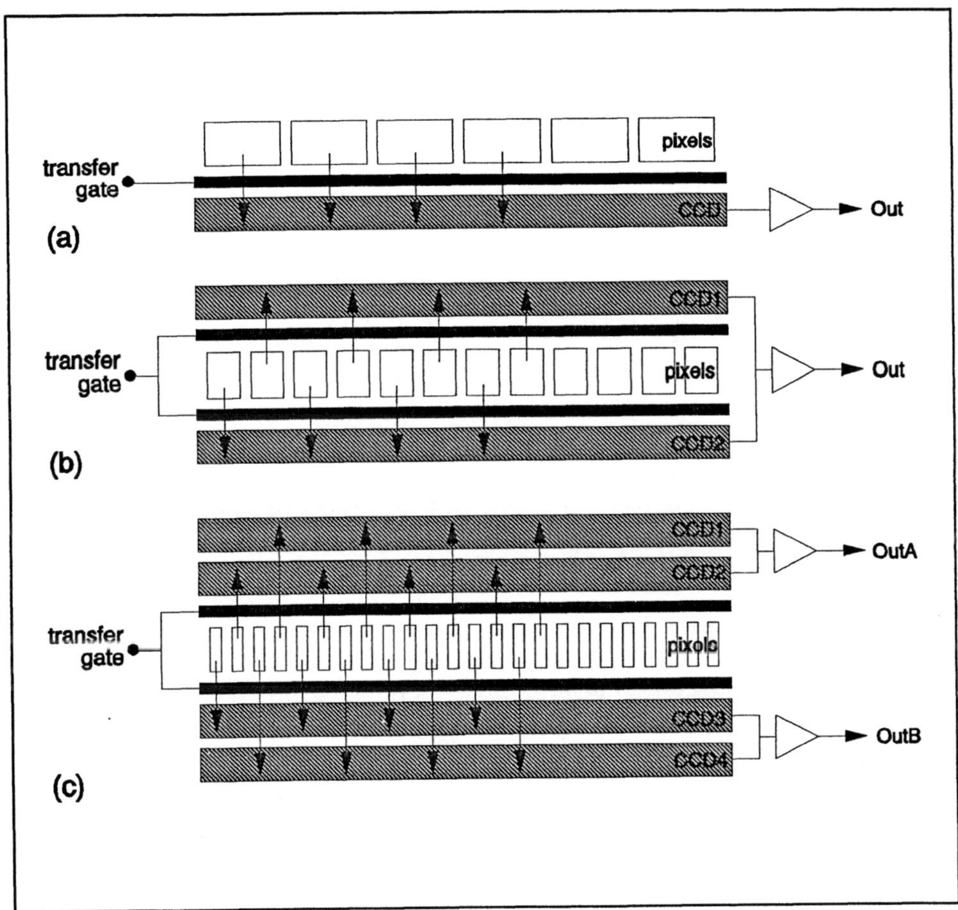

FIGURE 4.3. Device concepts of the different types of linear image sensor : a simple linear structure (a), a bilinear imager (b), and the quadrilinear concept (c).

If there is a need for a high-resolution imager with a higher number of pixels/mm, the length of the CCD unit cell can be a limitation, because it defines the smallest pixel pitch. In a bilinear imager (e.g. Goto 84) the pixel pitch can be decreased by a factor of two beyond the pitch of the CCD unit cell. As shown in Figure 4.3b, a bilinear device uses two CCD shift registers, one on each side of the single row of pixels. The basic operation of a bilinear imager is identical to that of a simple

linear one except that the charge packets are spread out after the integration over two CCD registers. Odd pixels are, for instance, moved to the bottom CCD line, even pixels to the top CCD line. Also this transport is indicated by arrows in Figure 4.3b. At the output stage of the device the charge packets are again multiplexed in such a way that the information at the output of the device is a successive stream of samples as they were taken by the photosensitive elements.
In addition to its increased pixel density, the bilinear CCD also has the advantage, compared to its linear counterpart with the same number of pixels, of a lower clocking frequency.

The demultiplexing technique characteristic of the bilinear imager can also be applied to make a quadrilinear imager (Herbst 76, Declerck 83, Sevenhans 83): see Figure 4.3c. Four CCD shift registers, two on both sides of the pixel row, are used to transport the charge packets to a double output structure (or even a single output stage). A possible way to split the information in four CCDs is shown in Figure 4.3c. In this quadrilinear imager the pixel pitch is equal to one fourth of the length of the CCD unit cell. If the partitioning across the various CCD registers is done properly, this configuration is highly suitable for high-resolution line-imager sensors : a total number of about 8,000 pixels on a single line is no longer an exception (Anagnostopoulos 93).

The principle of constructing more than one CCD shift register along the same side of a single pixel row is also applicable to high-density array imagers, as will be shown later in section 9.1.

4.2.2. ARRAY IMAGERS

In the case of two-dimensional solid-state imaging, several device architectures are possible and they are not just limited to CCDs : XY-addressed photodiodes (MOS-XY) and charge-injection devices (CID) can also be used. The CCDs themselves can be further subdivided into frame-transfer (FT) imagers, interline-transfer (IL) imagers, and frame-interline-transfer (FIT) devices. The basics of all these different types will be described, and a comparison of all the alternatives will follow at the end of this section.

Frame-transfer CCD

Basically all frame-transfer devices are built with MOS capacitors as sensing elements. On the other hand, all CCD cells needed to transport the charge packets are also composed of MOS capacitors, thus creating the possibility to combine both functions in a single structure using a string of CCD cells or a CCD shift register as a linear solid-state imager. By placing several of these linear devices alongside each other it is possible to build a two-dimensional-imaging array, of the kind illustrated in Figure 4.4. Furthermore, a CCD shift register of the same length, shielded from incoming light, can be added to each light-sensitive CCD register. These shielded CCD lines can act as a temporary memory for the information coming from the top light-sensitive lines and which is shifted into it.

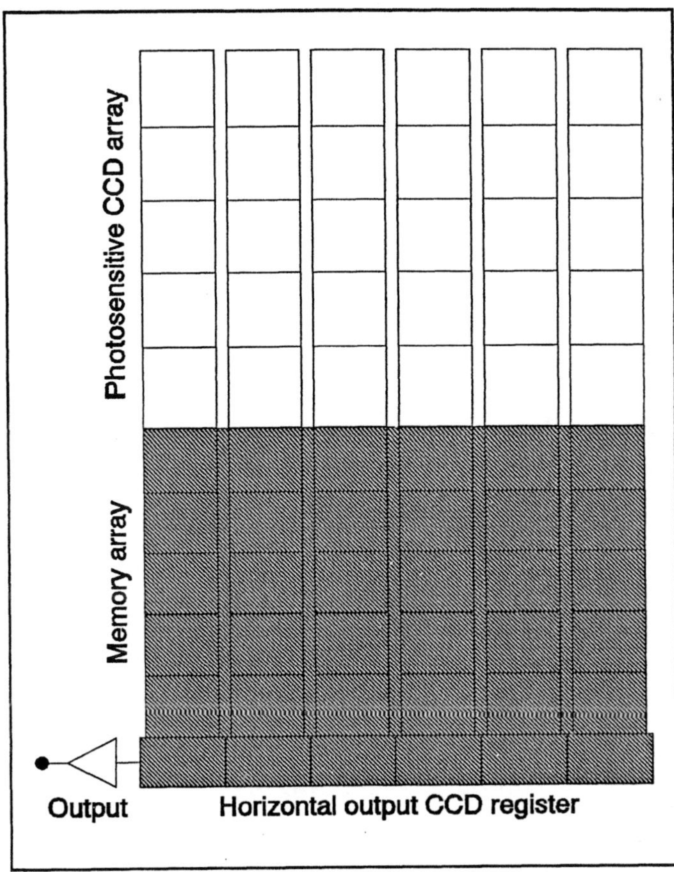

FIGURE 4.4. Device architecture of a frame-transfer image sensor.

The overall operation of a frame-transfer device can be described as illustrated in Figure 4.5 : in the light-sensitive top part of the device or image section, all CCD cells are biased into the integrating mode. Some of the CCD phases are connected to a high DC level and some to a low DC level. In this situation charges generated by impinging photons are collected in the induced potential wells (situation as it is shown in Figure 4.5a). At the end of a defined integration time the CCD shift registers transfer their charge packets to the corresponding CCD lines in the light-insensitive memory or storage section (see Figure 4.5b). This transfer from image to storage area has to be done as quickly as possible in order to avoid disturbance by extra charge generation of the information already available in the CCD shift registers of the imaging section.

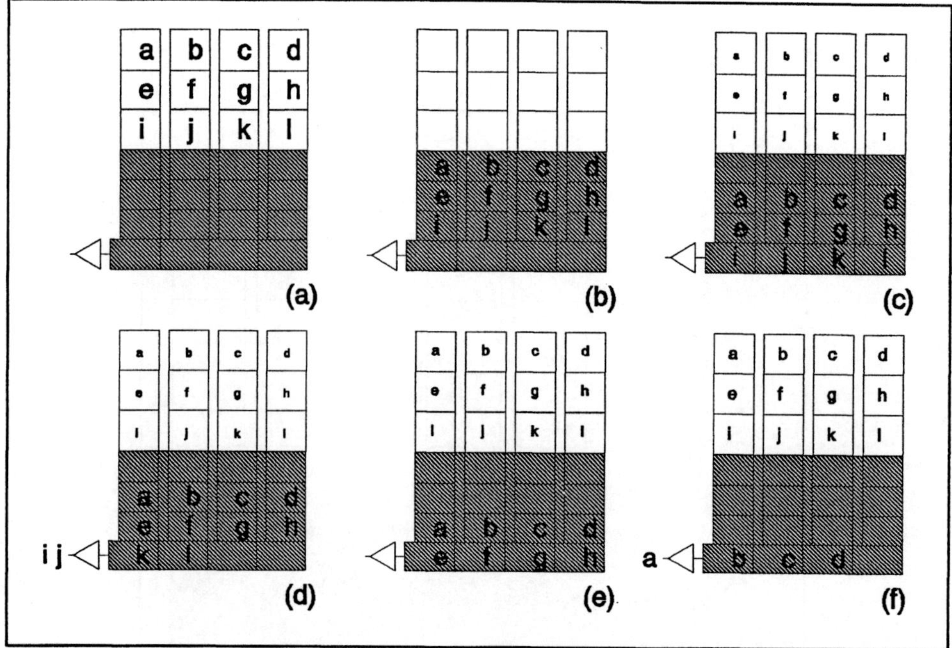

FIGURE 4.5. *Working principle of the frame-transfer CCD.*

Once the entire information from the image section has been transported to the storage section, the transport of the charge packets toward the output structure of the device can start. During this operation the charge packets located on a horizontal line (and thus belonging to different vertical CCD shift registers) are transferred to a single horizontal output CCD register (as shown in Figure 4.5c). This parallel-series transfer is similar to that also used in linear imagers. Once the charge packets have been shifted into the horizontal output register, the final transfer to the output stage can take place (see Figure 4.5d), charge packets can be converted to an electrical voltage, and the video signal can be processed further. The latter charge transport (parallel to series shift, horizontal charge transport) is repeated till the complete device has shifted its information to the output (see Figures 4.5e and 4.5f).

During the transport of all video information from the storage area to the output stage, the CCD cells of the image section are again biased into the integration mode, thus allowing the device to integrate the next image. Note that this causes a new image to build up in the image section while at the same time the preceding image is completely or partly located in the storage area. This effect is illustrated in Figures 4.5d, 4.5e and 4.5f by means of the "growing" information which is integrating in the image section.

In the approach of simultaneous video readout and integration, the timing of photosensing can be made fully independent of the timing of the video readout.

A simpler version of the frame-transfer imager is the full-frame device. Its image section and horizontal output register are similar to those of the frame-transfer CCD, but it has no storage area. The readout of the charge packets passes directly from the image section toward the horizontal output register. To avoid disturbance of the video image "on its way" to the output amplifier, the incoming light should be interrupted. This can be done by using a shutter, cutting the light path, or by means of a flashlight.
The full-frame imager is quite suitable for single shots or can be used in combination with an external memory.

Interline-transfer CCD

Like the frame-transfer device, the two-dimensional interline-transfer imager is constructed by placing several linear imagers alongside each other. In the interline-transfer device the linear array used is of the type shown in Figure 4.3a : the sensitive pixels are located near the shielded CCD transport register. The typical construction of the interline-transfer device is shown in Figure 4.6.

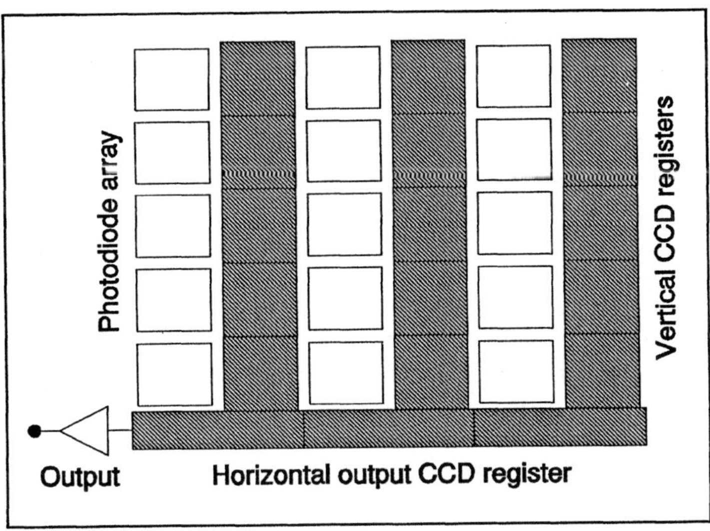

FIGURE 4.6. Device architecture of an interline-transfer imager.

Its basic operation can be as illustrated in Figure 4.7 : the integration of charge takes places in the pixels, which can be photodiodes or MOS capacitors. At the end of the integration time (see Figure 4.7a) the charge packets are shifted from the pixels into the vertical CCD shift register alongside them (see Figure 4.7b).

These shift registers are shielded from light and act as a memory for the information coming from the pixels.

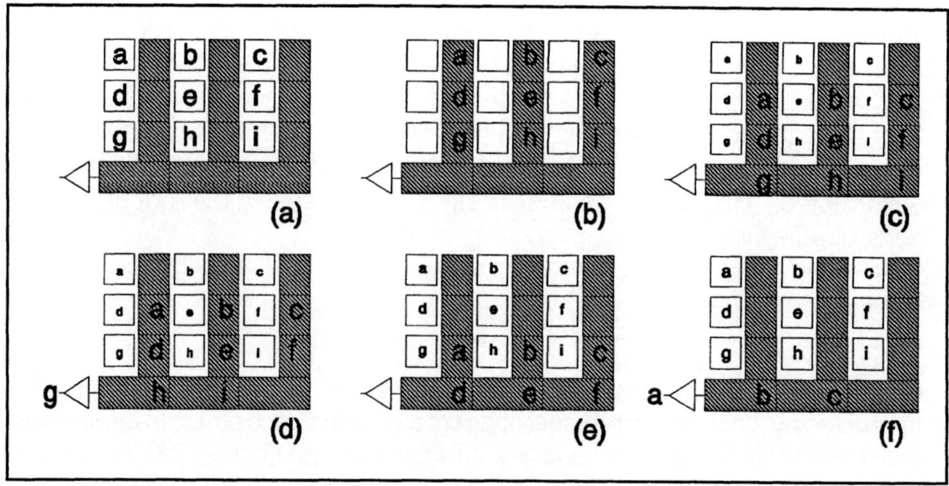

FIGURE 4.7. Working principle of the interline-transfer CCD.

After this transfer a new integration period can start and the transfer of the video information of the image just built up can also begin. The vertical shift registers are emptied into the horizontal-output register line by line in the same way as with the frame-transfer device (see Figure 4.7c). The serial information in the horizontal-output register is transferred to the output (see Figure 4.7d) and consequently converted into an electrical voltage or current. The sequence of parallel to series conversion and the horizontal readout is repeated till the complete device is empty (see Figures 4.7e and 4.7f).

Again as in the case of the frame-transfer device, the interline-transfer imager can integrate a new image during the storage of the previous image in its memory element : the vertical CCD shift registers.

An alternative to the IL concept, the resistive-gate sensor was introduced in the early days of interline-transfer CCDs. This type of imager is similar to the normal interline-transfer type as described here, but its vertical shift registers were built with a resistive gate instead of the normal clocking CCD gates (Heijns 78). Compared to the classical interline-transfer CCD, the resistive-gate sensor is limited as to speed and transfer efficiency.

Frame-interline-transfer CCD

A "marriage" between a frame-transfer imager and an interline-transfer device is
the frame-interline-transfer CCD : it has the light-sensitive area of the interline-transfer
device (photodiodes and vertical shift registers), combined with the storage area
of the frame-transfer device as illustrated in Figure 4.8. Its basic operation, which
is illustrated in Figure 4.9, combines the two previous devices : integration of the

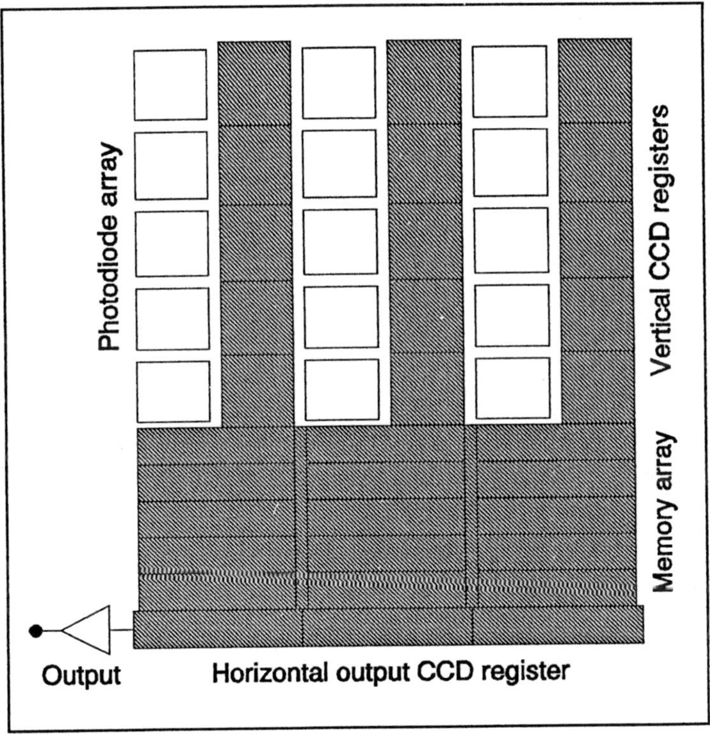

FIGURE 4.8. Device architecture of a frame-interline-transfer imager.

charges in the pixels (see Figure 4.9a), transferring the charge packets into the
vertical shift register (see Figure 4.9b), followed by fast vertical transport toward
the shielded memory part of the imager (see Figure 4.9c). This vertical transport
is not a line-by-line operation as in the interline-transfer structure instead, all the
charge packets are moved as quickly as possible, as was done in the frame-transfer
structure. The vertical shift registers act only as an intermediate memory location
and a transport channel between the pixels and the storage area. The conversion
of the parallel information in the storage area to the serial information in the horizontal
output register is similar to that already described for the frame-transfer and the
interline-transfer imagers (see Figures 4.9d, 4.9e and 4.9f).

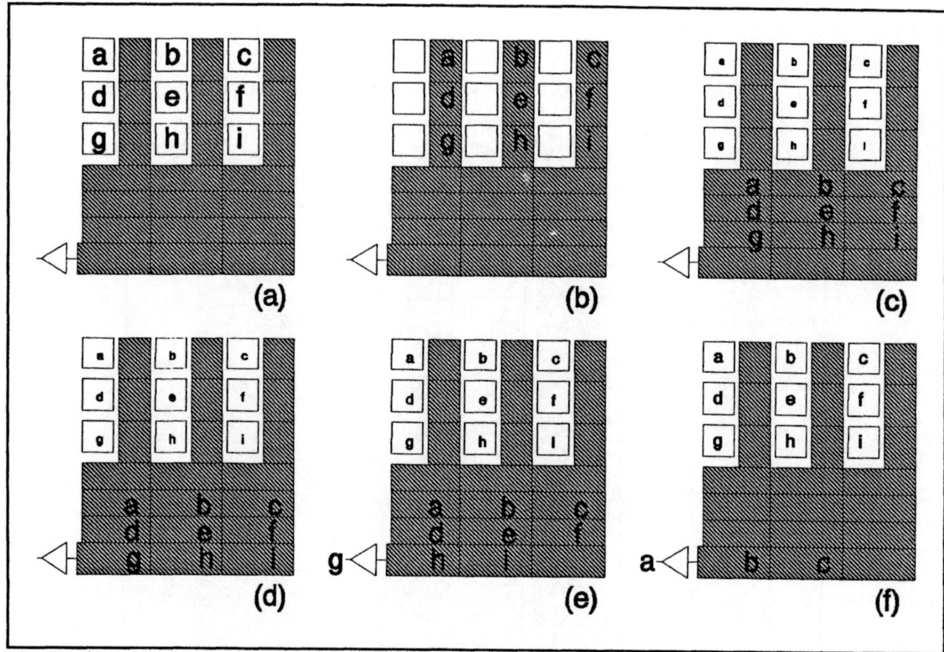

FIGURE 4.9. Working principle of the frame-interline-transfer CCD.

Although this is not its intended application, it is possible to have the video information of three images available in a frame-interline-transfer imager : one image built up in the light-sensitive photodiodes, a second stored in the vertical shift registers (as in the operation of the interline-transfer device), and a third stored in the memory area (as in the operation of the frame-transfer device).

Like the frame-transfer and the interline-transfer devices already described above, also the frame-interline-transfer CCD has an extension : the QFIT or the quasi-frame-interline-transfer device. The difference between the FIT and the QFIT is the storage area, the QFIT having only a fraction of the storage area of the FIT, which results in the distinct working principle of the device : after the frame shift its entire memory is filled with charge packets but the memory capacity is not enough to take the full video information. Part of it is still stored in the vertical shift registers of the imaging section. The most typical application of the QFIT is the charge-reset option of the device, which will be studied in chapter 6.

MOS-XY imager

An MOS-XY addressable imager is designed as a matrix of photodiodes each of which is provided with an MOS transistor acting as a switch. This setup is shown schematically in Figure 4.10. The basic structure of a unit cell is very similar to that illustrated in Figure 4.2b, where the photodiode is connected to the output by a sensing line. To convert the two-dimensional spatial information into a serial stream of the signals, electronic scanning circuits are added to the device to address all pixels sequentially and to read out the information they contain.

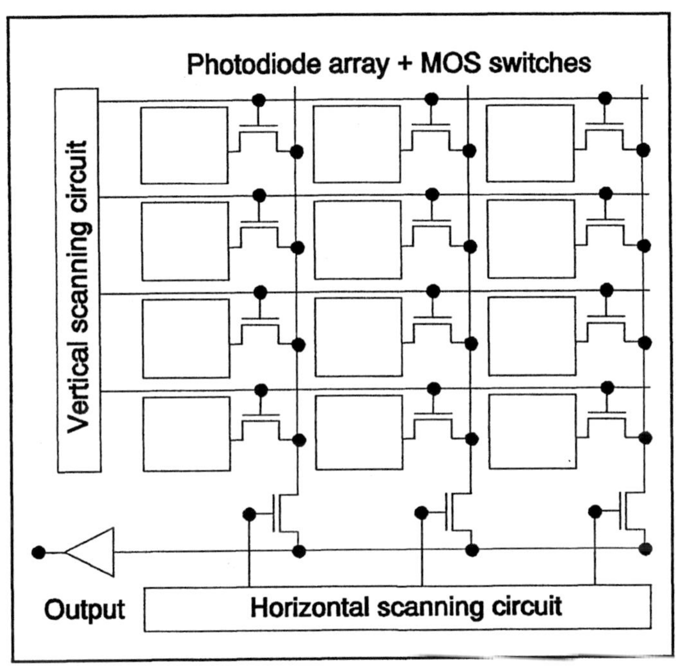

FIGURE 4.10. Device architecture of an MOS-XY addressable image sensor.

The essential operation of the device is based on this addressing feature, which is illustrated in Figure 4.11. At the start of a new field (see Figure 4.11a), the vertical scan circuit is activated and it selects, for instance, the first row of pixels by setting a high DC voltage at all gates of the MOS switches of this first row of pixels (see Figure 4.11b). Next, the horizontal scan circuit selects a pixel on one particular line by scanning its own outputs with a single high DC output while all the others remain at a low level (see Figures 4.11b and 4.11c). This combination of one output of the vertical scanner and one output of the horizontal scanner at a high level with all the others low, selects only one single pixel out of the two-dimensional matrix. This pixel can be emptied and dump its information in the output stage (see Figure 4.11b). Immediately after this event, the pixel can restart an integration and the next neighboring pixel will be addressed and read out. The overall sequence is further illustrated in Figures 4.11d, 4.11e and 4.11f.

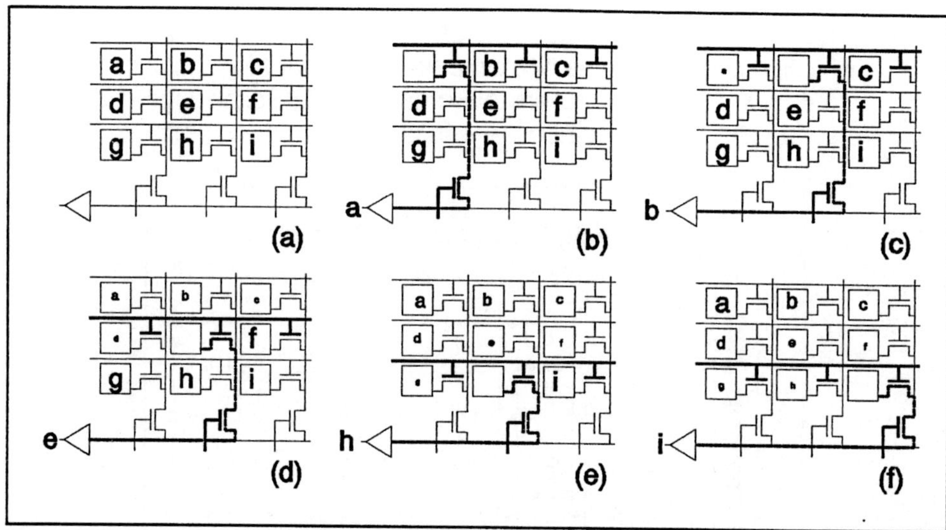

FIGURE 4.11. Working principle of the MOS-XY addressable imager.

In the MOS-XY addressable imager, every new integration period of a particular pixel starts at the moment it is read out. For this reason, the various pixels of the focal plane have all different starting and ending points of their integration time. This effect is also illustrated in Figures 4.11a through 4.11f by means of the "growing" information in the various pixels. With this concept in mind, it should be clear that the starting condition with all pixels having "equal amount of information" available in Figure 4.11a, is only for sake of simplicity to start the explanation concerning this working principle.

The MOS-XY addressable imager is limited in its performance due to the relatively high levels of noise including fixed-pattern noise. These are partly due to the fact that the low capacitance of the photodiode is connected to the high readout capacitance of the sensing line. The latter can be reduced by incorporating a horizontal CCD readout register instead of the horizontal scanning mechanism. This device concept is shown schematically in Figure 4.12. The imaging part of the sensor remains unmodified and is still scanned with a vertical scanner. The horizontal readout arrangement uses a buried-channel charge-coupled device and a suitable amplifier (Terakawa 80). The area in between the light-sensitive array and the CCD in Figure 4.12 is the charge-priming section which is used to supply an internal bias charge to the vertical transport lines just before the transfer of the signal charge. This charge priming makes the subsequent charge transfer from high-capacitance transport lines remarkably efficient. The signal charge transferred into the MOS capacitors can be simply "skimmed off" to the BCCD, leaving the bias charge behind. This mechanism of the charge-priming section resembles the fat-zero technique used in surface-channel CCDs to enhance their transport efficiency.

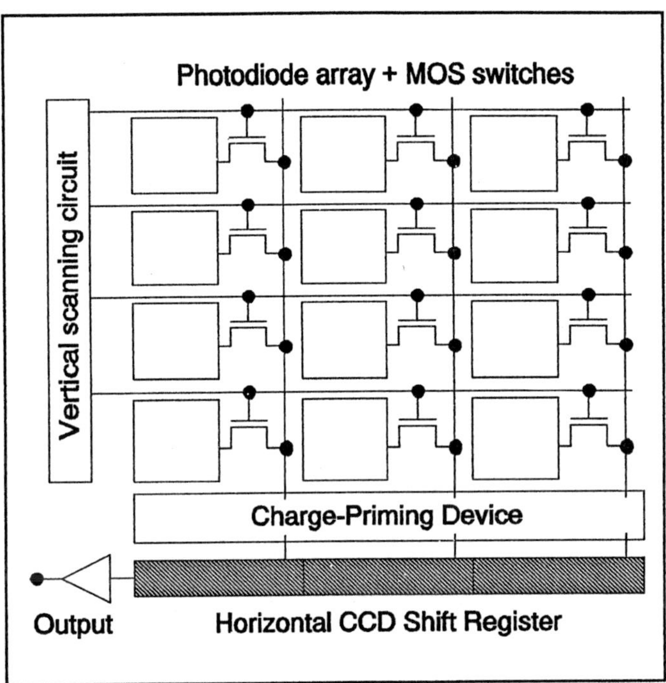

FIGURE 4.12. Device architecture of a MOS-XY imager with a horizontal CCD output register.

A more advanced imager, from the same MOS-XY family is the TSL (Transversal Signal Line) device. It is characterized mainly by a dual readout transistor in each pixel and horizontal instead of vertical readout lines (Noda 86). The TSL is more fully described in section 8.5.

<u>Charge-Injection Device</u>
The basic configuration of the charge-injection device (Burke 76) is similar to that of the MOS-XY addressable imager : it is constructed around a two-dimensional array of pixels, with two scanning circuits to address the pixels individually. A schematic setup is shown in Figure 4.13.

Basically different to the MOS-XY is the concept of the pixel itself. For sensing elements the MOS-XY uses photodiodes, while the CID imager utilizes MOS capacitances, typically with two overlapping gates. The basic working principle of the individual pixels is illustrated in Figure 4.14. The two gates forming such a pair can be differently biased; one being connected to a higher voltage (e.g. 10

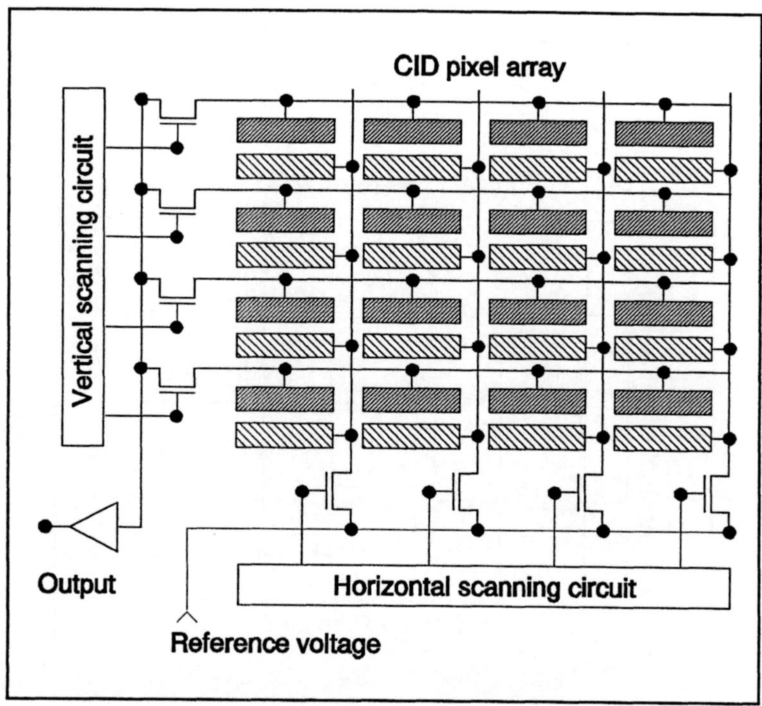

FIGURE 4.13. Device architecture of a charge-injection device.

V) the other to a lower voltage (e.g. 5 V), as shown in Figure 4.14a. Electrons generated can be collected in the potential well underneath the most positively-biased gate. By transporting the charges from the originally highest biased gate toward the second one by means of a clocking method similar to CCD transport (see Figure 4.14b), a displacement current is induced in the MOS gates. The quantity of integrated charge in the pixel, by measuring this displacement current, is sampled.

By restoring the original bias conditions, as shown in Figure 4.14c, the integration of charges can "continue", the previous readout process is a nondestructive readout. In the case when real-time video is needed (or when the readout should be destructive), the pixel is emptied by biasing both gates of the pixel to the low voltage and all previously integrated charges are dumped in the substrate. The clearing of the information is illustrated in Figure 4.14d. A displacement current is also generated in the negative-going MOS gate during the dumping operation. This can also be used to measure the charge collected in the potential well.

With reference to Figure 4.13, the operating principle of the two-dimensional CID can be described as follows : at the onset of the integration period, all pairs of gates are equally biased, one gate negatively (connected to a row electrode) and the

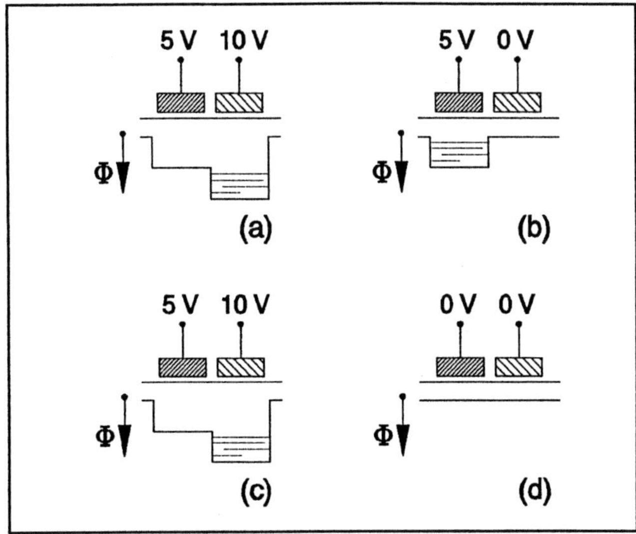

FIGURE 4.14. Working principle of the individual pixels of a charge-injection device.

other positively (connected to the column electrode) to integrate the charge. The pixels are scanned one after another by applying to a column electrode a low voltage which is even lower than the one applied to the row electrodes. This switching causes the charges of all pixels belonging to that particular column to shift their charges to the area beneath the row electrodes. A current amplifier at the end of the row senses the resulting displacement current that flows in the row electrode concerned. This current is proportional to the charge contained in the pixel and provides the sensor's output. At the end of each scan the row electrode switches to its minimum potential, causing the stored charge to inject into the underlying substrate and thus clearing the pixel for the next charge-integration period.

<u>Overview of Array Imagers</u>
If different device architectures are compared to each other, it is important to start from a common base, which in this section is an equally light-sensitive area (Collet 85, Theuwissen 90). Figures 4.4 to 4.9 are drawn in such a way that the imaging area of the frame-transfer imager, interline-transfer imager, frame-interline-transfer imager, MOS-XY sensor, and CID are all of the same size. Table 4.1. gives an overview of the pros and cons of the various device structures described.

By comparing Figures 4.4 and 4.5 it is clear that the frame-transfer imager has the largest chip size : besides its light-sensitive area it also contains a memory unit. On the other hand, the entire area which is defined as the imaging part of the frame-transfer device is light-sensitive. This is not true of the interline-transfer device : the image area also contains the vertical shift registers, which are light-insensitive. This effect is translated in the aperture ratio, which is defined as the area of light-sensitive silicon of one complete CCD cell divided by the total area of that particular CCD cell (= light-sensitive part plus the area of one stage from the vertical shift register). The interline-transfer CCD has a small chip size but also a small aperture ratio.

Although smear will be described later in Chapter 8, it may be pointed out here already that a characteristic of the frame-transfer device and the interline-transfer sensor is their relatively high smear level. The frame-interline-transfer CCD has been developed to overcome this problem. Unfortunately, it has the drawbacks of both the interline-transfer and the frame-transfer device : low aperture ratio and large chip size.

In-depth study of the shutter possibility for the frame-transfer device and the charge-reset operation for the interline-transfer CCD, will also be the subjects of section 6.4. Although the smear level is relatively high for frame-transfer imagers, these devices are the only CCD imagers which can be made fully smear-free if the camera in which the FT is used is provided with a shutter. Charge reset on the other hand is an interesting feature if the camera, built around the imagers, needs to have the option of shortening the integration time.

Comparing the device architecture of the frame-transfer and the (frame-)interline-transfer imagers, is should be no surprise to learn that the imagers of the frame-transfer type are much simpler in construction. "Only" CCD channels which are isolated from each other are used to build up the device. In the (frame-)interline-transfer technology, isolated CCD channels are needed with photodiodes in between. This complicates the production technology.

The first camcorders with a solid-state image sensor were built around an MOS-XY imager as its light-sensitive device. The fabrication technology of this type of sensor was very similar to the DRAM technology and it benefited from developments in this field of semiconductor application. Sensors of this type arrived on the market earlier and cost even less than CCDs did. However, MOS-XY devices are characterized by a relatively low signal-to-noise ratio, due to the high capacitance of the sensing line compared to the small capacitance of the photodiode.

Due to the addressable matrix, this type of sensor can be read out randomly, thus differing from standard serial video. The random access can be an advantage in certain applications such as, for instance, machine vision. This makes it is possible to address only part of the total imager array, which can be seen as a zooming function or windowing (Wadsworth 89).

This windowing function is also applicable to CIDs because they have the same addressability as MOS-XY imagers. For the same reason CIDs also have low signal-to-noise ratio in common with the MOS-XY sensors : a relatively large sensing-line capacitance is connected to the relatively small capacitance of the photosensitive cell.

Table 4.1. Overview of advantages and disadvantages of the different types of two-dimensional imagers under study. (All entries are explained in the text.)

DEVICE	PROs	CONs
FT CCD	Horizontal resolution High aperture ratio Shutter possibility	Large chip size High smear level
IL CCD	Small chip size	Small aperture ratio High smear level Charge reset Production technology
FIT CCD	Low smear level	Large chip size Small aperture ratio Production technology
MOS-XY	Production technology Windowing possibility	Signal-to-noise ratio
CID	Cell design Windowing possibility Nondestructive readout	Signal-to-noise ratio

Of the solid-state imagers the CID is the only one which offers the possibility of nondestructive readout. If the charge packets are not dumped into the substrate after measurement of the displacement current in the MOS gate, accumulation of extra charges in the potential well can continue, and extra readout cycles can take place.

WORTH MEMORIZING

Solid-state imagers can be configured in different organizations :
- one-dimensional devices, which can be subdivided into linear, bilinear and quadrilinear imagers;
- two-dimensional devices such as frame-transfer imagers, interline-transfer imagers and frame-interline-transfer

imagers as far as CCDs are concerned, and MOS-XY and charge-injection devices for XY-addressable architectures.

4.3. Conclusions

This first chapter on the subject of solid-state imagers describes the conversion of incoming photons into electrons and the subsequent readout of these electrons. For both processes, generation and readout, two options are specified. Photodiodes, made by a metallurgical n^+p junction or an np junction induced by means of an MOS capacitor, perform collection of the charges after generation of the electron-hole pairs by the absorption of the photons. The electric field across the np junction separates the electrons from the holes, and the minority carriers involved are captured in the potential well or depletion layer of the structure.

The next step in the photon-sensing process is the readout of the integrated charge packet. Two alternatives can be used : the sensing-line arrangement and the CCD readout. The former is a very simple construction but suffers from the fact that the small capacitance of the photodiode is connected to the large capacitance of the sensing line, which does not enhance its signal-to-noise performance. The latter alternative combines a more complicated structure with the small capacitance of a CCD cell and has superior signal-to-noise characteristics.

With these first fundamentals about solid-state imaging in mind, the various device architectures are listed, ranging from one-dimensional to two-dimensional devices. One-dimensional imagers can be further subdivided into linear, bilinear, and quadrilinear image sensors. Although the construction and driving of bilinear and quadrilinear sensors are slightly more complicated than those of linear arrays, they can have much higher numbers of light-sensitive cells per millimeter, which makes them very suitable for very high-resolution imaging.

Two-dimensional image sensors are divided into those with architectures relying on the CCD principle and those built around an XY-addressable matrix. The arrays described are :
- the frame-transfer imager with its separate image and storage sections, its high concentration of pixels, ensuring very high resolution but also resulting in relatively large chip size;
- the interline-transfer imager with its storage regions in between the light-sensitive pixels, resulting in a poor aperture ratio of the pixels but requiring only a small chip size;
- the frame-interline-transfer imager combining a standard interline-transfer image sensor with the storage area of the frame-transfer device. This architecture has been primarily developed to tackle the smear problem,

but unfortunately it combines the disadvantages of the interline- and frame-transfer devices;
- the MOS-XY addressable matrix of photodiodes which were about the first solid-state imagers to appear as a consumer product on the market, but with rather poor noise performance compared to CCDs;
- charge-injection devices, similarly with poor noise performance as well, but with the outstanding characteristic of being the only imager with a nondestructive readout mode.

Top left : a small frame-transfer CCD (2.7 mm image diagonal) containing 200,000 pixels intended for low-cost applications. The imager is mounted in a plastic package (courtesy of Sanyo). Top right : a 768 x 484 interline-transfer imager which is compatible with a 2/3 inch optical format (courtesy of Kodak). Middle : two cross sections through the CCD structure along the direction of charge transport are shown. In the bottom cross section the overlapping poly-Si gates can be recognized, in the top cross section an optimized technology is applied to minimize the overlap capacitances of the various CCD gates. The latter have a length of 3 μm (courtesy of Philips). Bottom left : top view of a frame-interline-transfer device which uses the M-FIT architecture to capture the images in a progressive mode (courtesy of Matsushita Electric Company). Bottom right : the application of solid-state imagers in a portable HDTV camera : the camera contains three chips, each having 2.2M pixels (courtesy of BTS).

Chapter 5

FUNDAMENTALS OF SOLID-STATE IMAGING

In this section further attention will be paid to the fundamentals of solid-state imaging but not as far as the architecture of imager types is concerned. This subject has been discussed in the preceding chapter. The main accent will now be put on the device physics. As already stated in section 4.1, the basis of solid-state imaging can be divided into two parts : first, the absorption of photons in the substrate of the device and, second, the collection of the generated charge carriers. Both are described here with some additional fundamental details.

The most important optical properties such as the spectral response, the quantum efficiency, spatial resolution, and aliasing will also be studied. The spectral response is a measure of the output voltage or current as a function of the incoming light energy. As will be illustrated, the spectral response is extremely wavelength-dependent. The same is true of the quantum efficiency, which is defined as the number of electrons generated for each incoming photon. Although the first two parameters in fact depend on the wavelength or energy of the photons, the spectral response of solid-state imagers is directly proportional to the number of photons impinging on the device. Resolution will be dealt with using the modulation transfer function, which is a very powerful aid in describing the ability of the devices to resolve spatial information. As will be seen, however, the "integration in time" and the "averaging through a window" of the image sensors to capture their information, can give rise to aliasing problems. Although present in an imager in the form of spatial information, the aliasing components are fully similar in nature to those already described during the study of electrical-signal sampling (see section 3.5).

5.1. Absorption of photons

Figure 5.1 shows schematically what happens in the bulk of the semiconducting substrate of the imager when the surface is struck by a photon flux ϕ_0. If the energy of the photons (h.ν) is higher than the bandgap E_g of the semiconductor, photons will be absorbed and the actual flux $\phi(x)$ at a depth x in the substrate will differ from the incoming flux ϕ_0. The photon flux $\phi(x)$ as a function of the depth x can be written as :

$$\phi(x) \ = \ \phi_0 \cdot e^{-\alpha \cdot x} \ , \tag{5.1}$$

where α is the absorption coefficient of the substrate material.

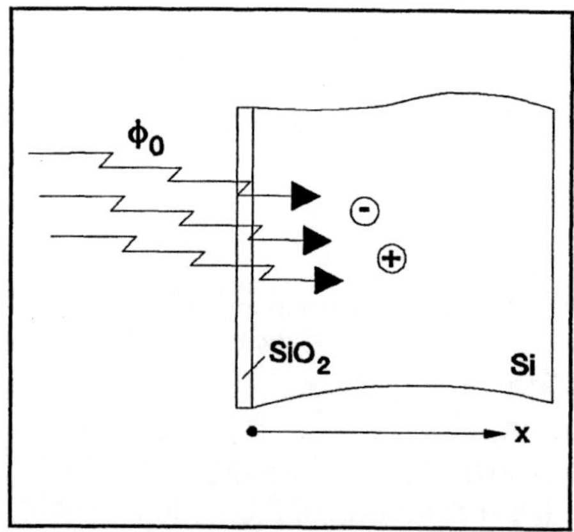

FIGURE 5.1. Schematic illustration of the generation of an electron-hole pair, due to the impinging photons, in the bulk of the silicon.

The decrement of the photon flux dϕ within a piece of silicon of thickness dx, due to photon absorption in the silicon layer of thickness dx, can be deduced as :

$$d\phi(x) = -\phi_0.\alpha.e^{-\alpha.x}.dx \ . \qquad [5.2]$$

On the assumption that each absorbed photon will generate a single electron-hole pair, the generation velocity g(x) of the electrons or the holes at a certain depth x in the substrate will equal the local decrement of the flux dϕ, divided by the energy of a single photon h.ν :

$$g(x).dx = -\frac{d\phi}{h.\nu} = \frac{\phi_0.\alpha.e^{-\alpha.x}}{h.\nu}.dx = \frac{\phi_0.\alpha.e^{-\alpha.x}.\lambda}{h.c}.dx \ , \qquad [5.3]$$

where λ is the wavelength of the incoming light and c the velocity of the light.

It is important to note that the generation velocity g(x) depends closely on the wavelength λ of the incoming light and on the incoming flux ϕ_0, but also via the absorption coefficient α on the characteristics of the substrate material used. A parameter of interest with respect to this phenomenon is the penetration depth x^*, defined as the depth at which the remaining flux $\phi(x^*)$ is equal to :

$$\phi(x^*) = \phi_0.e^{-1} \qquad [5.4]$$

or :

$$x^* = \alpha^{-1} .$$

[5.5]

Characteristic of the remaining photon flux at the penetration depth x^* is its value, namely 37 % of the incoming flux. The penetration depth x^* is a measure for the piece of silicon on which a fixed amount of light energy is absorbed.

The dependence of the absorption coefficient α on the wavelength λ for a silicon substrate (White 76) and the corresponding curve for the penetration depth are shown in Figure 5.2. If the wavelength of the incoming light is 500 nm (green light), $\alpha = 10^4/$cm and $x^* = 1$ μm, for a wavelength of the incoming light equal to 1000

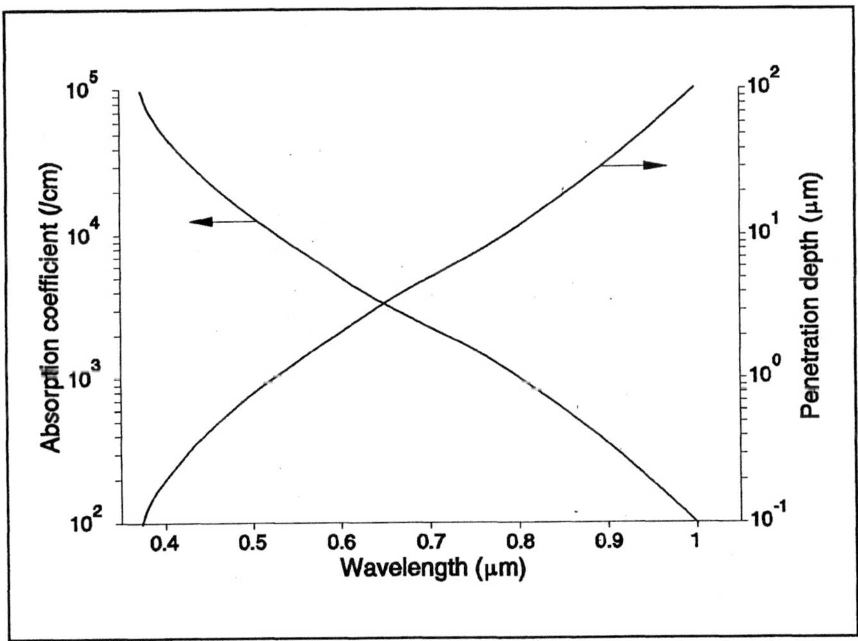

FIGURE 5.2. The absorption coefficient of silicon together with its corresponding penetration depth as a function of the wavelength of the incident light.

nm (infrared light), $\alpha = 100/$cm and the penetration depth is of the order of 100 μm. Figure 5.2 shows clearly that infrared light can penetrate much more deeply into silicon before it is absorbed than green light. For even shorter wavelengths (blue light, for example) the absorption coefficient is higher still, and the penetration depth still smaller, while the photon flux is absorbed in a much thinner layer of silicon.

WORTH MEMORIZING

Absorption of photons in silicon results in the generation of electron-hole pairs. The amount of charge generated depends on the incoming flux, the wavelength of the incoming light, and the absorption coefficient of the semiconducting substrate. Depending on the wavelength, the depth at which a fixed amount of photon flux is absorbed, is very shallow (e.g. blue light) or quite considerable (e.g. infrared light).

5.2. Collection of generated carriers

After the absorption of photons and generation of electron-hole pairs, the charge carriers will be separated by means of an electric field, as already described in section 4.1. This electric field is the result of the np junction used. The reverse-biased np junction can be induced by an external gate voltage or can be defined by a metallurgical junction, as it is in photodiodes. The depletion region spanned by the electric field plays an important role in the description of the collection of charge carriers. Because all minority carriers generated in this depletion region will be captured, the collection efficiency for these charge carriers in this region will always be 100 %.

On the other hand, not all photons will be absorbed in the depletion layer and some of them will generate electron-hole pairs in the neutral bulk of the semiconductor material. To be collected in the depletion region of the np junction or in the potential well of a CCD cell, the minority carriers generated in the neutral bulk have to diffuse toward the collection site.

This situation is shown in Figure 5.3, where a p-type silicon substrate of thickness t_{si} is shown, with a depletion region which has a width x_d, induced by a positive voltage V_G on the gate. During the diffusion process the charge carriers can also be lost by recombination if the diffusion length of the minority carriers is too small.

The total efficiency of this collection process η_c can be described by :

$$\eta_c = \eta_{dl} + \eta_{bulk} , \qquad\qquad [5.6]$$

where the collection in the depletion layer is described by η_{dl} and its counterpart from the neutral bulk by η_{bulk}.
These two components η_{dl} and η_{bulk} can be written as (Van de Wiele 76) :

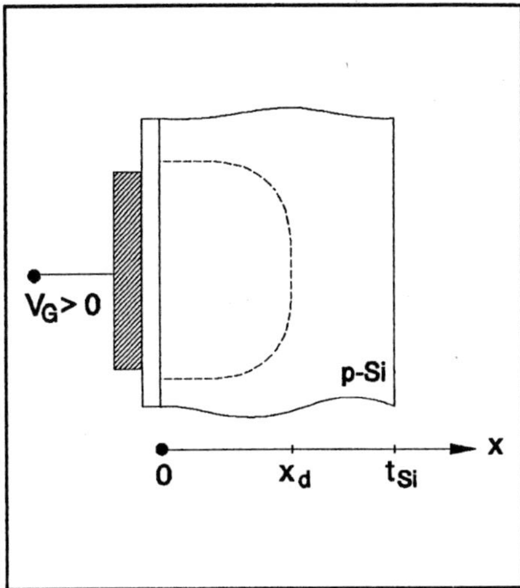

FIGURE 5.3. Schematic cross section of an MOS capacitor on a p-type silicon substrate, with a positive voltage on the gate. The depletion region is indicated by the dotted line.

$$\eta_{dl} = 1 - e^{-\alpha . x_d} \tag{5.7}$$

and :

$$\eta_{bulk} = \frac{\alpha . L_n^2}{\alpha^2 . L_n^2 - 1} \cdot \left(\alpha . e^{-\alpha . x_d} + \frac{e^{-\alpha . t_{Si}} - e^{-\alpha . x_d} . \cosh\left[(t_{Si} - x_d)/L_n\right]}{L_n - \sinh\left[(t_{Si} - x_d)/L_n\right]} \right) , \tag{5.8}$$

where L_n is the diffusion length of the minority carriers.

The formula for η_{dl} can be seen as a 100 % collection efficiency for those carriers which are generated within the depletion layer thickness w_d. The relation which represents η_{bulk} is somewhat more complicated but takes into account the recombination phenomenon with the parameters L_n and t_{Si}, which are respectively the diffusion length and the thickness of the substrate.

Note the dependence of η_c on the various parameters : longer wavelength of the incoming light in combination with its lower absorption coefficient results in less absorption in the depletion region and the overall collection efficiency depends to a great extent on the bulk characteristics of the silicon. A larger diffusion length L_n results in a higher collection efficiency.

On the other hand, if the incoming light has a shorter wavelength, it will be much more absorbed in the depletion region itself, and the recombination characteristics of the bulk material will play a much smaller role in the collection of the minority carriers.

WORTH MEMORIZING

After generation of the electron-hole pairs the minority charges have to be collected in the potential wells of the CCD. This process can be divided into two separate mechanisms :
- collection of charges generated in the depletion region. This process is highly efficient because the charges are collected at the place where they are generated;
- collection of charges generated outside the depletion region. This process relies on the diffusion of the charges to the potential wells and has a lower efficiency because the charges can recombine during their diffusion process.
Due to the combination of the two collection phenomena, the overall collection efficiency is highly wavelength-dependent.

5.3. Spectral response

A possible measure for the output response of the imager due to an optical input signal is the spectral response R (A/W) of the device, defined as the ratio of the output current of the sensor to the incoming light power. The output current of a CCD imager I_{out} (A/cm^2) can be written as the number of minority carriers Q_n (C/cm^2) collected divided by the integration time T_{int} :

$$I_{out} = \frac{Q_n}{T_{int}} \, . \tag{5.9}$$

(This study is concerned with n-channel devices transferring electrons but the theory naturally also applies to p-channel CCDs.)

If the output structure of the CCD imager under study is a floating-diffusion node connected to a source follower, the relation between the number of carriers Q_n and the output voltage measured at the output amplifier, V_{out} (like formula [2.6]) can be written :

$$Q_n = \frac{C_{FD} \cdot V_{out}}{A_{SF} \cdot A_{cell}} \qquad [5.10]$$

where :
- C_{FD} : floating-diffusion node capacitance,
- A_{SF} : gain of the source-follower structure,
- A_{cell} : area of a single cell or pixel.

Putting all these parameters together in the definition of the spectral response results in :

$$R = \frac{I_{out}}{\Phi_0} = \frac{C_{FD} \cdot V_{out}}{A_{SF} \cdot A_{cell} \cdot T_{int} \cdot \Phi_0} \; . \qquad [5.11]$$

Although not explicitly mentioned in the preceding formula, R is closely dependent on the wavelength of the incoming light because, as discussed earlier, the absorption and the collection of charges depend on the wavelength or the energy of the photons impinging the imager surface. The relation between the spectral response R and the wavelength of the incoming light is shown in Figure 5.4 (Dyck 82). The data illustrated is substrate-material dependent and, of course, holds only for silicon. On the basis of this figure a number of interesting observations can be made, bearing in mind that the curves are drawn for a constant incoming light flux Φ_0 :
- the parameter in this figure is the diffusion length L_n of the electrons in the p-type substrate. The curve for $L_n = 0$ gives the response of only the depletion region because with $L_n = 0$, all electrons generated outside the depletion region will recombine immediately. The curve for $L_n = \infty$ gives the response when each absorbed photon delivers an electron to the charge packet because none will recombine. Of course, both situations are theoretical, and the reality lies somewhere inbetween. Curves for different realistic values of L_n are shown : the smaller L_n is, the greater the recombination effects and the smaller the response at the imager output will be;
- the level of response due to collection in the depletion region is shown by the curve $L_n = 0$, while the level of response due to collection in the neutral bulk can be found between curve $L_n = 0$ and the curve for the actual value of L_n;
- the almost linear part common to all curves at their left-hand side can be explained as follows : with a constant input of light energy, the number of photons is a linear function of wavelength because the energy of the photons is wavelength-dependent. This explains the linear character of the curves : longer wavelengths incorporate more photons. For smaller wavelengths on the other hand, the absorption coefficient is so high that

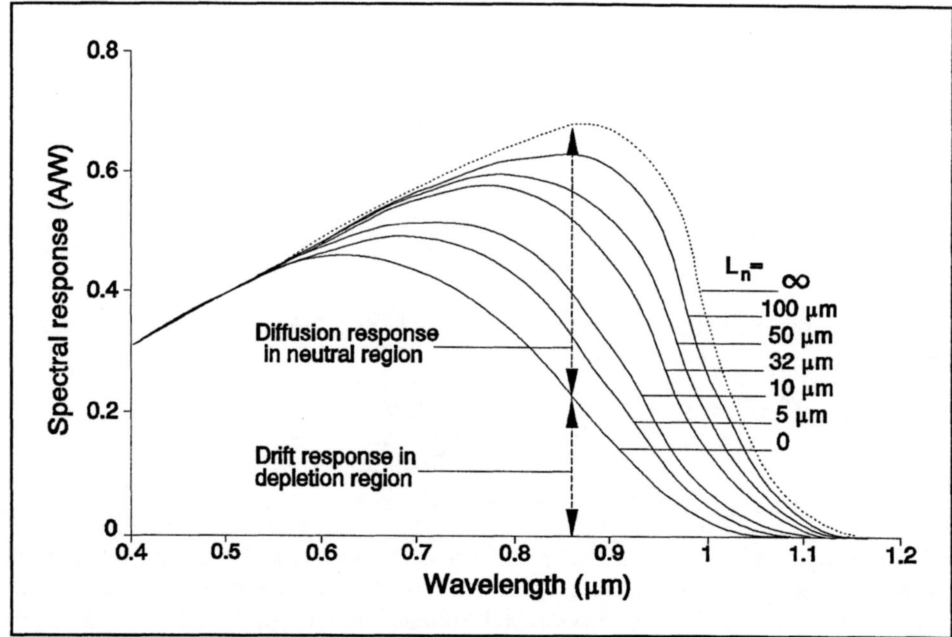

FIGURE 5.4. The graphs show the spectral response of an imager as a function of the wavelength of the incident light, parameter is the diffusion length.

almost all photons are absorbed in the depletion layer. For these reasons the curves all coincide initially (all photons being absorbed in the depletion region) and are linear (the number of photons is proportional to the wavelength);

- because of the limited thickness of the substrate itself and the lower absorption coefficient for longer wavelengths, not all the photons with these wavelengths penetrating the substrate are absorbed. A proportion of them simply passes through the complete semiconductor and do not generate any electron-hole pair. The thinner the substrate or the longer the wavelength of the light, the smaller the spectral response of the imager for longer wavelengths will be. For these reasons, the spectral response becomes small and, possibly as low as zero, for long wavelengths. Even if the diffusion length of the minority carriers is infinity, the imager will not deliver any output signal for these kinds of optical inputs.

A practical value for L_n is in the range around 100 μm if the devices are made on standard p-type silicon. However, the value of L_n can be shaped artificially : the top part of the substrate can be engineered to have a high value for its diffusion length, with a extremely small value for L_n in the rest of the substrate. This abrupt changeover can be introduced by choice of the substrate material or doping.

Fabricating the CCD on a p-p$^+$ substrate (e.g. an epitaxial p-type top layer of 10 μm on a thick silicon substrate, highly p$^+$ doped) results in a high value of L_n in the p-layer and a very small value of L_n in the p$^+$-bulk. Another alternative for fixing the diffusion length by means of doping profiles in the silicon substrate is a vertical-npn structure. The precise effect of this on L_n and the background for applying this npn-doping combination will be explained later in section 6.3.

If relation [5.11] is rewritten, the output voltage V_{out} can be expressed as :

$$V_{out} = \frac{R \cdot A_{SF} \cdot A_{cell} \cdot T_{int} \cdot \Phi_0}{C_{FD}} \qquad [5.12]$$

and V_{out} is a linear function of the integration time T_{int} and the incoming photon flux Φ_0. This relation is shown in Figure 5.5, where V_{out} is plotted as a function of T_{int}, with the input light flux Φ_0 as a parameter. Note the slow increase of V_{out}, even for $\Phi_0 = 0$ μW/cm^2. This corresponds to the conditions in dark. The measure of V_{out} as a function of the integration time T_{int} with $\Phi_0 = 0$ μW/cm^2 equals the dark current. This latter effect is not included in [5.12].

WORTH MEMORIZING

The spectral response of a solid-state image sensor such as CCD depends very greatly on the wavelength of the incoming light and on the characteristics of the substrate material in terms of its diffusion length. On the other hand, the spectral response for a given wavelength is completely linear in relation to the integration time and the incoming photon flux.

5.4. Quantum efficiency

Although the spectral response R can be a parameter helping to define an imager's output, a more practical value for comparing different sensors with each other is the quantum efficiency η. This parameter is defined as : the number of collected electrons divided by the number of photons impinging on the device.

The number of electrons can be expressed as :

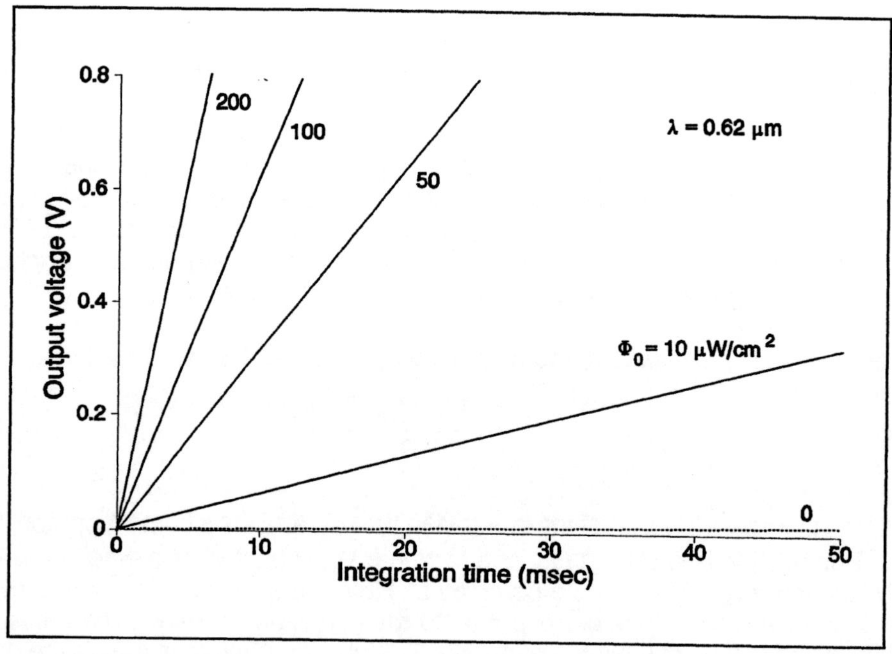

FIGURE 5.5. Relation between the output voltage and the integration time with the incident light flux as parameter.

$$\text{number of electrons} = \frac{Q_n}{q.T_{int}} \,, \qquad [5.13]$$

while the numbers of photons can be written as :

$$\text{number of photons} = \frac{\Phi_0}{h.\nu} \,. \qquad [5.14]$$

Replacing ν by its equivalent λ/c, the quantum efficiency can by expressed as :

$$\eta = \frac{Q_n.h.c}{q.T_{int}.\Phi_0.\lambda} \,. \qquad [5.15]$$

Taking into account the expression [5.11] for R, it is easy to define the spectral response R as a function of the quantum efficiency η :

$$R = \frac{\eta \cdot q \cdot \lambda}{h \cdot c} \, , \qquad\qquad [5.16]$$

and, vice versa :

$$\eta = \frac{R \cdot h \cdot c}{q \cdot \lambda} \, . \qquad\qquad [5.17]$$

Figure 5.6 shows a typical spectral response of a linear CCD imager which has MOS capacitors as its sensing elements. The spectral response R is plotted along

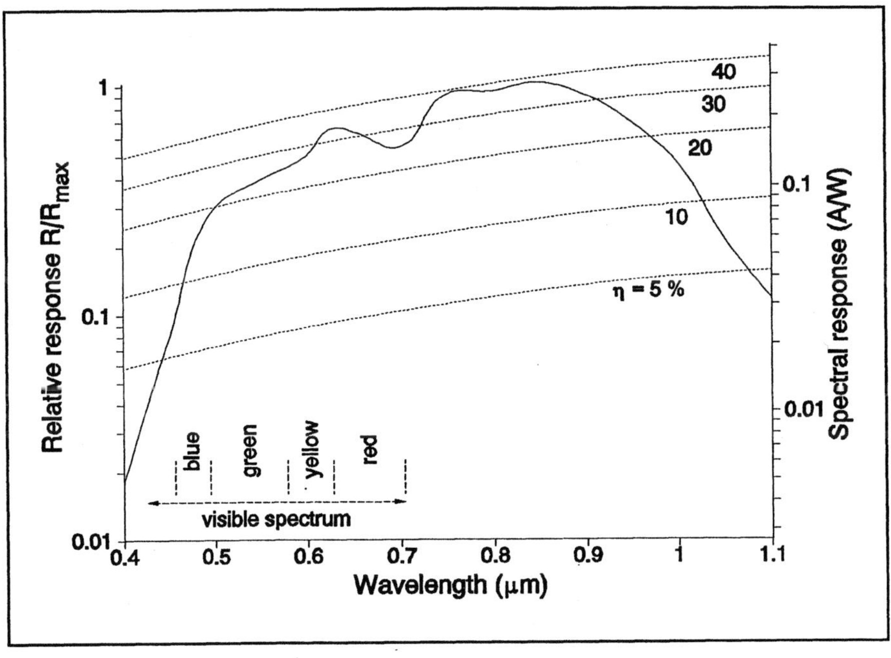

FIGURE 5.6. A typical spectral response of a linear CCD imager with MOS capacitors as sensing elements. Lines of constant quantum efficiency are included in the plot.

both Y axes, normalized at the left and, in absolute values at the right. Figure 5.6 also includes lines of constant quantum efficiency. Observe that these lines run parallel to each other but not parallel to the horizontal axis. This is due to the already described effect of increasing the number of photons with increasing wavelength of the incident light for a constant input light flux Φ_0, as indicated by [5.14].

So far as the response of the CCD is concerned, three different regions can be distinguished in Figure 5.6 :

- for shorter wavelengths, the device is somewhat insensitive. This is due to the absorption of photons in the polysilicon gates which completely cover the pixels of the CCD imager;
- for longer wavelengths, the device is again somewhat insensitive, but now the photons are hardly absorbed at all, because the absorption coefficient of the substrate material is greatly reduced;
- the region in between, where the quantum efficiency reaches its maximum or about 30 % to 40 % of the impinging photons give rise to an electron which is sensed at the output of the CCD. Notice the peaks and valleys in the response curve. These can be explained by multiple reflections of the light in the multilayer structure on top of the silicon substrate.

WORTH MEMORIZING

A parameter commonly used to describe the response of an imager to an optical stimulus is the quantum efficiency. An analytical expression can be derived which links the quantum efficiency and the spectral response to each other.

5.5. Resolution

Image sensors are used not only to detect light-intensity variations in time, but also to detect light variations in the spatial domain. Here, too, the devices have their limits : not all of them can detect the same high spatial frequency. This detection limit depends on the total number of pixels available. The length or the area of the device also plays a crucial role in detecting spatial information. A parameter describing the detection capabilities of an imager as regards spatial information is its resolution. Resolution is a measure for the highest spatial frequency which can be resolved by the device taking a specific contrast in account. These spatial frequencies are stated in line-pairs/mm (lp/mm).

The response of an image sensor to changing spatial frequencies is characterized by its modulation transfer function (MTF). As will be shown, the MTF curve of the device makes it easy to specify its resolution. The MTF itself is defined as the response of the complete optical system to a sinusoidally changing spatial frequency of the input signal with frequency f_{sig}. Both values, the response of the imager R and the input frequency f_{sig}, are normalized respectively to the response of the imager at zero spatial frequency and to the sampling frequency of the CCD.

The MTF of the complete optical system can be expressed as :

$$MTF = MTF_D \cdot MTF_{\varepsilon} \cdot MTF_G \cdot MTF_L \qquad [5.18]$$

with :
- MTF_D : diffusion MTF of the image sensor,
- MTF_{ε} : MTF as a result of the incomplete charge transport,
- MTF_G : geometric MTF of the image sensor,
- MTF_L : MTF of the lens of the optical system.

Discussion of the lens characteristics MTF_L is beyond the scope of this study, because it is a parameter of the optical system used and not a sensor parameter.

5.5.1. DIFFUSION MTF

Portion of the minority carriers generated by photons have to diffuse to the potential well in order to be collected. Because the diffusion of these charges takes place in the quasi-neutral bulk, no predefined direction for this diffusion exists. Many charge carriers will be not collected at all because they recombine or diffuse in any other than those directions of the potential wells. Another possibility of "misdiffusion" is shown in Figure 5.7, where a generated electron (in the quasineutral p-type bulk) diffuses toward a pixel which is not located exactly above the generation site and actually contributes to the charge packet adjacent to the pixel to which it was supposed to diffuse.

For example, the simplified situation in Figure 5.7 has two electrons in the upper potential well, one which in fact does not belong where it was collected, and another which ultimately arrived in the correct potential well. The lower potential well is empty, although it had normally to contain also a single electron. Instead of having one electron in each potential well, the end result is two electrons in the upper well and no electrons at all in the lower well. The original input signal is "contaminated" by the diffusion of the charge carriers through a region where they are not guided toward the correct bucket.

This effect of misdiffusion is described by the diffusion modulation transfer function : MTF_D. By solving the continuity and diffusion equation it is possible to derive the following equation for MTF_D (White 76) :

$$MTF_D = \frac{1 - \dfrac{e^{-\alpha . x_d}}{1 + \alpha . L_K}}{1 - \dfrac{e^{-\alpha . x_d}}{1 + \alpha . L_n}}, \qquad [5.19]$$

the parameter L_K being defined by :

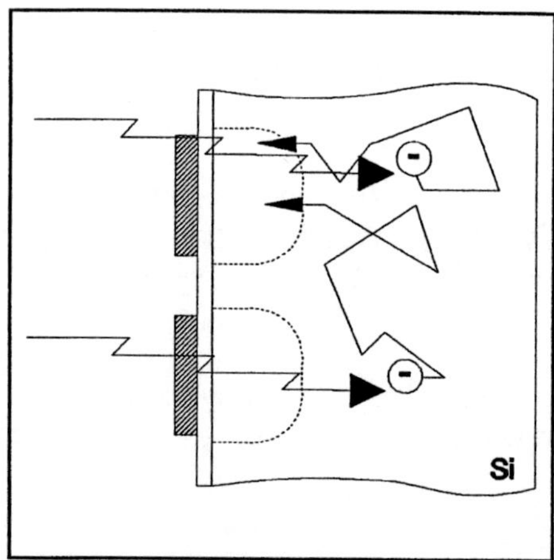

FIGURE 5.7. Degradation of the MTF due to "misdiffusion" of generated charge carriers.

$$L_K = -\frac{1}{L_n^{-2} + (2\pi . f_{sig})^2} \, , \qquad\qquad [5.20]$$

where f_{sig} represents the frequency of the spatial input signal.

Figure 5.8 shows the behavior of MTF_D as a function of the input signal frequency, with the thickness of the depletion region x_d (= 10 μm) and the diffusion length L_n (= 100 μm) fixed, and the wavelength λ of the light as a parameter.
Referring to Figure 5.8, the following observations can be made :

- for higher values of the wavelength λ of the light, the MTF_D decreases due to the fact that, corresponding to higher values of λ, the absorption coefficient of silicon α is decreasing, the light is effectively penetrating deeper into the substrate before getting absorbed, and consequently the charge carriers have to diffuse over a longer distance, which increases the chance of a misdiffusion;
- for lower values of the input frequency, a misdiffusion over a small number of pixels is a smaller problem than it is at higher frequencies. For example, if the mean misdiffusion is over a distance comparable to three pixels, input frequencies considerably lower than a repetition frequency corresponding to these three pixels are almost unaffected by the diffusion MTF_G;
- although not shown, an increase in L_n will increase the diffusion possibility and will consequently increase the chance of misdiffusion and adversely affect the MTF_D;

- increasing x_d will decrease the number photons which pass through the depletion region without being absorbed and consequently the MTF_D will increase.

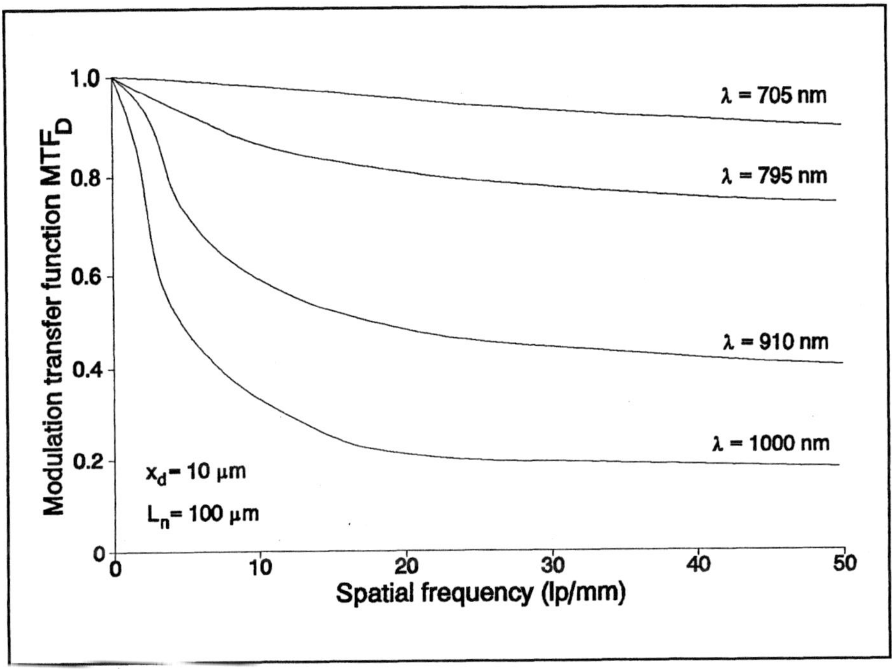

FIGURE 5.8. The behavior of the diffusion MTF as a function of the spatial frequency. The wavelength of the incident light is used as a parameter.

An interesting conflict can be seen with regards to the diffusion length L_n in the discussion of the light sensitivity and the diffusion MTF. A high value of L_n will make the sensor more sensitive to light with higher wavelengths, but on the other hand it will have a negative effect on MTF_D. Vice versa, all measures to optimize MTF_D via the parameter L_n will adversely affect the spectral sensitivity.

5.5.2. TRANSPORT MTF

As discussed in an earlier section (3.1), the transport inefficiency ϵ will degrade the frequency response of a CCD delay line. Of course, the same holds for the CCD imagers. In this section the same expressions can be handled as derived previously, except for the conversion of the spatial frequency f_{sig} instead of the frequency f of the electrical input signal. The conversion to and from f_{sig} and f_s can be written as :

$$f = \frac{f_{sig}}{1/CL}.f_c \; , \qquad\qquad [5.21]$$

f_c and CL being respectively the CCD clock frequency and CCD cell length.

The generalized expression for the transport MTF_ε of a CCD with N cells and m gates per cell can, with the aid of relation [3.22], be written as :

$$MTF_\varepsilon = e^{-N\varepsilon_m.[1-\cos(2\pi\,f/f_o)]} \; . \qquad\qquad [5.22]$$

By translating this expression to the spatial-frequency plane and bearing in mind that the Nyquist frequency f_N equals $1/2P_{pix}$, with P_{pix} the pixel pitch, it becomes
- for a CCD sensor with CL equal to the pixel pitch P_{pix} (e.g. a simple CCD line sensor) :

$$f = \frac{f_{sig}}{2.f_N}.f_c \qquad\qquad [5.23]$$

$$MTF_\varepsilon = e^{-N\varepsilon_m.[1-\cos(\pi\,f_{sig}/f_N)]} \; , \qquad\qquad [5.24]$$

- for a CCD sensor with CL equal to 2 pixel pitches (e.g. a bilinear CCD line sensor) :

$$f = \frac{f_{sig}}{f_N}.f_c \qquad\qquad [5.25]$$

$$MTF_\varepsilon = e^{-(N/2)\varepsilon_m.[1-\cos(2\pi\,f_{sig}/f_N)]} \; , \qquad\qquad [5.26]$$

- for a CCD sensor with CL equal to 4 pixel pitches (e.g. a quadrilinear CCD line sensor) :

$$f = \frac{2.f_{sig}}{f_N}.f_c \qquad\qquad [5.27]$$

$$MTF_\varepsilon = e^{-(N/4)\varepsilon_m.[1-\cos(4\pi\,f_{sig}/f_N)]} \; . \qquad\qquad [5.28]$$

Note that the number of transports in the expressions for MTF_ε changes from N in [5.24] to N/2 in [5.26] and to N/4 in [5.28] in accordance with the linear, bilinear and quadrilinear nature of the devices.

It is interesting to observe the nature of the curves representing the aforementioned MTF$_e$ expressions, as illustrated in Figure 5.9. In the case of the linear sensor

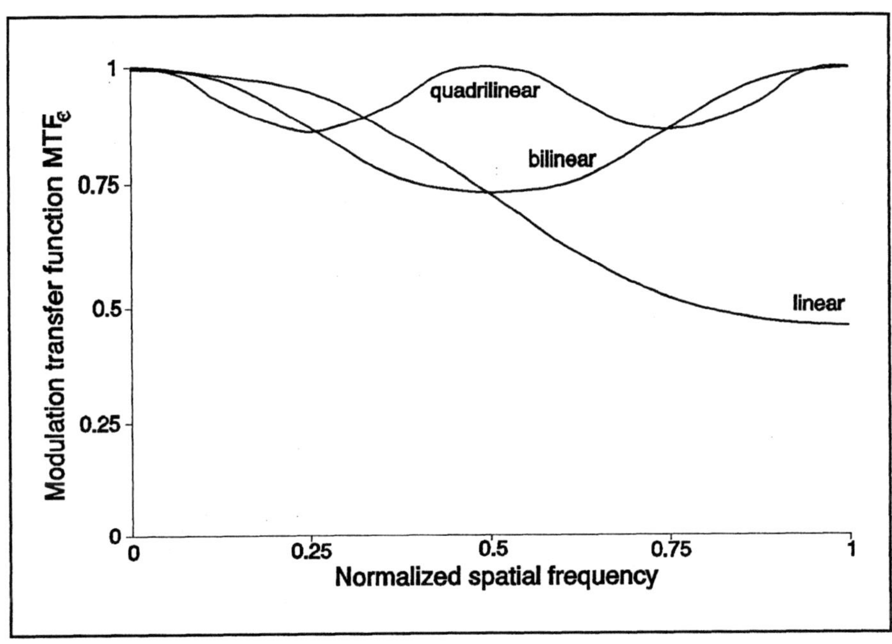

FIGURE 5.9. MTF degradation due to the transfer inefficiency, for a simple linear, a bilinear and a quadrilinear imager with equal f_i.

the MTF$_e$ decreases continuously with increasing spatial input frequency f_{sig}. This contrasts with the bilinear sensor : MTF$_e$ reaches a minimum value at $f_{sig} = f_N/2$ and returns to MTF$_e$ = 1 for $f_{sig} = f_N$. This behavior can be simply explained by the fact that the input signal corresponding to the Nyquist limit is a block pattern with alternately a pixel white and a pixel black. All odd pixels (e.g. all white ones) are shifted in one shift register of the CCD and all even pixels (e.g. the black ones) in the other shift register. Each individual shift register seems to contain DC information. All packets in a particular shift register have the same number of charge carriers in their potential wells.

The worst case for the bilinear device is found at $f_{sig} = f_N/2$, when the shift registers each contain alternating black and white information, and any transport inefficiency affects the difference between black and white most.

A similar statement is true of the quadrilinear device, but now there are two minima in the MTF$_e$ curve, for $f_{sig} = f_N/4$ and for $f_{sig} = 3.f_N/4$. Also at $f_{sig} = f_N/2$ and at $f_{sig} = f_N$, MTF$_e$ becomes equal to unity, because in both cases the four shift registers contain charge packets representing DC information, mutually different in amplitude.

Thus far only the effect of transport inefficiency on the MTF in a single CCD register has been involved. This applies unaltered to one-dimensional imagers. In the case of a two-dimensional device the behavior of the MTF$_e$ becomes complicated. With such image sensors and during the readout process, the charge packets have all to be transported in a combination of a vertical CCD register and a horizontal output register, irrespective of the architecture of the imager. In most devices the horizontal register is optimized for high-speed operation, and the vertical one for charge-handling characteristics, which makes the transport inefficiency parameters for both registers different in most applications.

Nevertheless, the aforementioned theory on MTF$_e$ remains valid, although the effect on the characteristics of a two-dimensional imager will be a combination of two independent MTF curves. On the other hand, what makes the transport MTF study of a two-dimensional imager complicated is the fact that not all charge packets have to travel the same distance before they come to the output. CCD cells located closer to the output amplifier will suffer less from the transport MTF than CCD cells located further away from the output node.

5.5.3. GEOMETRIC MTF

A solid-state image sensor uses its pixels to sample the optical information projected onto the device. But the pixels do not have an infinitely small window with which to sample the input signal. The pixels have a certain width and the sampling of the optical input signal is averaged across the pixel's width. Physical observation of the input information through a 'window' can be mathematically expressed by taking the convolution of the input signal together, with the windowing function. Taking the convolution in the spatial domain is the same as multiplying the input in the frequency domain with the Fourier transform of the window.

Figure 5.10 shows a single pixel row, with a pixel width Δx, a pixel pitch P_{plx}, and its center at x_0. Any arbitrary optical input signal $H_{in}(x)$ can be written as :

$$H_{in}(x) = H_0.[1 + A_0.\cos(2\pi f_{sig} x)] \qquad [5.29]$$

with H_0 and A_0 respectively its DC and its AC component.

The output signal $H_{out}(x_0)$ of the pixel at location x_0 with respect to this input signal $H_{in}(x)$ can be written as the integral of $H_{in}(x)$ between the boundaries $(x_0 - \Delta x/2)$ and $(x_0 + \Delta x/2)$, or :

$$H_{out}(x_0) = \frac{1}{\Delta x}.\int_{x_0-\frac{\Delta x}{2}}^{x_0+\frac{\Delta x}{2}} H_0.[1 + A_0.\cos(2\pi f_{sig} x)].dx , \qquad [5.30]$$

which can be written in the shortened form:

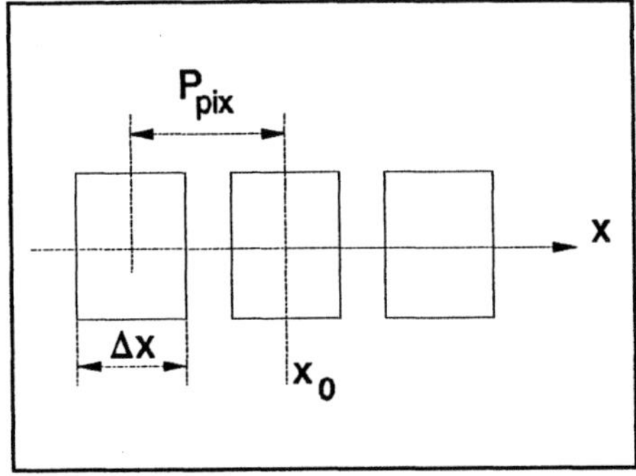

FIGURE 5.10. Representation of a single pixel row with a pixel pitch equal to P_{pix}, a pixel width Δx and with its center at x_o.

$$H_{out}(x_0) = H_0.[1 + M.\cos(2\pi f_{sig}\Delta x)] \, , \qquad [5.31]$$

the parameter M being defined as :

$$M = A_0 . \frac{\sin(\pi f_{sig}\Delta x)}{\pi f_{sig}\Delta x} \, . \qquad [5.32]$$

The geometric modulation transfer function MTF_G can be defined as :

$$MTF_G = \frac{M(f_{sig})}{M(f_{sig}=0)} = \frac{\sin(\pi f_{sig}\Delta x)}{\pi f_{sig}\Delta x} \, . \qquad [5.33]$$

Normalizing f_{sig} by f_N, and bearing in mind that :

$$\frac{1}{2.P_{pix}} = \frac{f_s}{2} = f_N \, , \qquad [5.34]$$

the expression for MTF_G can be transformed to :

$$MTF_G = \frac{\sin\left(\dfrac{\pi \Delta x}{2 P_{pix}}.\dfrac{f_{sig}}{f_N}\right)}{\dfrac{\pi \Delta x}{2 P_{pix}}.\dfrac{f_{sig}}{f_N}} \, . \qquad [5.35]$$

Figure 5.11 shows the geometric MTF for three different pixel organizations :
- curve (1) for noncontiguous pixels : $\Delta x < P_{pix}$ (e.g. $\Delta x = P_{pix}/2$) which is

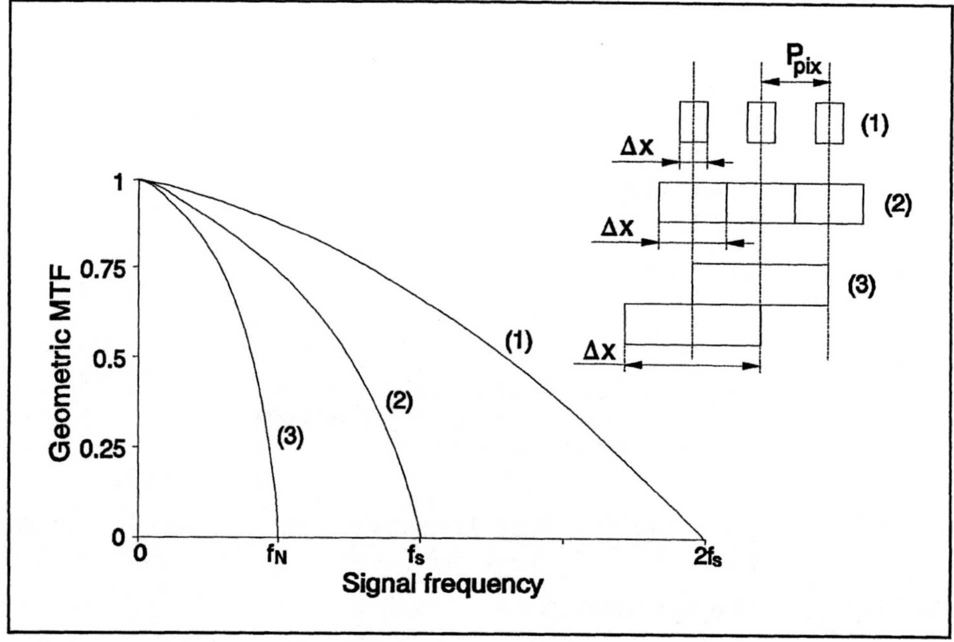

FIGURE 5.11. Illustration of the geometric MTF curves for three types of pixel organizations : (1) noncontiguous pixels, (2) contiguous pixels, (3) overlapping pixels.

the case for the pixels on one horizontal line of an interline-transfer CCD. The MTF_G becomes equal to :

$$MTF_G\big|_{\Delta x = P_{pix}/2} = \frac{\sin\left(\dfrac{\pi}{4}\cdot\dfrac{f_{sig}}{f_N}\right)}{\dfrac{\pi}{4}\cdot\dfrac{f_{sig}}{f_N}} \ . \qquad [5.36]$$

This curve has $MTF_G = 0$, for $f_{sig} = 2.f_s = 4.f_N$;
- curve (2), for contiguous pixels : $\Delta x = P_{pix}$, which can be the case for the pixels on one horizontal line of a frame-transfer CCD. The MTF_G becomes equal to :

$$MTF_G\big|_{\Delta x = P_{pix}} = \frac{\sin\left(\frac{\pi}{2}\cdot\frac{f_{sig}}{f_N}\right)}{\frac{\pi}{2}\cdot\frac{f_{sig}}{f_N}}. \tag{5.37}$$

This curve has $MTF_G = 0$, for $f_{sig} = f_s = 2.f_N$;
- curve (3), for overlapping pixels : $\Delta x > P_{pix}$, (e.g. $\Delta x = 2.P_{pix}$), which can be the case for vertically interlaced pixels (see section 6.2 below). The MTF_G becomes equal to :

$$MTF_G\big|_{\Delta x = 2P_{pix}} = \frac{\sin\left(\pi.\frac{f_{sig}}{f_N}\right)}{\pi.\frac{f_{sig}}{f_N}}. \tag{5.38}$$

This curve has $MTF_G = 0$, for $f_{sig} = f_s/2 = f_N$.

Note that the noncontiguous pixels show the highest value of MTF_G or that an interline-transfer CCD has inherently a higher geometric modulation transfer function than any other CCD imager. This feature explains why of devices with equal numbers of pixels on a line, this type of imager has potentially the highest resolution. However, as is the case for L_n with respect to MTF_D, it is not an advantage in all situations to have a high value for MTF_G. The reason why will be explained in the section which follows.

So far, the effect of misdiffusion, transport inefficiency, and the geometrical pixel aspects have been described. However, these parameters are not the only ones determining the MTF of an imager. Other sources can be :
    - the uniformity of the light transmission across the device;
    - the uniformity of the photosensitivity of the device itself;
    - multiple reflections of the incoming photons;
    - nonperpendicularly impinging photons.

WORTH MEMORIZING

The resolution of an image sensor, which is its ability to resolve spatial variations in the incoming signal, is, of course, one of its prime characteristics. This parameter can be aptly defined as the modulation transfer function or MTF. The modulation transfer functions is a combination of different modulation

transfer functions each which can independently degrade the overall performance of the device. The main components are :
> **- the diffusion MTF which describes the degradation of the device's MTF due to the possibility of misdiffusion of the electrons generated outside the depletion layer;**
> **- the transfer MTF which includes the nonideal charge-transfer effects;**
> **- the geometric MTF taking into account the influence of the geometrical features of the pixels themselves.**

5.6. Aliasing and Moiré effects

The sampling nature of a solid-state imager, in combination with the MTF_G behavior, can result in a very strange interference signal known as aliasing or Moiré effect.

If a signal with a well-defined and limited frequency spectrum is sampled with a frequency f_s, the output signal has an unlimited frequency spectrum in which the original frequency spectrum of the input signal is repeated and mirrored around all multiples of the sampling frequency f_s. This is one of the main characteristics of the sampling theorem : the frequency characteristic of the input signal is repeated and mirrored at $f = 0$, f_s, $2.f_s$, $3.f_s$, ... as gone into the study of sampling in section 3.5.

As long as the basic frequency spectrum of the input signal is limited to the Nyquist frequency $f_N = f_s/2$, there will be no interference between the basic spectrum around $f = 0$ and the one repeated around $f = f_s$. In all other cases frequencies will interfere, as can be seen in Figures 3.7 and 3.8

If this sampling theorem is applied to the solid-state imager, the basic frequency characteristic can be found in the MTF_G of the sensor. The MTF_G curve will be repeated and mirrored around f_s (and all subsequent multiples of f_s). This is shown in Figure 5.12 for an MTF_G curve of an imager with contiguous pixels : the solid line is the baseband and the dashed curve is the one mirrored and repeated around f_s. The interference between the two curves will be readily observed : if an input signal with $f_{in} > f_N$ is presented to the sampling system, it will be recognized in the output signal as a signal with a frequency equal to :

$$f_{out} = f_s - f_{in} \, , \qquad\qquad [5.39]$$

which is completely different from the original signal, or the original high-frequency input signal is reflected in the output signal as a low-frequency signal.

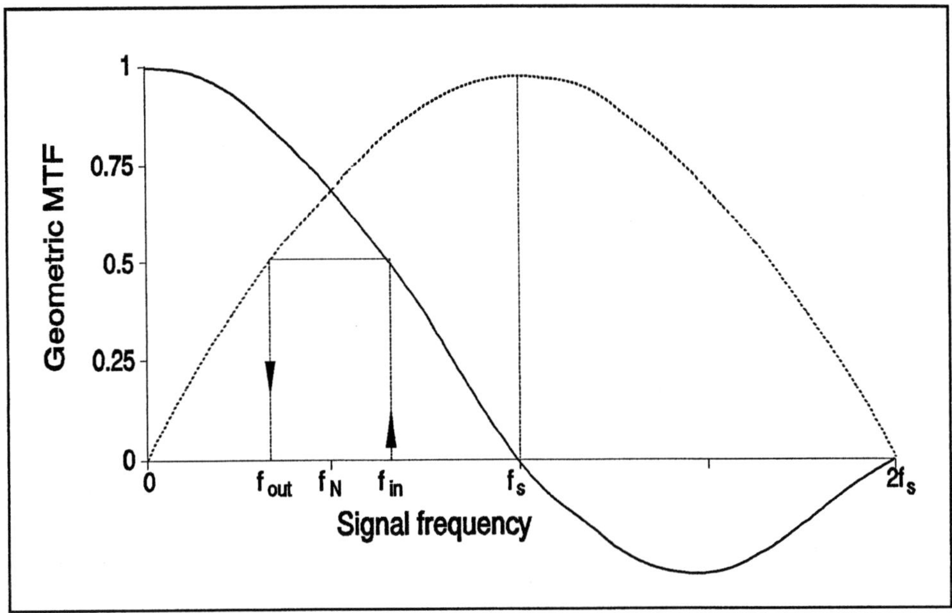

Figure 5.12. Illustration of the origin of the Moiré effect or aliasing.

From the system point of view it is impossible to distinguish this "aliasing signal" from any original low-frequency signal, and the interference cannot be compensated or corrected. The only way to avoid the Moiré effects is to limit the bandwidth of the input signal to frequencies lower than f_N (e.g. by means of an optical low-pass filter). Because of this limitation, a limit is set to the usable bandwidth of the imaging system (in the case of vertical interlaced TV systems (see section 6.1.1) this limitation is known as the Kell factor) (Hsu 86) :

$$\frac{\textit{usable bandwidth}}{\textit{Nyquist frequency}} \approx 0.7 \ldots 0.9 \, , \qquad [5.40]$$

which means that the usable bandwidth of the system is only 70 % to 90 % of that predicted by the Nyquist theorem.

It is interesting to note the fact that a high-resolution device or an imager with a high $\mathrm{MTF_G}$ curve (with constant f_N and f_s) is more susceptible to Moiré effects than a device with a lower $\mathrm{MTF_G}$. A higher $\mathrm{MTF_G}$ gives rise to a mirrored and repeated curve which is also greater in amplitude, and consequently the signals interfered with, are greater in amplitude. For this reason there has to be a trade-off between resolution and Moiré effects : for instance, on one hand the interline-transfer imager has a high $\mathrm{MTF_G}$, but the same holds for the aliasing components, while on the other hand the frame-transfer imager has intrinsically a lower MTF_G, but suffers less from Moiré effects.

WORTH MEMORIZING

The windowing action of a solid-state imager's pixels in combination with the sampling mechanism gives rise to the aliasing effect. Aliasing or Moiré effects cannot be corrected once they have appeared in the output signal. The only way to avoid them is to limit the spatial frequency of the input signal.
Imagers with the highest resolution potential always suffer most from aliasing effects.

5.7. Conclusions

In this chapter dealing with the fundamentals of solid-state imaging, several items are reported whose chief importance lies in the optical specification of imaging devices. The spectral response, quantum efficiency, resolution, and the sensitivity to aliasing all go to define the ability of the solid-state imager to convert the light information into electrical signals. The information carried by the photons can be offered to the imager as time varying or spatial variations. To promote study of the aforementioned characteristics of the imaging devices in greater depth, a more or less fundamental description of the processes involving the absorption of photons and the collection of the generated carriers is given.

Photons with a different energy content or of a different wavelength are absorbed at different depths in the silicon substrate. This effect is due to the wavelength-dependent absorption coefficient of the silicon substrate. This last parameter is relatively high for short wavelengths (e.g. UV and blue light) and relatively low for longer wavelengths (e.g. red and IR light). In practice this means that blue light is absorbed almost completely in a very thin layer of silicon, while red light penetrates quite deeply into the silicon before being absorbed. In combination with the collection of generated charge carriers, this effect of a differentiated absorption coefficient can give rise to the typical spectral response curve of a solid-state image sensor. Charges are only collected if they are generated into the depletion layer or potential wells of the CCD. If not, they first have to diffuse to the potential wells. On their passage to the potential wells, however, they can recombine and get lost. So the overall spectral response for a charge-coupled device can be summarized as follows :
    - for "short" wavelengths the photons are absorbed in the potential wells themselves and basically a relatively high response to photons of this

kind can be expected. But before generating electron-hole pairs, these photons have, of course, to reach the substrate itself and this problem limits the spectral response of the devices. Many short-wavelength photons have already been lost in the top layers above the silicon substrate. For this reason, and completely contrary to what the theoretical absorption curve of the CCD would predict, charge-coupled devices have only a limited response to blue light and are almost completely insensitive to UV light (unless special precautions are taken);

- for "long" wavelengths the photons are absorbed quite deeply into the silicon and almost completely outside the depletion regions. The diffusion process is not 100 % efficient and not all the generated electrons will reach the potential wells. In view of this physical effect, charge-coupled devices are not very sensitive to near-infrared light and are completely insensitive to an infrared stimulus (again unless special precautions are taken);

- for "medium" wavelengths the spectral response of the CCDs is a combination of the foregoing effects. For instance, "green" photons can easily find their way to the silicon substrate, are absorbed almost completely in the depletion layer, and result in a high spectral response and a high quantum efficiency.

The resolution or the ability of the devices to resolve spatial information is completely accounted for the modulation transfer function or MTF. This parameter can be divided into different parts : the diffusion MTF, the transport MTF, and the geometric MTF. All have been described extensively in this chapter in the form of a theoretical analysis and their impact on practical applications.

The diffusion of the charge carriers after their generation by the photons has to overcome the problem of misdiffusion besides the already mentioned point of recombination. Diffusing carriers are not directed toward the correct pixel of the imaging device, and the chance that they arrive at a wrong location is quite high, especially in the case of carriers generated deep in the silicon bulk (by long-wavelength photons). This effect can severely degrade the resolution characteristic of the image sensor.

As already described during study of the delay lines, transport inefficiencies can influence the amplitude and phase characteristics of the signal passed through the CCD. Of course, this statement is still valid for imaging applications of charge-coupled devices.

The pixels of the image sensors are of finite width and integrate the information over a finite time. Or they may act as an "averager" in both the spatial and the time domain. This complex behavior of the device has its consequences for the resolution characteristic owing to the geometry of the pixel. In section 5.5.3 analytical expressions have been derived to illustrate the influence of the pixel configuration on the MTF curve.

In combination with the sampling theory, which is also valid for imagers, the specific geometric MTF curve can explain the theoretical background for the aliasing or Moiré effects.   When present, these well-known artifacts are introduced via the nonlimited frequency spectrum of the spatial input signal. Moiré effects can never be eliminated when present in the electrical video signal because they cannot be distinguished from original low-frequency components.   The only way to get rid of these spurious signals is to limit the spatial input frequency.   This can be done by using optical low-pass filters, for instance.

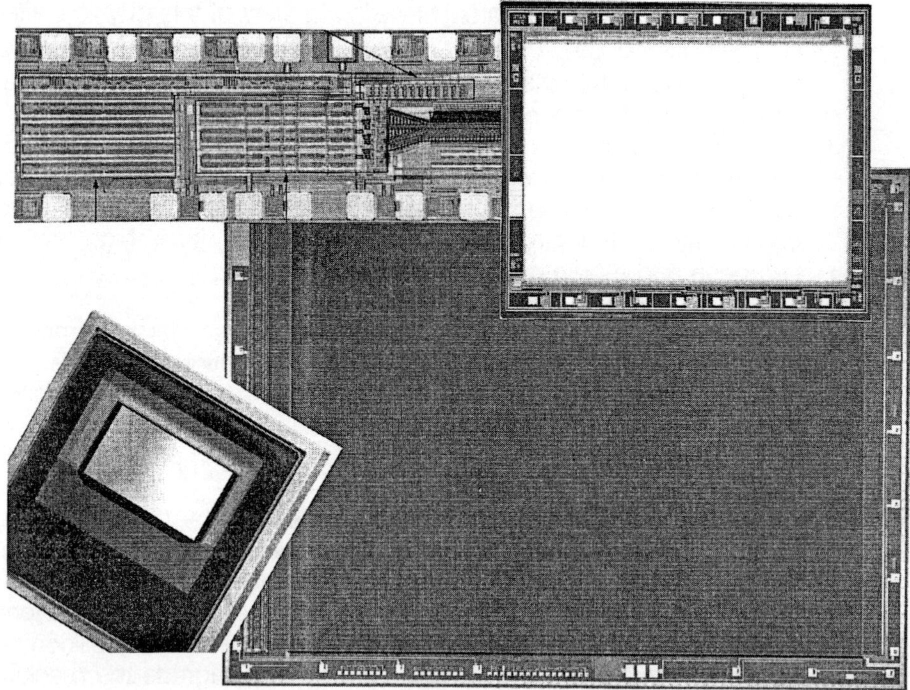

Top left : a linear array having three parallel pixel rows for color applications (R, G, B). The chip contains part of the driving and processing electronics (courtesy of Sony). Top right : a small interline-transfer image sensor with 270,000 pixels and an image diagonal compatible with a 1/3 inch optical format. The device is intended for camcorder-type applications (courtesy of Matsushita Electric Company). Bottom left : a frame-transfer image sensor with 2.2M pixels dedicated to HDTV applications. Within the package of the device, special precautions are taken to minimize optical reflections (courtesy of Philips). Bottom right : a top view of a Charge-Sweep Device with 512 x 512 pixels. The pixels are of the PtSi Schottky-barrier type optimized for IR imaging (courtesy of Mitsubishi).

Chapter 6

SOLID-STATE IMAGING FOR TELEVISION APPLICATIONS

For many years the application of solid-state imaging for television applications was exclusively limited to imaging tubes, but after the introduction of the first video camera with a solid-state imager as its image pickup element (Hanmo 83), everything changed very rapidly. Nowadays, nearly the opposite is true : imaging for TV applications is almost exclusively the domain of solid-state imagers and in particular charge-coupled devices.

Quite a long period elapsed between the introduction of charge-coupled devices in 1969 and the introduction of the first solid-state camera. The main reason for this was the complexity of imager-fabrication technology. Not surprisingly, the first camera was built around an MOS imager because this device could be produced using a somewhat simpler technology which was particularly pushed by the DRAMs. Nevertheless, CCD cameras came soon after cameras with MOS imagers. Table 6.1 gives an overview of the announcements up to 1980 of two-dimensional imaging arrays with the characteristic of being TV compatible (Koch 83). After the end date shown in Table 6.1 almost every company involved in television and video electronics announced new devices.

From Table 6.1. it will be seen that U.S. companies were first, followed by the Europeans. In these two continents the CCD was invented at two different locations but almost at the same time, so the introduction of the first imagers from these laboratories was not surprising. In 1979 however, the Japanese also showed their interest in solid-state imaging. This interest never slowed down and even almost pushed almost all imaging-tube activities from the scene.

The reasons for the success of solid-state imaging at the expense of imaging tubes are the many advantages of CCDs over classical imaging devices, namely :
- solid-state imagers have many mechanical characteristics which imaging tubes lack : they are much smaller in volume and weight, and no vacuum is required;
- CCDs have a fixed geometry which has proved to be very reliable and robust, and need almost no tuning of voltage settings;
- the cost of solid-state devices has been reduced drastically by the technology push and, nowadays, CCDs cost much less than any tube;
- compared to tubes, CCDs are operated at a much lower supply voltage and consume almost no power. Solid-state devices are not sensitive to external electromagnetic fields whereas tubes with an electron beam inside are;

- as far as the optical characteristics of the two challengers are concerned,
solid-state devices are almost as sensitive as tubes and present no image
lag or burn-in phenomena, but tubes are the equal of CCDs if resolution
is compared.

Table 6.1. : Overview of the announcements of two-dimensional, TV-compatible imaging arrays from the invention of the CCDs until 1980.

Year	Company	Array (H x V)	Type
1973	Fairchild	100 x 100	CCD (interline transfer)
1974	Bell	496 x 475	CCD (frame transfer)
	GE	244 x 248	CID
	RCA	512 x 320	CCD (frame transfer)
1978	Philips	300 x 200	CCD (resistive gate)
	Siemens	100 x 100	CID
1979	Hitachi	484 x 384	MOS XY
	Sony	492 x 226	CCD (interline transfer)
	Matsushita	512 x 383	CCD (frame transfer)
	Hitachi	320 x 244	MOS XY
	Fairchild	488 x 380	CCD (interline transfer
	Sony	490 x 570	CCD (frame transfer)
	TI	800 x 800	CCD (full frame)
	NEC	490 x 384	CCD (interline transfer)
	Toshiba	512 x 340	CCD (frame transfer)
	Siemens	100 x 100	MOS XY
1980	Matsushita	506 x 413	CCD (interline transfer)
	Hitachi	484 x 384	MOS XY
	Siemens	100 x 100	MOS XY
	Matsushita	492 x 404	MOS XY
	Sharp	488 x 385	CCD (interline transfer)
	Toshiba	492 x 400	CCD (interline transfer)

The only field in which a tube can beat a solid-state imager is picture quality : due
to the two-dimensional sampling process of a solid-state imager, aliasing problems
occur and the disturbing Moiré effects can be present in the displayed picture in
both the horizontal and in the vertical direction. Imaging tubes suffer only from
vertical aliasing components.

All the above-mentioned advantages of solid-state devices in general and charge-
coupled devices in particular have pushed imaging tubes almost completely out
of the field. Today, only CCDs are used as image sensors where TV applications
are concerned. In particular, their favorable signal-to-noise ratio compared to these

of MOS-XY and CID devices makes them attractive for these kinds of applications.

On the subject of imaging for TV applications, a few typical features of the imagers will be described : for instance, how a sensor can be operated in the interlaced and the progressive mode, how color can be reproduced, how spurious signals can be avoided upon overexposure of the imaging device, and shortening of the integration time by charge reset or an electronic shutter function.

6.1. Scanning modes

In the world of TV standards, different scanning modes exist, of which interlaced scanning and progressive scanning are the two best known. This discussion of scanning modes will be limited to interlaced and progressive scans. Both have their pros and cons. The solid-state imagers are designed dedicated to specific scanning mode, although with a progressive-scanned image sensor it is also possible to deliver interlaced video information. In the latter case, however, the image sensor is not employed in its most optimized design.
The formerly used imaging tubes do not suffer from this dedicated scan design. The imaging tubes can be scanned fully custom defined and with the same tube interlaced and progressive scans are possible.

Interlacing is a technique used in TV systems to enhance the vertical resolution of the displayed picture without increasing the bandwidth of the transmission system. The field frequency on a TV monitor is 50 Hz for the CCIR standard (Europe) and 60 Hz for the EIA standard (US and Far East). All these 50 or 60 pictures/sec or fields are not displayed one after another in the same way, but there is a spatial offset on the display system between the odd fields and the even fields. This spacial offset is equal to half the vertical pitch of a video line and in this way the video lines of the even fields are displayed exactly in between the video lines of the odd fields.

This principle is illustrated in Figure 6.1a : the odd lines of the TV monitor are activated during the first field (their numbers are indicated left-hand side of Figure 6.1a) and the even lines are displayed in the second field (their numbers are indicated at the right-hand side of Figure 6.1a). The sequence in which the information is displayed on the monitor (which is sketched in Figure 6a) is as follows : 1, 3, 5, 7, 9, 11, 13, 2, 4, 6, 8, 10, 12, 14. With this display technique the vertical resolution is increased compared to the straightforward display of all fields one on top of another. By definition, two interlaced fields make up one frame or :

$$2 \; fields \equiv 1 \; frame \; . \qquad\qquad [6.1]$$

The opposite of interlacing as a display technique is the progressive scan. Figure 6.1b shows a monitor screen which is activated in the progressive-scanning mode.

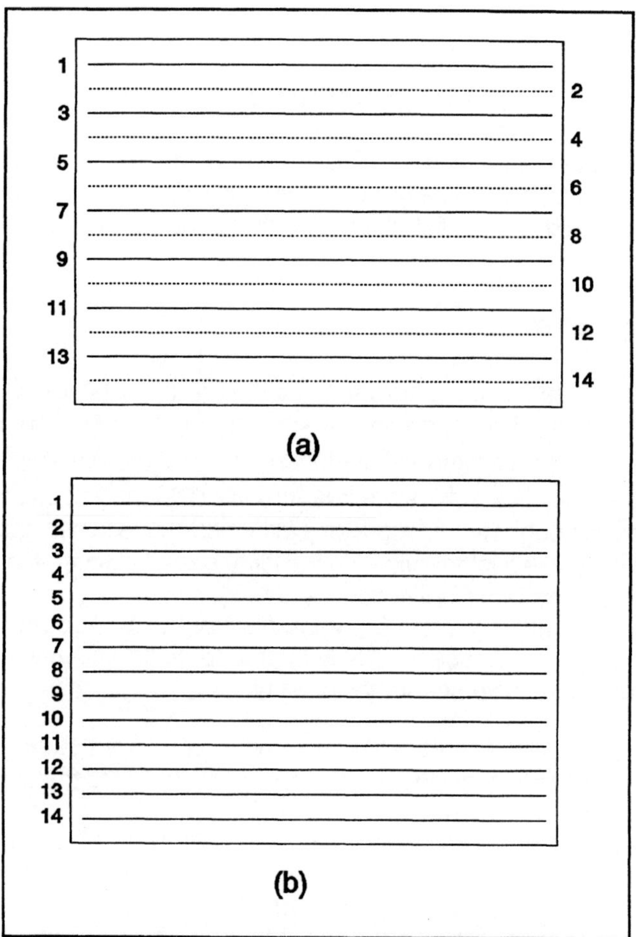

FIGURE 6.1. *Various display modes for displaying the video information on the TV monitor : (a) interlaced scanning, (b) progressive scanning.*

All lines are written during every pass, and each field is exactly the same. Note that the following equality applies to the progressive-scanning mode :

$$1 \; field \equiv 1 \; frame \; . \qquad\qquad [6.2]$$

6.1.1. INTERLACED SCANNING

For consumer-type applications, interlaced scanning is of fundamental importance because all classical TV monitors use interlaced video information. The image sensor which will be coupled to the display has to provide video information to the display in such a way that interlacing is still possible, or the image sensor has also to capture its information by a kind of interlacing process. There has to be an offset equal to half the vertical pitch of a video line between two consecutive fields. The way to construct these interlaced fields by means of the CCDs is quite simple, but different for the frame-transfer and (frame-) interline-transfer devices.

Frame-transfer CCD
Integrating charge packets in a frame-transfer CCD is simply done by applying DC voltages to the gates of the imaging part of the device. This is shown in Figure 6.2a, where a cross section of a four-phase device during an integration period is depicted. Of the four phases only Φ_1 is set to a low voltage (e.g. 0 V), and all the others are connected to a high voltage (e.g. 10 V). In Figure 6.2 low voltages are indicated by (-), high voltages by (+). These DC settings define the pixel areas : the potential wells are defined between two Φ_1 gates. A single pixel is indicated in Figure 6.2a by the solid rectangle : its width is equal to a single CCD channel, its length is equal to four gates.
But it is also possible, with the same flexibility, to define the potential wells between two Φ_3 gates, as is done in Figure 6.2b. The corresponding gate voltages are indicated as well as its unit cell. By comparing Figures 6.2a and 6.2b it should be clear that the pixels are shifted over half the vertical pixel pitch. This is how interlacing is provided in a frame-transfer CCD : defining the pixel separation by a low DC voltage, the center of the pixel can be shifted easily. Note that perfect interlacing with this technique is only possible if the number of phases is even because the pixel shift is limited to a multiple of the gate length.

(Frame-) Interline-transfer CCD
The photodiodes of a (frame-) interline-transfer imager are fixed and do not have the flexibility to be shifted according to the interlacing mechanism. This problem is solved by providing the imager with a double quantity of photodiodes, but the number of charge packets shifted towards the output stage is kept constant. The way this is done is illustrated in Figures 6.3a and 6.3b (Takizawa 83). These show three vertical CCD registers, each with two CCD cells, together with two photodiodes for each CCD cell. For the odd-field read out, the odd-numbered pixels are shifted out (Figure 6.3a); for the even-field read out, the even-numbered pixels are shifted out (Figure 6.3b). Interlacing is relatively straightforward, but this system has integration times of 40 msec for each single pixel instead of the classical 20 msec for the CCIR standard (33.4 msec instead of the 16.7 msec of the EIA standard). This causes some image lag and some disturbance from moving objects in the

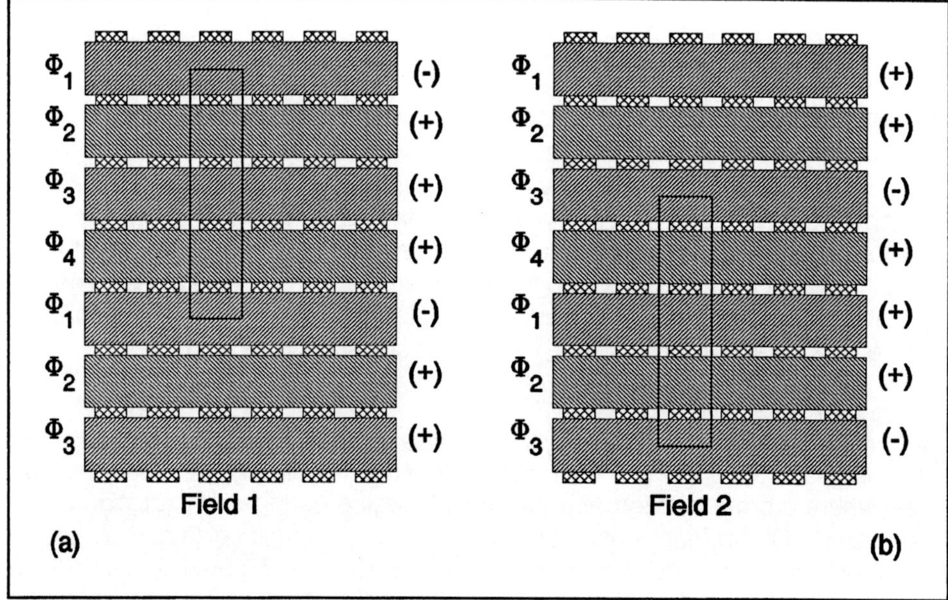

FIGURE 6.2. *Interlacing in a frame-transfer imager by correct definition of the gate voltages : (a) for the odd field, (b) for the even field.*

scene owing to the overlapping of integration periods in time.

To obviate this drawback, the frame-integration mode is changed to the field-integration mode shown in Figures 6.4a and 6.4b. The same device as in Figures 6.3a and 6.3b is shown but in the field-integration mode all photodiodes are emptied after each integration period. They are summated two by two. Interlacing is achieved by choosing each time another neighbor to be added to, as illustrated in Figures 6.4a and 6.4b. In the field-integration mode the integration time is again limited to the original one, and all integration periods are equally long for all pixels. For these reasons field-integration-mode interlacing is applied in (frame-) interline-transfer devices in almost all situations.

Interlacing with an MOS-XY imager or a charge-injection device is similar to that with interline-transfer CCDs in frame-integration or field-integration mode : the number of pixels is doubled, but the number of charge packets fed to the output during one field period remains unaltered.

6.1.2. PROGRESSIVE SCANNING

Progressive scanning is a technique used in applications where high(er) vertical resolution is of prime importance and where, for instance, the bandwidth of the

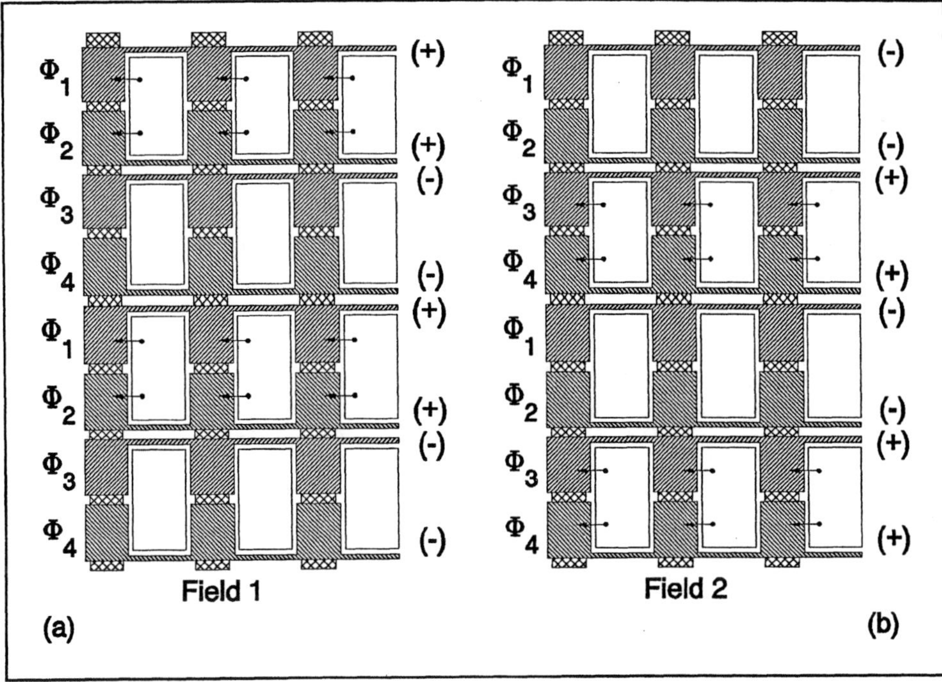

FIGURE 6.3. Interlacing in (frame-) interline-transfer imagers by reading out different lines in the frame-integration mode alternately for the odd (a) and the even (b) field.

transmission medium imposes no limit on the bandwidth of the video information. Typical examples of progressively scanned monitors are professional apparatus and computer monitors. In situations where an image-sensing system has to be coupled to a progressive-scanned monitor, the video information delivered by the image sensor has also to adhere to the monitor format. Unlike interlace-scanned image sensors, the construction of progressive-scanned devices is straightforward.

A consequence for the frame-transfer CCDs is that the number of lines in the image section and in the storage section has to be doubled compared to the interlaced-scanned alternatives. For a standard image format, the area of the image section remains the same and, only the number of vertically defined lines doubles.

In (frame-) interline-transfer devices, the total number of photodiodes is unaltered compared to an interlace-scanned device, but the vertical shift registers have to be adapted in such a way that they can handle twice the number of charge packets. (In the case of a frame-interline-transfer device, also its memory area has to be doubled to store twice as much charge packets than its interlaced-scanned version.) This idea is illustrated in Figure 6.5, where the construction of a single vertical shift

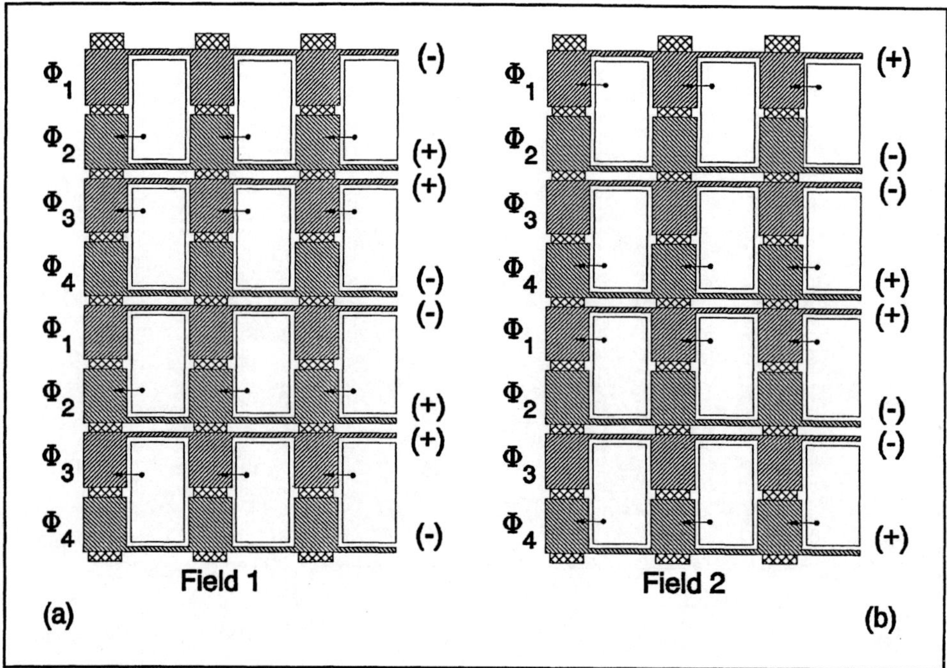

FIGURE 6.4. Interlacing in (frame-) interline-transfer type imagers by adding two different lines together in the field-integration mode for the odd (a) and the even (b) field.

register together with its corresponding photodiodes is shown for interlaced- and progressive-scanned imagers. Figure 6.5 shows from left to right the situation for a frame-interlacing device (a), a field-interlacing imager (b) and two progressive-scanned arrays (c-d). The two interlacing schemes are equivalent to those shown in Figures 6.3 and 6.4. Both progressive-scanned imagers make use of a three-phase vertical CCD register. Note the difference in construction of the second clock Φ_2 : in the first situation the gate is running horizontally (Abe 90) (the interconnection to the right between two photodiodes is not shown) and in the second situation the gate is running vertically. The latter construction gives some more flexibility in connecting the polysilicon gate and in designing the interconnection from one pixel to the other (Kobayashi 93). This idea of vertically running gates can also be applied to frame-transfer imagers with high pixel densities (Beck 82, Bosiers 91).

WORTH MEMORIZING

Interlaced scanning is implemented with a frame-transfer imager using an appropriate DC setting on the CCD gates during the

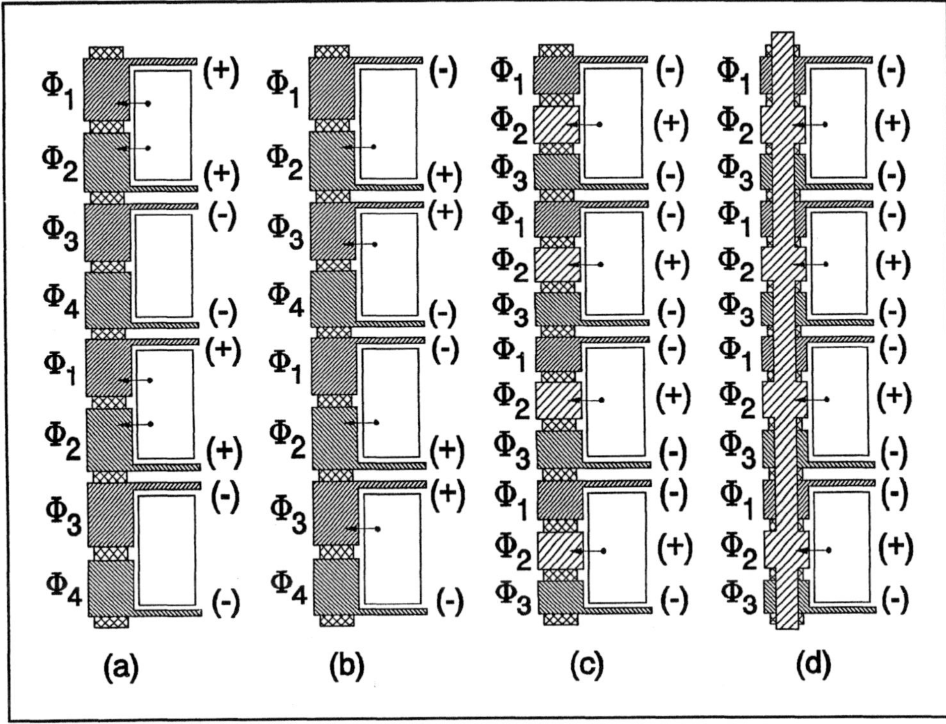

FIGURE 6.5. Progressive versus interlaced scanning for (frame-) interline imagers : field interlacing (a), frame interlacing (b), and two progressively scanned devices (c-d).

integration periods. (Frame-) Interline-transfer CCDs have doubled their number of photodiodes to obtain interlacing. The contents of each pair of photodiodes are added to get field interlacing. In the case of frame interlacing the two consecutive fields are defined by all odd rows of photodiodes (for field 1) and all even rows of photodiodes (for field 2).

Progressive scanning necessitates doubling the number of lines in the image section of a frame-transfer imager, but needs only an optimized structure for the vertical registers and memory zone of the FIT and IL sensors. In the latter case the full number of photodiodes needed for a progressive-scanned picture is also already available in an interlaced version of the FIT or IL imager.

6.2. Color imaging

In the case of imaging for consumer applications, the CCD camera should be able to capture color images and to deliver a color-coded signal to the video system.

The most straightforward way to do this is to provide the camera with three sensing devices, one for each main color : red, green and blue. This technique is only applied in systems where very high camera performance is needed, but in most cases of consumer applications, an optical system built around three sensing devices is too expensive. To overcome this drawback the color separation of the incoming optical information is done with a single chip and a color filter applied to it. Previously, the color filter was fabricated on a separate glass plate and glued to the CCD (Ishikawa 81), but nowadays, all single-chip color cameras are provided with an imager which has the color filter on-chip processed (Dillon 78) and not as a hybrid. The color filters for a single-chip color camera can be organized in different structures, and even the colors themselves can differ from manufacturer to manufacturer. The filters can be defined on the chip in stripes or as a mosaic while, the colors can be primary or complementary ones.

6.2.1. STRIPE FILTERS

The organization for a stripe filter, built with primary colors (Lee 83), is shown in Figure 6.6a by means of a top view, together with a horizontal cross section of

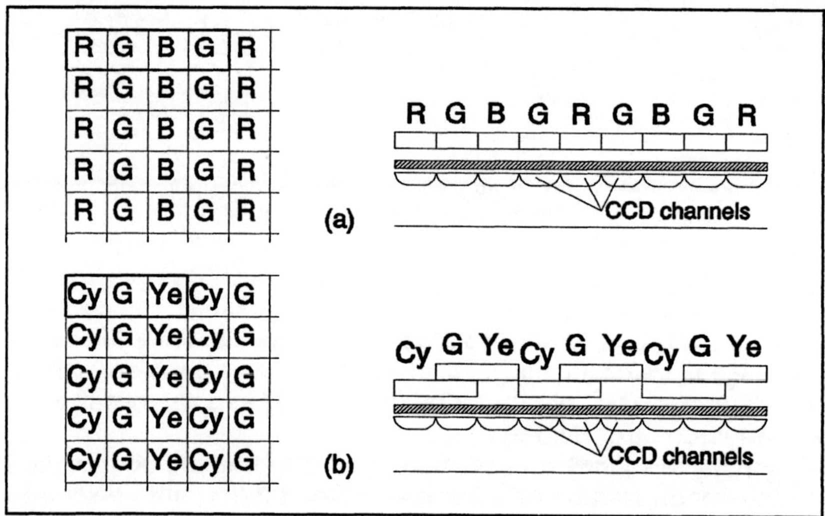

FIGURE 6.6. Color-filter arrangement in stripes with (a) primary and (b) complementary colors.

the pixels. Each single pixel is covered by a red, green or blue filter. In the example shown, the repetition rate is equal to four pixels : in each group of four, two filters are green, one red, and one blue. (The basic filter configuration which is repeated

horizontally and vertically is indicated by an extra rectangle in Figure 6.6.) The reason for covering more pixels with a green filter has to do with :
- the sensitivity of the human eye, which is most sensitive to green light;
- the current coding for color TV systems, where both the sensitivity and the resolution of the camera are defined by a combination of the red, blue and green signals in which the amount of green is dominant (see [6.13]).

Thus the G signal is the most important input signal. Reading out the CCD gives successively a red, a green, a blue and again a green pixel on each video line. Restoring the RGB signal for a monitor is straightforward.

The alternatives to the primary colors are the complementary ones shown in Figure 6.6b; the accompanying cross section illustrates how three colors are generated from only two filter layers. The cyan and yellow layers each cover two adjacent pixels, of which one common to both is covered by cyan and yellow. Bear in mind that a cyan filter absorbs red light and a yellow one blue light, or :

$$Cy = B + G \qquad\qquad [6.3]$$

and :

$$Ye = R + G . \qquad\qquad [6.4]$$

The overlap of the two filters is then only sensitive to green light. Reading out the CCD does not give the RGB signal, but to rectify this, the RGB signal has to be generated from the complementary colors by means of a matrixing operation :

$$R = Ye - G \qquad\qquad [6.5]$$

$$G = G \qquad\qquad [6.6]$$

$$B = Cy - G . \qquad\qquad [6.7]$$

The main difference between the using of the CyGYe filters and the RGB is the simpler fabrication technology involved. Only two layers have to be deposited instead of three. But there is another interesting advantage in using complementary colors : the electrical signals R, G and B extracted from the Cy, G and Ye video levels contain all frequency-limited components. The bandwidth of these signals is limited to one third of the bandwidth of the signal in the case of a black and white imager. But when, for instance, CyGYe stripe filters are used, in addition to these low-frequency R (available in Ye), G (available in G), and B (available in Cy) components, a high-frequency G^* (available in Ye, G and Cy) can also be obtained. Because each pixel contains the high-frequency green information, the signal G^* can simply be extracted from the CCD output signal by passing its complete CyGYe signal through a high-pass filter. The luminance signal intended for the intensity signal on the monitor can be reconstructed from R, G, B, and G^*, including some of the high frequency information. In other words, the information obtained from a sensor with complementary colors can contain more details than the information obtained from the same sensor covered with primary colors, and its resolution will be higher.

6.2.2. MOSAIC FILTERS

Mosaic-filter architectures are typically used with (frame-) interline-transfer CCDs (Manabe 83) and MOS-XY addressed imagers (Aoki 82). The reason why (frame-) interline-transfer devices make use of mosaic filter architectures has to be found in the fact that even the simplest stripe filter organization reduces the bandwidth of the color signals to at least one third of the bandwidth of the imager without color filters. To maintain high resolution in the horizontal direction for color sensors with stripe filters, the number of pixels per line has to be increased considerably. This can be an acute problem with (frame-) interline-transfer devices, but can be implemented more easily with a frame-transfer CCD thanks to the less complex structure of the CCD cell. To overcome this disadvantage, (frame-) interline-transfer imagers employ mosaic configurations. It will be clear that mosaic organizations reduce the horizontal color resolution to only half the black and white resolution, but they also do the same to the vertical resolution. Stripe filters do not alter the vertical resolution.

An example of complementary colors in a mosaic organization on an interline-transfer CCD in the frame-integration mode is shown in Figure 6.7a (Takizawa 83). (The basic filter configuration which is repeated horizontally and vertically is indicated by an extra rectangle on Figure 6.7.) The different pixels belonging to lines n and n+1 contain the following information :

$$W \qquad G \qquad W \qquad G$$

and

$$Ye \qquad Cy \qquad Ye \qquad Cy$$

in which the white signal W is composed of :

$$W = R+G+B . \eqno{[6.8]}$$

By combining these pixels in a two-by-two matrix, the color information can be reconstructed using the relations [6.5], [6.6] and [6.7].

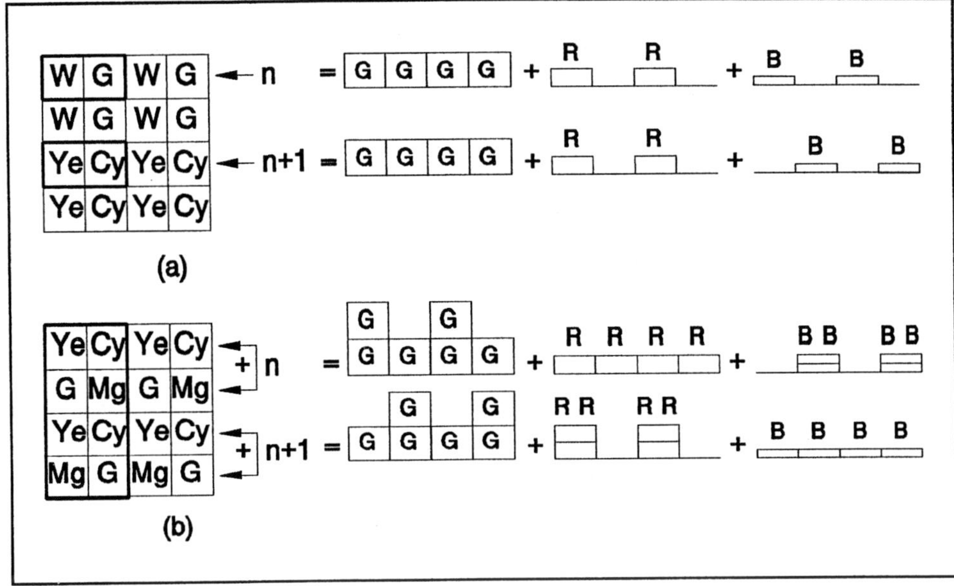

FIGURE 6.7. Color filter arrangement for a mosaic filter with complementary colors in the frame-integration mode (a) and the field-integration mode (b).

The luminance information Y is available in each group of 2 horizontal neighbors :

$$Y = W + G = Ye + Cy = R + 2G + B . \qquad [6.9]$$

For the interlaced lines n' and (n+1)', the story can be repeated : a matrix of two-by-two pixels provides all the information necessary to restore the color information needed by the monitor.

If the video signal is no longer coded as RGB but with the two color-difference signals R-Y and B-Y, line-sequentially superimposed on the luminance signal Y, the generation of these two signals from the sensor (with mosaic filters) is much simpler if the mosaic-filter configuration is adapted and if a magenta (Mg) pigment is added. An example for the field-interlaced scanned device is included in Figure 6.7b (Kohno 85, Tanaka 90). Bear in mind that, with the interlacing concept, two vertically adjacent colors are always mixed together. Subtracting two horizontal neighbors yields :

$$(Mg + Cy) - (G + Ye) = 2(B - G/2) \simeq 2(B - Y) \qquad [6.10]$$

and

$$(Mg+Ye)-(G+Cy) = 2(R-G/2) \approx 2(R-Y) \,. \qquad [6.11]$$

To have some insight into these relations, it has to be remembered :
    - that magenta contains the blue and the red information, i.e. :

$$Mg = R + B \qquad [6.12]$$

    - that the luminance signal Y is about half the value of the green information added by part of the red and part of the blue signal, i.e. :

$$Y \approx G/2 \,. \qquad [6.13]$$

Taking these relations into account, generation of the corresponding video signals R-Y and B-Y is quite straightforward.

This color-filter configuration and its ease of video processing make this overall technique very attractive for use in almost all up-to-date consumer camcorders.

6.2.3. PRIMARY OR COMPLEMENTARY, STRIPE OR MOSAIC

Designing an image sensor for a single-chip color camera always involves resolving the problem of primary or complementary colors and of a stripe or a mosaic organization. Some of the pros and cons of both systems already have been mentioned, and some others are included in the following list comparing :

    - primary colors and complementary colors :
        + from a technology point of view, a complementary filter in its simplest configuration is easier to process : only two layers (for cyan and yellow) instead of the three layers required when primary colors are used;
        + more light is lost with primary filters (only one pixel blue, two pixels green, one pixel red, as shown in Figure 6.6a), compared to complementary filters (one pixel blue, three pixels green, one pixel red as in Figure 6.6b); this can be a drawback for the primary colors when sensitivity is concerned;
        + with complementary filters, each pixel contains green information (when Mg is not used); this characteristic can be exploited to generate an extra high-frequency component in the luminance signal;
        + matrixing is needed with complementary color filters, which is not optimum for the signal-to-noise ratio of the system; this drawback is absent with primary colors.

    - stripe filters and mosaic filters :

+ stripe filters retain the color information from pixels all of which lie on the same horizontal line; mosaic filters retain their information from a two-by-two matrix : stripe filters need more pixels on a line if they have to demonstrate the same horizontal resolution as a mosaic filter, but mosaic filters show only half the vertical resolution in chrominance compared to a stripe filter;

+ signal processing with mosaic filters is more complicated because at least one horizontal delay line is needed (Knop 85);

+ the signal-to-noise ratio of the luminescence signal is comparable for the two systems.

In today's devices intended for real-time video, complementary colors are used almost exclusively to arrange on-chip color filters. All (frame-) interline-transfer devices and MOS-XY imagers use the mosaic configuration, while stripe filters are often applied on frame-transfer devices. Frame-transfer CCDs have the capability to provide a higher horizontal resolution compared to (frame-) interline-transfer devices, and this makes them basically more suited to stripe filters.

An interesting trend which is developing, however, is that frame-transfer devices also being used in combination with mosaic filters (Okada 91). The reason for this nonconventional change cannot be found in any special function and characteristic of the frame-transfer sensor in connection with the color-filter array. However, the availability of video-processing ICs is also driving the frame-transfer imagers in the direction of mosaic color filters. All video processing ICs on the market are designed by manufacturers who previously produced only (frame-) interline-transfer imagers. Consequently, all matrixing and color processing is being dedicated to mosaic filters.

Opposite to real-time video, still-video or electronic photography relies on imagers with primary colors, mainly to benefit from the absence of color matrixing and the high signal-to-noise ration of the system.

6.2.4. COLOR SEPARATION BY A PRISM

As already mentioned in the introduction to this section on color imaging, the ultimate solution is an optical system with three image sensors, namely one for each primary color. Splitting of the incoming optical information into the three components is done by means of a color-splitting prism, both for classical imaging tubes (de Lang 65) and for solid-state image sensors (Murata 83). The camera head of a professional color camera is shown in Figure 6.8 : the prism is placed behind the lens of the camera, and glued to it are the three imaging devices for the blue (CCD (B)), green (CCD(G)), and red (CCD(R)) information. These image sensors are electronically driven fully in parallel, and the complete black and white resolution of each device is available in each color component. After signal preprocessing of the CCD output signals, the signals are fed into a signal processor to generate the chrominance and luminance signals. Luminance can be generated as :

$$Y = (0.11*B) + (0.59*G) + (0.30*R) \ . \qquad\qquad [6.14]$$

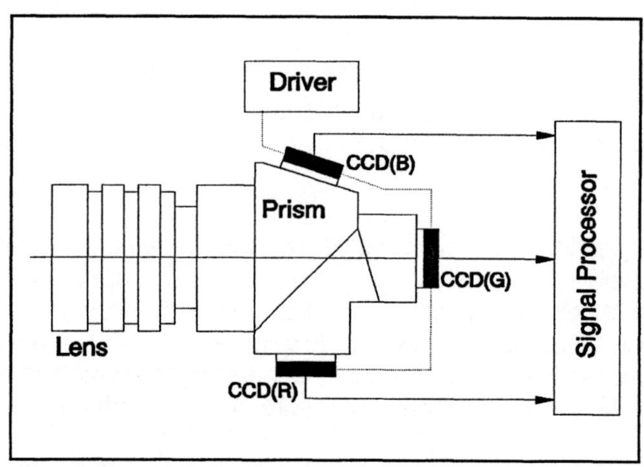

FIGURE 6.8. Color separation by means of a prism as used in a professional camera head.

A simple trick to enhance the resolution of the camera shown in Figure 6.8 is to give the "green" CCD a horizontal offset, equal to half the pixel pitch P_{pix}, in relation to the "blue" and "red" CCD. This construction with offset is shown schematically in Figure 6.9 (Shinoda 87).

Compared to stripe filters with primary colors in single-chip color cameras, this professional color-splitting technique faces some related problems : the luminance signal of the video apparatus has the same bandwidth as the color components, while some extra bandwidth in luminance is an advantage. This is based on the facts that the human eye can resolve more luminance details than chrominance details.

Nevertheless just a few exceptions, all cameras for professional applications use a color-splitting prism with three image sensors, one for each primary color. Among the few exceptions, however, some utilize a dichroic prism with three imagers, two of them for the green channel, and one for the red and blue channels (Ikeda 85, Okano 91). The two "green" imagers are spatially offset on two (green) sides of the prism. The third imager, with an on-chip red and blue stripe filter, for instance, is placed on the third side of the prism. The advantage of this construction is the high resolution in the luminance but, unfortunately, it is accompanied by low resolution in the chrominance signal.

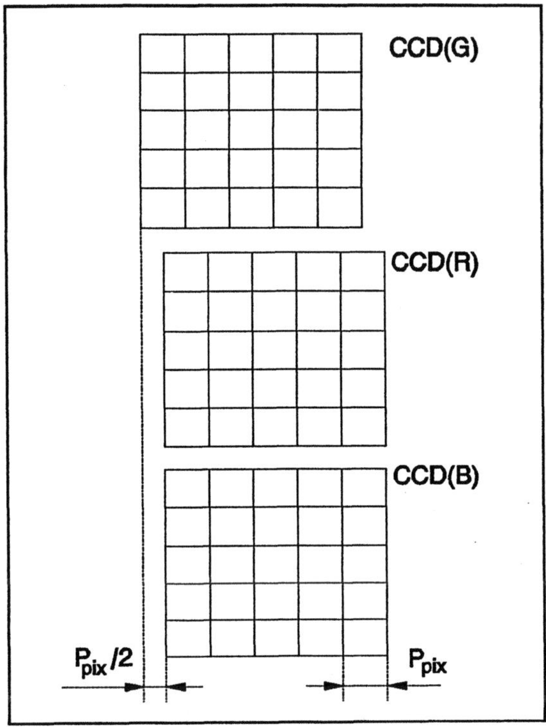

FIGURE 6.9. Spatial offset of the three imagers in the camera head of Figure 6.8, used to increase the horizontal resolution of the camera system.

Compared to tubes, solid-state imagers have, as already mentioned, an extremely stable geometry. As regards the professional imaging with three devices, the stable geometry of the solid-state imagers can be interpreted as an advantage, but also as a disadvantage. On the one hand, once the three devices are aligned and glued to the optical prism, the complete optical block can be considered as a geometrically very stable device. On the other hand, alignment has to be done extremely precisely with CCDs and the pixels are fixed on the sensor surface, whereas in the case of tubes the light-sensitive surface can be shifted : further alignment of the tubes after mechanical fixing can be done electrically. In practice, however, the advantages of solid-state image sensors prove to far exceed the disadvantages with respect to alignment : in practical situations aligning problems are no longer obvious.

6.2.5. COLOR IMAGING WITH LINEAR ARRAYS

The discussion on color imaging with two-dimensional imagers is equally applicable to linear devices. Primary and complementary on-chip color filters can be used. They can be detailed on a separate glass plate and glued to the imager. Worth mentioning in the context of color imaging with linear arrays are the various sensor architectures which are possible, proceeding from three linear arrays on a single chip to a single line sensor with individual filter dots on every pixel. The various filter-sensor organizations are shown in Figure 6.10 (Kawamoto 91). They are from top to bottom :

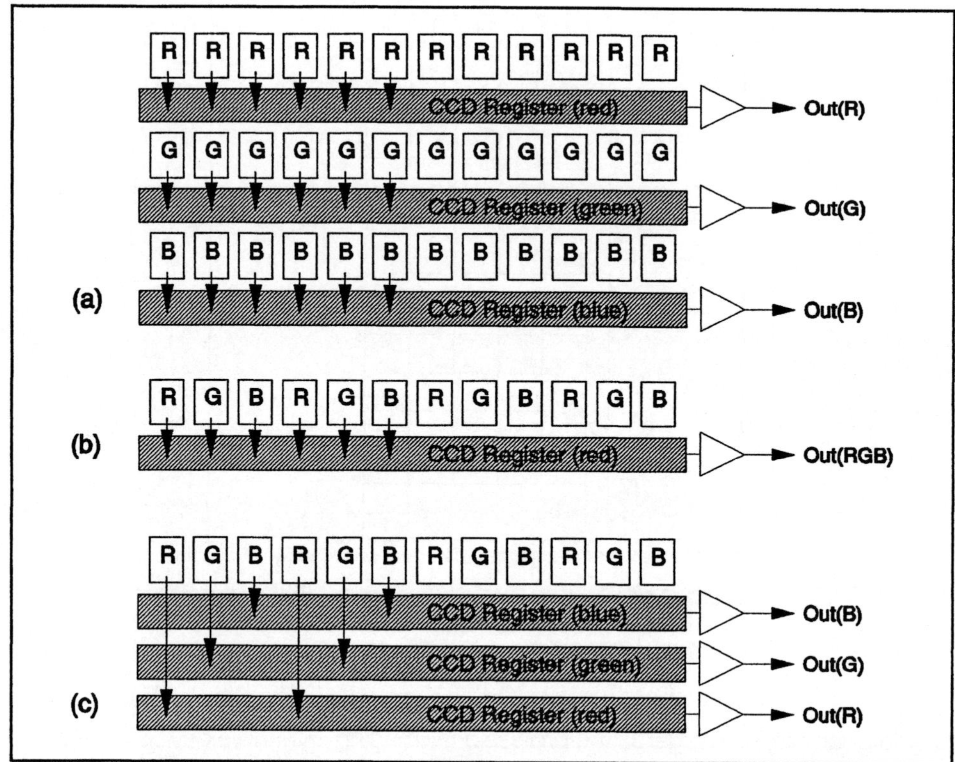

FIGURE 6.10. Illustration of three different sensor configurations used for color imaging with linear arrays.

- a complete triplet structure (Figure 6.10a) : a separate line array is included for every color and all three image-sensor lines are located on the same chip. A stable mutual geometry is maintained, but the different image-sensor lines capture different lines of the object. In reconstructing the video signal, this shift between the three different colors has to be compensated for, which costs extra hardware and/or software. However, full resolution is maintained for all three colors;
- a single line array with a single CCD shift register (Figure 6.10b) : this is the most elementary line array provided with a color filter. Every pixel is covered with an appropriate filter dot. Compared to the preceding

triplet array, this sensor has only one third of the resolution when the pixel pitch is kept constant. The single CCD register contains all the color information of all three primary components;
- a single line of light-sensitive pixels in combination with three CCD registers, one for each primary color component (Figure 6.10c). This concept has also lower resolution, but all pixels are on the same line. The three horizontal CCDs can be driven with a lower clock frequency (: 3) and all color components are automatically multiplexed and fed to the outside by their own output amplifier.

Every architecture from Figure 6.10 has its specific application fields. Where high resolution is of prime importance, the first example with the triplet line array is preferable to the other two. To avoid complex matrixing, even an additional array to generate the luminance signal can be incorporated. It can be built without a color filter (Lees 90). An image-sensor chip of this kind (with different chrominance arrays and an additional luminance array) can be used in telecine machines in which classical movies are converted into video signals by scanning the celluloid film mechanically in the transverse direction and electronically in the longitudinal direction. Correction for the offset between the different arrays is usually done electronically off-chip. For situations where ease of use is more important, the second or more especially the third solution, depicted in Figure 6.10c, is attractive.

WORTH MEMORIZING

Color imaging can be implemented in several ways :
- with a color-splitting prism and three chips, one for each primary color;
- with a color filter pattern on a glass plate and glued to the CCD;
- with a color filter array deposited directly on-chip.
The color filter array can be configured as a stripe or as a mosaic filter pattern. The colors themselves can be the primary colors (R, G, B) or the complementary components (Cy, G, Ye, Mg). Choice from all the options mentioned is very largely application-dependent :
- professional applications : three chips glued to a color-splitting prism;
- consumer applications : on-chip mosaic filter array with the Cy, G, Ye and Mg components.

6.3. Blooming and antiblooming

An attractive feature of solid-state imaging devices compared to imaging tubes is their insensitivity to burn-in effects and the absence of image lag. Solid-state imagers empty their potential wells after each integration period, and once this operation is over, the potential wells are free of any information from the previous integration. Or the new image to be integrated contains no information from the preceding image : no lag, no burn-in.

Another effect also associated with excessive light input is blooming. If the amount of light is so great that the potential wells are full of electrons even before the end of an integration period, any additional photon can still give rise to an extra electron. The result of this is illustrated in Figure 6.11 : the already full well will spill these electrons over into the neighboring wells, and the displayed image will show artifacts in the scene as column or line defects around the highlight. This spillover of excess electrons to other pixels is called blooming. Solid-state imagers for TV applications are all provided with means to counteract blooming or with an antiblooming structure. Depending on the architecture of the image sensor and on its fabrication technology, different kinds of antiblooming mechanisms are used : a lateral or horizontal overflow drain, a clocked or charge-pumped antiblooming mechanism or a vertical draining system which is different for (frame-) interline-transfer and frame-transfer devices. All these structures will be described in this section.

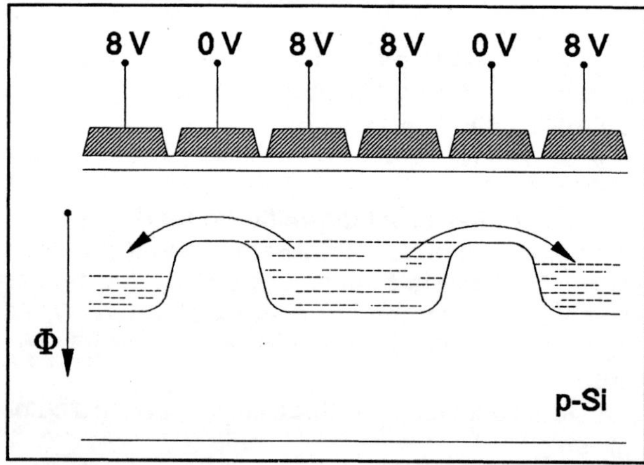

FIGURE 6.11. The origin of the blooming effect : excess electrons spill over to neighboring pixels and cause some typical artifacts on the monitor.

6.3.1. LATERAL OR HORIZONTAL ANTIBLOOMING

The actual blooming is caused by charges spilling to neighboring pixels. This can be avoided by creating a site to which charges are preferentially spilled instead of spilling over to other pixels across the barrier gates. An antiblooming (AB) drain is located near each pixel and the separation between this drain and the pixel itself is effected by means of a separate antiblooming gate (Levine 84). This construction is shown in cross section in Figure 6.12, together with a sketch of the surface potential. From left to right the n^+ AB drain, the AB gate, the CCD gate, and the p^+ channel-stop diffusion can be recognized. The potential barrier underneath the AB gate can be influenced by the voltage on the gate and can be set to such a voltage that the surface potential underneath the AB gate is the highest of all pixels' surroundings. Excess electrons which cannot be handled in the full well of the pixel will spill over the lowest barrier (= highest in electrostatic potential) and will arrive in the AB drain. This n^+ diffusion is externally biased to a high DC voltage and will drain off all the electrons which flow to it.

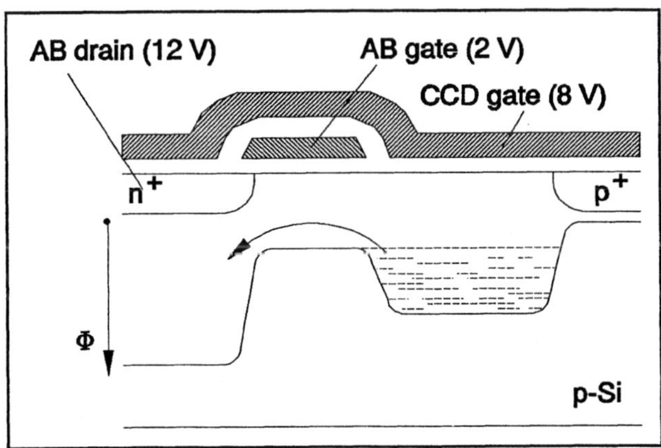

FIGURE 6.12. Anti-blooming by means of a horizontal gate and drain construction.

This lateral overflow drain or horizontal antiblooming structure has the advantage of being easily adjustable by the external voltages on the AB gate and the AB drain. The main drawback of this structure, however, is its occupation of silicon area which normally belongs to the light-sensitive pixel. The incorporation of a horizontal anti-blooming drain is paid for by a decrease in light sensitivity due to this loss of sensitive pixel area.

6.3.2. CHARGE-PUMPED OR CLOCKED ANTIBLOOMING

Charge-pumped antiblooming or clocked antiblooming makes use of a mechanism whereby electrons and holes are recombined (Beck 82, Hynecek 85). Prior to spilling over to other pixels, the electrons are given the opportunity to recombine with holes. It is essential to provide an overexposed pixel with enough holes, so that they can effectively recombine with the excess electrons before blooming can occur and artifacts are created on the displayed picture. The provision of holes can be effected in the following way : the holes necessary for the recombination are stored in the interface states of the MOS structure. In a buried-channel CCD the implantation profiles are optimized in such a way that there is no interaction between the electrons in the CCD channel and the interface states, but in the case of charge-pumped antiblooming this condition is set to its actual limit. There is indeed no interaction between the electrons from the CCD channel and the interface states as long as the CCD well is below saturation. At the onset of full-well situation, however, the restriction on interaction no longer applies and the electrons from the charge packet can communicate with the hole-filled interface states, or any excess of electrons will interact with the interface states. During interaction these electrons will find holes to recombine with, and will disappear without giving rise to any problem of blooming.

A cross section of the buried-channel CCD is shown in Figure 6.13b : an n^--type top layer on a p-type substrate. The potential diagram in depth through the center of the pixel is shown in Figure 6.13a for three different gate voltages. Before the start of an integration period, the voltage applied is - 5 V (curve 1) to pin the surface potential in the CCD channel to the p-substrate voltage and/or the p^+ channel-stop implant. In this situation the silicon surface is inverted, its surface potential is equal to the potential of the p^+-regions and all interface states are filled with holes. (This action is performed by the charge-pumping mechanism.) During the integration the gate voltage is set to +5 V; curve 2 shows this situation for an empty well, curve 3 for the full well situation. Note that "full well" corresponds to the case when there is practically no barrier between the interface potential and the channel potential. With a full well any additional electron can interact with the interface, and will find there a hole to recombine with. To complete the potential diagram, curve 4 is added, which is the situation with the barrier gates set to 0 V gate potential.

One important condition for operating this mode of antiblooming is the availability of enough holes to recombine. Under extreme overexposure conditions, the stock of holes will decrease rapidly. However, to refill the interface states with holes, the gates which are biased positively during the integration can be pumped one after the other during each line-blanking period. In this way the stock of holes can be replenished.

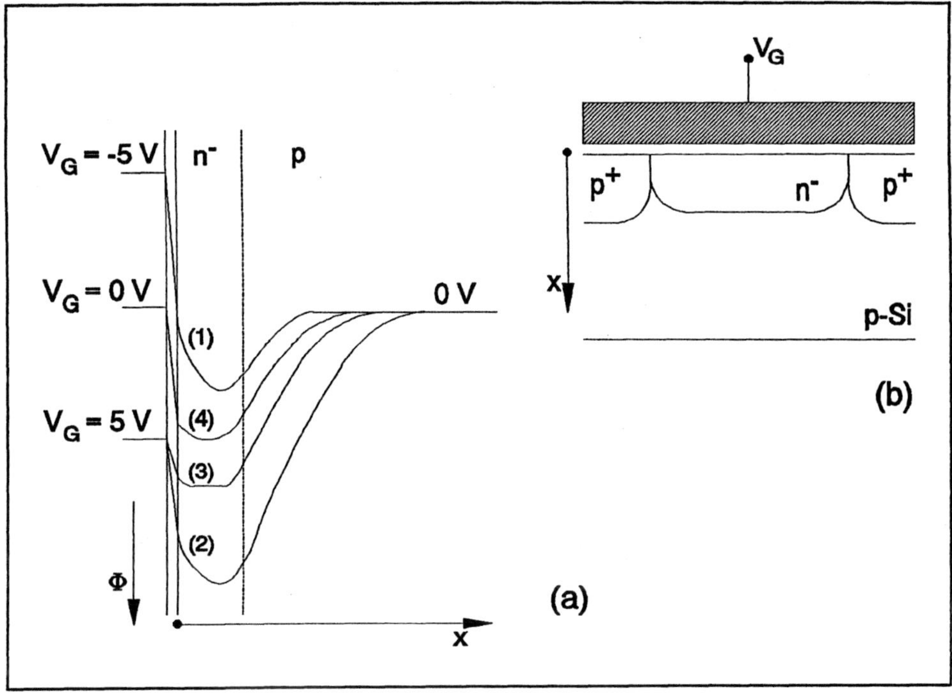

FIGURE 6.13. Antiblooming by a dynamic charge-pumping mechanism. A cross section of the image cell (b) and a potential profile in-depth (a) through the center of the pixel are shown.

Although the charge-pumped or clocked antiblooming takes up no light-sensitive silicon area, the antiblooming mode needs a rather complicated clocking scheme (three levels, clocking in the line-blanking period) and is limited to MOS capacitors with an np-silicon sandwich. Additionally, there is the need for enough interface states, whereas processing in all CCD technologies is optimized to keep the number of interface states as low as possible. The maximum capacity of charge-pumped antiblooming is about 50 to 100 times overexposure with modern technology.

6.3.3. VERTICAL ANTIBLOOMING

Like horizontal antiblooming, vertical antiblooming has a preference for a spillover site but, while the antiblooming means are located laterally near the pixels in the horizontal case, the drain is located underneath the pixels in the vertical antiblooming mode.

(The meaning of "vertical" for the type of antiblooming is different from that of "vertical" as applied to the vertical shift registers of an interline-transfer CCD, for instance. In the case of the antiblooming structure "vertical" should be interpreted as "in-depth". Despite the ambiguity of the term, the word "vertical" is still used in both meanings.)

The vertical antiblooming structure for a (frame-) interline-transfer CCD is located underneath the photodiode (Ishihara 82, Oda 83), as shown in Figure 6.14b, which is a cross section of an interline-transfer CCD unit cell. From left to right is drawn the p^+-channel stopper, the n^+p photodiode, the transfer region between the photodiode and the (vertical !) CCD shift register, the n-buried channel of the CCD shift register, and again the p^+-channel stopper. These structures are all diffused in a p-type well on top of a n-type silicon substrate. Note the "weak" spot in the p well exactly underneath the photodiode.

In the situation considered, the p well is biased to a low DC voltage (e.g. 0 V) and

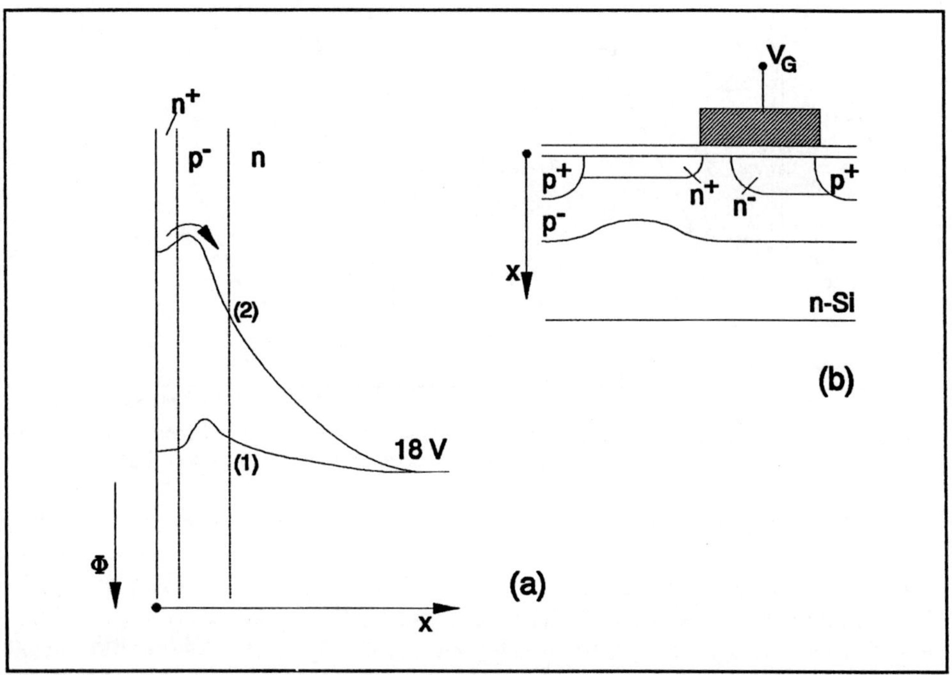

FIGURE 6.14. Antiblooming by a vertical n-p-n photodiode of a (frame-) interline-transfer cell : a cross section (b) and a potential profile in depth (a) through the center of the pixel.

the n substrate to a high DC voltage (e.g. 18 V). Figure 6.14b presents the electrostatic potential diagrams in the middle of the photodiode, namely the fully depleted n^+ top layer and the fully depleted p well. Curve 1 is for an empty well, e.g. at the onset of an integration period. If the well starts filling up, the potential in the n^+-top layer and in the p well will decrease. At the point of saturation, shown by curve 2, any additional electron in the potential well will be attracted to the high potential of the n substrate. It will pass across the small remaining barrier of the p well towards the n substrate which is acting as a drain. Steady state is reached and the vertical overflow drain is doing its job.

The vertical antiblooming structure in a MOS XY-sensor is very similar to the construction used in an interline-transfer device. The vertical n-p-n structure is optimized underneath the photodiode, almost independently from the rest of the pixel (Koike 80).

In the case of a frame-transfer imager, a slightly different vertical antiblooming structure has to be used, because in this case the vertical overflow drain has to be located underneath the CCD channel itself (van de Steeg 85, Nichols 87). A cross section of a unit cell is shown in Figure 6.15b, depicting from left to right : the p$^+$-channel stopper, the n-buried channel of the (vertical !) CCD shift register,

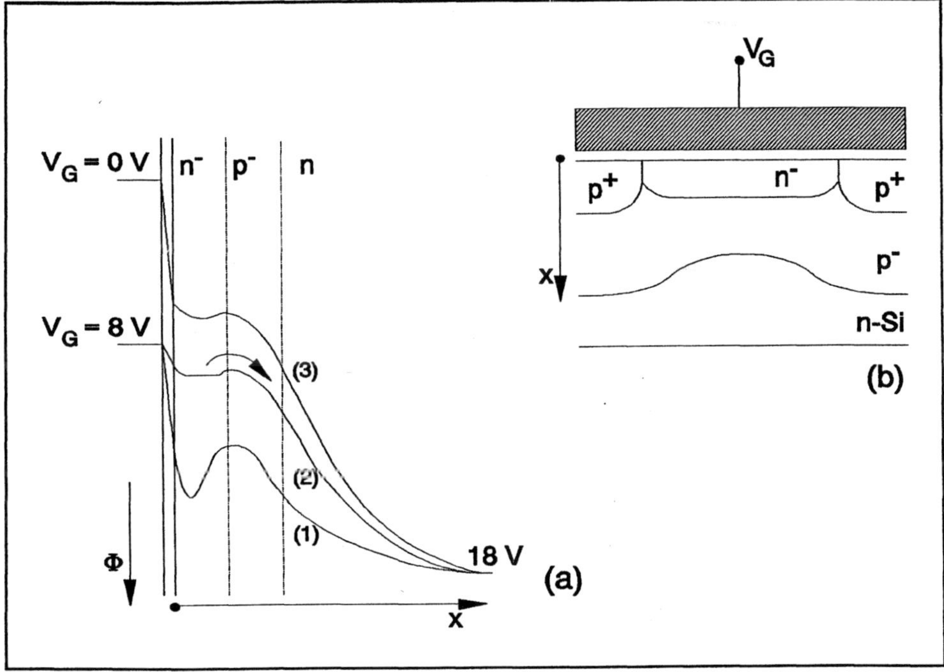

FIGURE 6.15. Antiblooming by a vertical n-p-n structure of a frame-transfer cell : a cross section (b) and a potential profile in-depth (a) through the center of the pixel.

and again the p$^+$-channel stopper. In the frame-transfer CCD also, all structures are diffused in a p well on top of an n-type silicon substrate. Note again that the "weak" spot in the p well is exactly in the middle of the CCD channel. In the situation being examined the p well is biased to a low DC voltage (e.g. 0 V) and the n substrate to a high DC voltage (e.g. 18 V). Figure 6.15a illustrates the electrostatic potential diagrams in the middle of the CCD channel, namely the fully depleted n layer and the fully depleted p well. Curve 1 is for an empty well. If the well starts filling up, the potential in the n-type buried channel and in the p well will decrease. At the point of saturation shown in curve 2, any additional electron in the potential well, will be attracted to the high potential of the n substrate. It will pass across the small remaining barrier of the p well toward the n substrate which is acting as a drain. Steady state is reached and the n-type substrate acts as a vertical overflow drain. Curve 3 in Figure 6.15a depicts the situation underneath a barrier gate biased, for instance, to 0 V. Note that excess electrons will spill across the lowest barrier toward the n-type substrate and not toward neighboring pixels because in this case the barrier to be overcome is much higher.

Vertical antiblooming structures for all categories of imagers are characterized by some typical features :
- they are compact and take up no extra silicon;
- they require complicated numerical simulations to design the pixels of the CCDs : two- and three-dimensional electrostatic potential simulations are needed to optimize the overflow mechanism;
- all electrons generated deeper in the silicon than the point where the p well has its minimum potential will be attracted by the n-type substrate, and will never reach the CCD channel. This cutoff action considerably limits the (red and infrared) sensitivity of the device but it eliminates the problem of the diffusion MTF and minimizes the dark current collection;
- the "weak" point in the p well can be generated in two ways : either by a double p-well construction (the first shallow, the second deeper), or by a single p-well diffusion where the "weak" point is created by means of two sideways diffusions meeting in the "weak" point.

In commercially available cameras for consumer and broadcast applications, the imagers are almost exclusively fitted with vertical anti-blooming drains because of their compactness and effectiveness. Ability to handle overexposure levels exceeding 10^4 is not exceptional in solid-state imagers incorporating a vertical antiblooming construction.

WORTH MEMORIZING

Antiblooming is a means to protect against overexposure and can be easily incorporated in CCD architectures. Its use means that artifacts in displayed images can be avoided. Several antiblooming options are possible :
> **- horizontal antiblooming characterized by simple tuning of the structure but taking up quite a lot of the active pixel's silicon area;**
> **- charge-pumped or clocked antiblooming based on the presence of interface states and activated by a complicated way of clocking;**

- vertical antiblooming featuring great compactness, high efficiency, low dark current and high diffusion MTF at the expense of lower red and infrared sensitivity and complicated cell optimization.

6.4. Charge reset or electronic shutter

Due to the low level of image lag with interline-transfer and frame-interline-transfer CCDs, or even its complete absence with frame-transfer devices when using solid-state image sensors instead of imaging tubes, the sharpness of the pictures is much higher with present-day cameras compared to the previous generation. Nevertheless, the integration of an image still takes 20 msec (CCIR standard) or 16.7 msec (EIA standard). During this time any (fast-) moving object can give rise to some additional unsharpness.

To overcome this effect due to the relative long integration time, a technique called charge reset or electronic shuttering is applied. The method can also be defined in photographic terms as electronic exposure timing. The integration time of the imager is shortened to 1/100 sec, even down to 1/4000 sec, without changing the field rate of 50 fields/sec or 60 fields/sec. This is achieved by splitting the classical integration time of 20 msec or 16.7 msec into two parts of which the second part is equal to the desired integration time (e.g. 1/250 sec), while the first part is the original minus the desired integration time (20 msec - 1/250 sec, or 16.7 msec - 1/250 sec). This splitting effect is shown schematically in Figure 6.16.

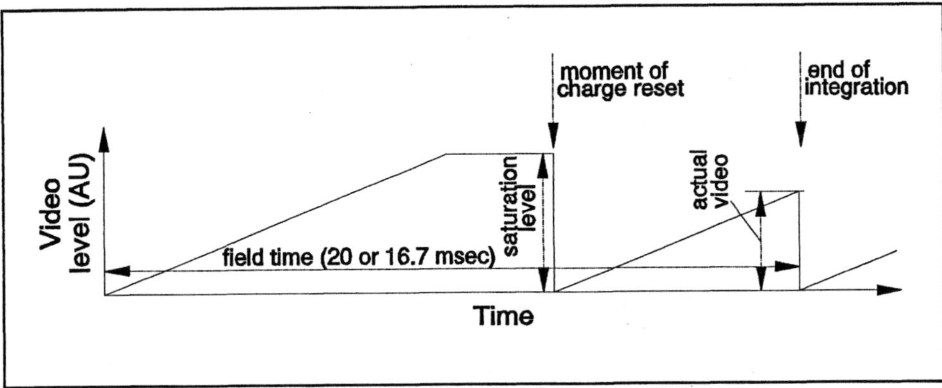

FIGURE 6.16. Splitting the active integration time into two parts to shorten the ultimate integration time.

Although charge reset or electronic shuttering was originally introduced to enhance the sharpness of images, it can also be used to provide the imagers with an electronic iris function. If the amount of light impinging on the sensor is too high, the integration time can be shortened to collect only part of the charge generated during the normal integration period. Unlike a mechanical iris, an electronic one, of course, does not affect the optical depth of focus.

The question is how to get rid of the charges which are integrated during the first part of the classical integration time, to restart the integration during the shortened period, and to read out only the charge packets integrated during this second integration period. The measures taken to empty all potential wells at the charge-reset point depend on the type of sensor used.

6.4.1. CHARGE RESET IN FRAME-TRANSFER CCDS

In a frame-transfer CCD the charge-reset operation can be carried out quite simply : if the CCD gates of the imaging section are connected all together to a low DC voltage, the potential wells can all be emptied by the vertical antiblooming mechanism. The potential diagrams for this operation mode are shown in Figure 6.17, the corresponding cross section of the imager being the same as that included in Figure 6.15b.

The electrostatic potential curves shown in Figure 6.17a are respectively for an empty potential well at the start of integration (curve 1), a full well at the onset of vertical antiblooming (curve 2), a barrier gate (curve 3), and the situation for charge reset (curve 4). In this last case all the gates are biased especially negative (e.g. - 4 V), no potential well remains to catch electrons and all charges are drained to the n-type substrate. Restoring the DC voltages in accordance with the integration mode, restarts the integration process.

6.4.2. CHARGE RESET IN (FRAME-) INTERLINE-TRANSFER CCDS

Electronic charge reset in (frame-) interline-transfer CCDs is somewhat complicated. There is no direct means of influencing the potential in the photodiodes themselves, and they cannot be emptied in the same way as for the frame-transfer devices. For this reason, the photodiodes are emptied in the classical way at the point of charge reset, into the vertical shift registers. These CCDs subsequently transport their information, not to the output amplifier, but to an extra drain on top of the device. This is shown schematically in Figure 6.18. The drain can be a simple n^+ diffusion connected to a high DC voltage.
Although this operational principle of charge reset with frame-interline-transfer devices seems to be straightforward and applicable to IL sensors, there is one severe restriction : at the charge-reset point, the vertical CCD channels must no longer contain any video information from the previously integrated image. This means

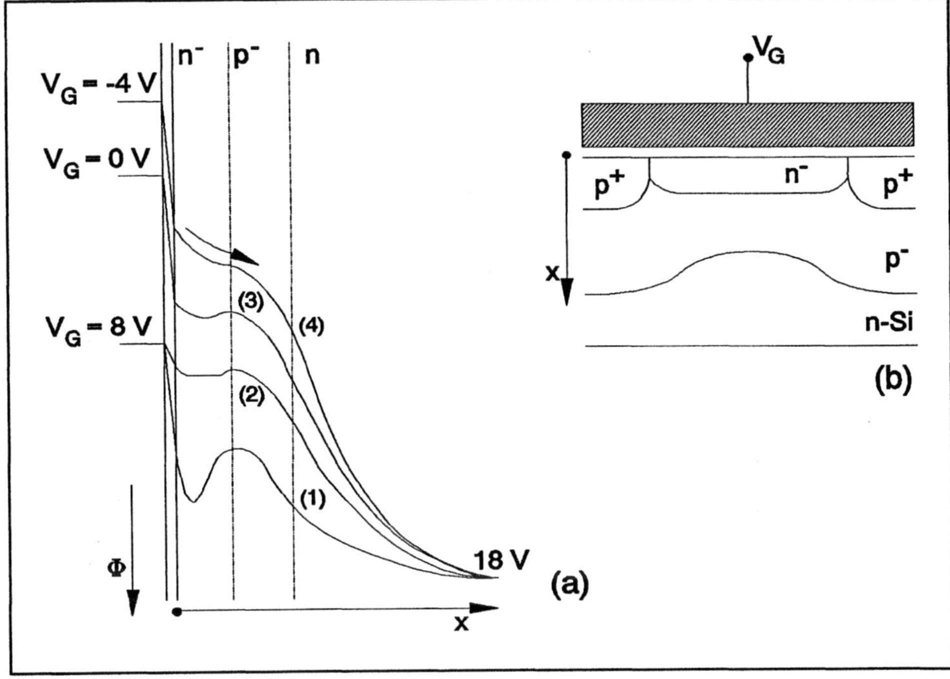

FIGURE 6.17. Potential diagrams through the center of a frame-transfer pixel during charge reset (a) together with the cross section of the structure (b).

that the charge-reset point for an interline-transfer device can only be located at the onset of the vertical blanking period and that the shortening of the integration time is limited to an actual integration time not exceeding the vertical blanking period, which is the same as : the longest integration time in the charge-reset mode is about 1/1000 sec. Intermediate values between 1/1000 sec and 1/50 sec or 1/60 sec are not possible with interline-transfer devices.

The situation changes completely if the interline-transfer imager is replaced by a frame-interline-transfer device. The vertical CCD registers of the imaging section of the FIT device are empty at almost every moment, and are available to receive charges from the photodiodes. In the event of charge reset, they can be passed on to a drain at the top of the sensor simply by reversing the clocking sequence of the vertical CCD clocks. On the other hand, the choice of integration-time shortening is completely free with a frame-interline-transfer device. This flexibility, absent when an interline-transfer imager is used, was the basis for the introduction of the frame-interline-transfer CCD. The additional advantage of reduced smear has accelerated this introduction process considerably.

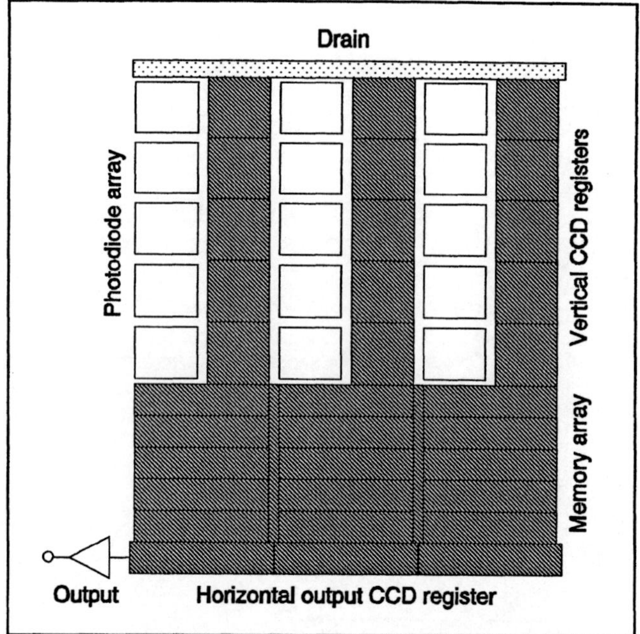

FIGURE 6.18. Charge reset in a frame-interline-transfer type imager by means of an extra drain at the top of the device.

The quite simple solution for FIT did not lift the restriction on interline-transfer devices. Integration-time shortening is limited to the vertical blanking time. A completely new technique for charge reset in IL and FIT devices has been introduced a couple of years ago. The photodiodes are no longer emptied via the vertical CCD registers, but extra means are provided for the photodiodes to empty them directly toward the n-type substrate. The idea is based on the same principle already mentioned during the discussion of frame-transfer imagers : the potential curve running through the center of the pixel is stretched by externally applied voltages, so that the pixel can no longer contain any charge carrier. All minority charges are immediately drained off to the n-type substrate. This process is shown for the IL or FIT photodiode in Figure 6.19.

Comparison of the cross sections in Figures 6.14b and 6.19b shows that the main difference between the structures is in the construction of the photodiodes : a p^+-top layer has been added in the new diode concept. This shallow implantation connects the upper side of the photodiode to the potential set at the channel stoppers. This has the effect of fixing or pinning the potential at the top side of the photodiode. The incorporation of this p^+-top layer has several advantages, of which some will be explained later. With reference to the study of the charge reset, the fact that the upper side of the photodiode is pinned is most important.

The working principle of the pinned photodiode is explained by the various curves

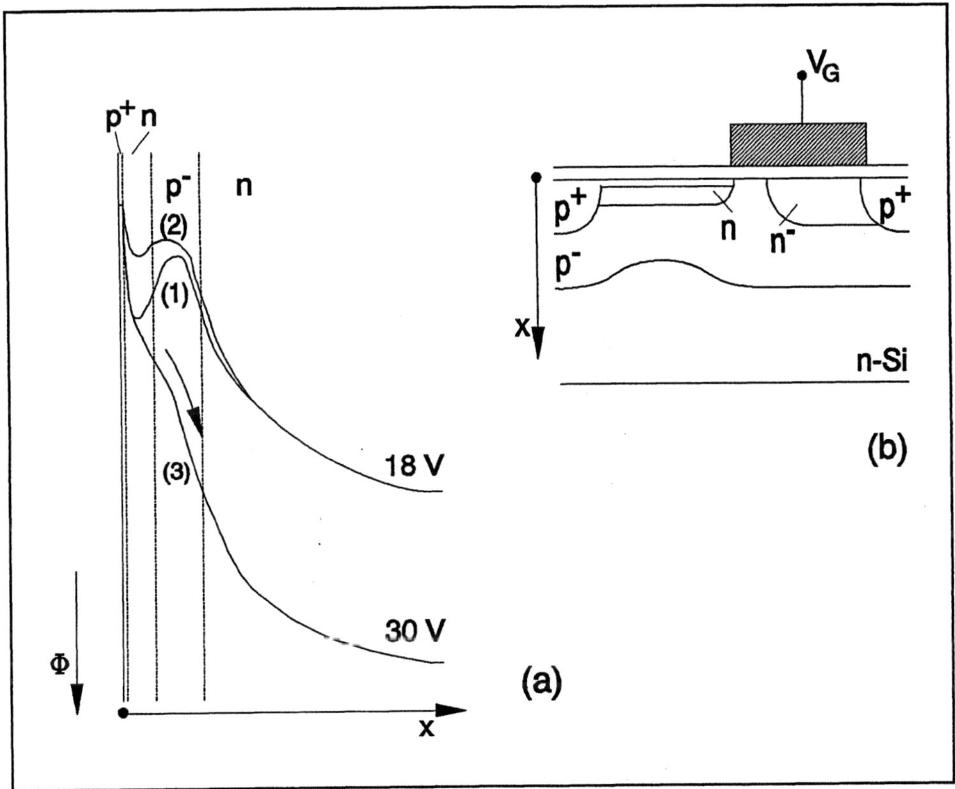

FIGURE 6.19. Potential diagrams through the center of an interline- or frame-interline-transfer pixel during charge reset (a) and its cross section (b).

representing the electrostatic potential through the center of the photodiode toward the substrate. These curves are valid for the following situations (Kuriyama 91, Hojo 91) :

- the onset of an integration period with empty photodiodes is illustrated by curve 1. The potential well in the np⁻ region is capable of storing the minority carriers generated;
- curve 2 illustrates the situation at the point of antiblooming. The remaining potential barrier toward the substrate is so small, that additional electrons can no longer be stored and will flow immediately to the substrate;
- curve 3 shows the potential profile during the charge-reset process. The potential curve is pinned at the photodiode side by the p⁺ layer and is

stretched into the substrate by a high voltage on the n-type substrate. Note that there is no longer any potential well between the p^+-top layer and the n-type substrate. Electrons can no longer be handled and will be drained directly toward the n-type substrate. This is how the charge-reset process operates.

The technique of applying a high-voltage pulse to the n-type substrate to empty the photodiodes gives the IL and FIT devices the same flexibility as the FT imagers so far as charge resetting is concerned. In all cases the integration time can be established with an extra driving pulse to the substrate (for the IL and FIT) or to the gates (for the FT).

Because of the capacitive coupling between the n-type substrate and the floating-diffusion capacitance, there will always be some interference from the high-voltage pulse on the substrate toward the output signal. For typical applications subject to classical TV standards, charge-reset pulses are applied to the device during the horizontal blanking periods, so that the interference is not visible to the end user. When the application is not subject to any TV standard, the interference cannot always be suppressed completely and might degrade the quality of the image displayed on the screen.

6.4.3. CHARGE RESET IN MOS-XY AND CID IMAGERS

A control of the photodiodes belonging to an MOS-XY imager is not available as it was the case for (frame-) interline-transfer CCDs. Emptying of the light-sensitive cells has to be done by a normal readout process. This complicates the design of the imager because, fully parallel to the readout facility of the classical MOS-XY device, a second drive and shift circuit has to be added to the chip to perform the "dummy" readout necessary for charge reset (Nishizawa 87). Figure 6.20 illustrates the MOS-XY sensor provided with the means for electronic shuttering. The effective integration period is defined as the time between the dummy readout and the effective readout cycle.

Note that charge reset with an FT, IL or FIT sensor is done simultaneously for all pixels, unlike the way it is done with the MOS-XY device : during the complete field time, charge reset is processed, pixel after pixel, fully synchronously with the normal pixel readout. In fact the charge-reset technique for all devices is fully in line with the end of the integration time. In the case of CCDs, integration stops at the moment the charge transfer takes place to the vertical registers or the storage area. In the case of MOS-XY sensors, the end of the integration time is defined by the readout of the pixels. This readout moment is different for all pixels.
The aforementioned theory makes the integration time for every pixel in the focal plane, irrespective of sensor type, exactly equal, with or without the electronic-shuttering function.

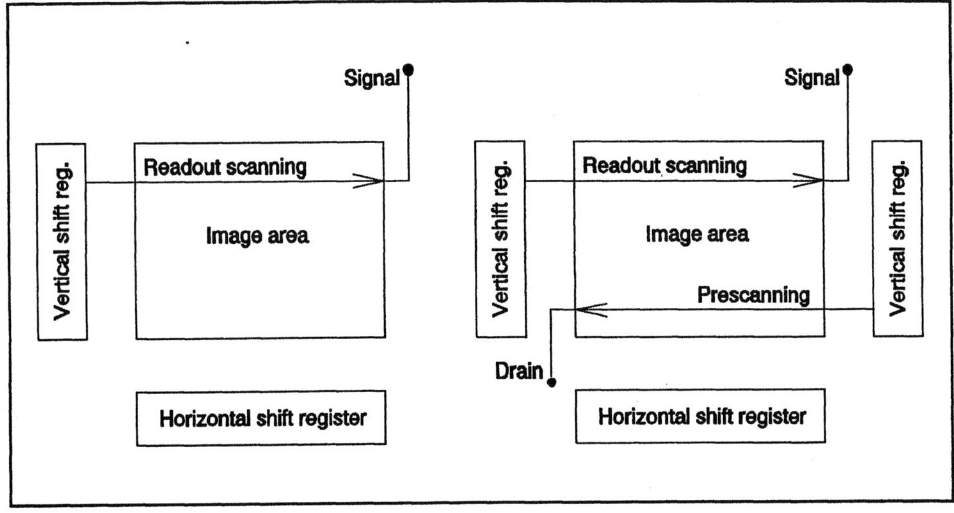

FIGURE 6.20. Charge reset in an MOS-XY imager by means of an extra dummy read-out "in parallel" to the normal operation of the device.

The same technique of electronic shuttering by means of a "dummy" readout has to be applied to the CID sensor. Although control of the MOS capacitors can empty all potential wells at once for all pixels, this will generate different integration times for the various pixels on the sensor area. On one hand integration will start for all pixels at the same time but, on the other, integration will end for each pixel with its readout, and this is completely different for all pixels. In the end also this means incorporating of extra driving electronics and a dummy readout stage in order to incorporate the charge-reset option into a CID chip.

WORTH MEMORIZING

Adding a charge-reset or an electronic-shuttering mode to a solid-state image sensor makes it possible for the device to completely empty its imaging section without destroying the charge content in the storage area or in the vertical CCD registers. With this technique the integration time can be shortened, resulting in sharper pictures or an electronic iris control on the sensor chip.
Charge reset or electronic shuttering benefits from the inclusion of the anti-blooming structure when CCDs are used. Appropriate pulsing of the gates (for an FT) or of the substrate (for an IL or FIT) drains all charge carriers out of the pixels into the substrate. With MOS-XY and CID sensors an extra dummy read-out mode is needed to provide the device with an electronic shutter.

6.5. Conclusions

The main application of solid-state image sensors is in consumer-type camcorders. Once CCDs found their way into these products, the production boom in solid-state imagers started. In this new application field, it is not surprising that considerable effort was put into developing CCDs to make them TV-compatible. This means in the first place compatible with the aspect ratio, the number of lines and the number of pixels on a line, so as to fill the TV screen completely. However, compatibility with the TV standard also relies on ability to use the same scanning mode. TV monitors use interlaced scanning and solid-state imagers have consequently to be operated on the same interlacing principle. When frame-transfer CCDs are used, the pixels are defined simply by the DC voltages on the CCD gates during the integration time. This flexibility makes frame-transfer imagers very suitable for interlaced scanning. From field to field the definition of the exact pixel geometry is shifted one line height simply by using the appropriate DC settings on the gates.

Interline- and frame-interline-transfer devices behave differently. They make use of photodiodes fixed in the silicon surface. To shift the video lines on the sensor from field to field each line is defined by two vertical adjacent photodiodes. In the frame-interlacing mode, the odd-numbered lines of photodiodes are read out during the first field, and the even-numbered lines of photodiodes are shifted to the output during the second field. In the field-interlacing mode, the contents of two diodes are added together in the vertical registers before the readout mode starts. Definition of the different interlaced fields depends on the choice of the added vertical neighbors. The alternative to the interlaced scanning is the progressive scan. In this scanning mode, the CCD designer has to make sure that every pixel needed on the monitor is also available in a single image section of the imager.

A typical characteristic of imaging for consumer applications is color imaging. This feature is implemented by splitting the incoming information up into different colors : according to whether they are primary or complementary components. Color splitting can be done by an appropriate color filter on top of the imager. Depending on the filter organization, the various color components can be arranged into a stripe pattern or a mosaic pattern. Both have their advantages and their disadvantages, as also has the use of primary of complementary pigments. For consumer-type camcorder applications, however, all devices are provided with a mosaic color filter, based on a combination of cyan, green, yellow, and magenta components.

In this chapter attention has been paid not only to color imaging for consumer-type applications, but also to color imaging for professional use. The convenience of the on-chip color filter is obtained at the cost of resolution per color. This effect is not acceptable for broadcast applications, for instance, and is circumvented by the introduction of a color-splitting prism whereby full resolution for each color component is maintained.

A very marked characteristic of the solid-state imager by comparison with image tubes is its invulnerability to overexposure. Charge-coupled devices have built-in protection against blooming so that excess charge carriers generated are drained off. This means that full potential wells do not spill their extra charge over in the event of overexposure. After readout of the charge packets, all potential wells are completely empty again and the devices do not suffer from image lag as tubes do.
In most of the devices used today antiblooming protection is incorporated underneath the pixels in the bulk of the silicon. Although not simple to integrate in the processing this vertical anti-blooming construction has several very attractive advantages : it protects the device very efficiently from overexposure, it decreases the collection of dark current, and enhances the diffusion MTF.

As already mentioned, solid-state image sensors have a much lower image lag than image tubes. Image lag limits the temporal resolution of the imaging element. Although the temporal resolution of solid-state imagers is already quite high, it can still be enhanced by electronic shuttering or charge reset. This feature reduces the integration time to almost any arbitrary value. Instead of integrating during the standard field time of 20 msec or 16.7 msec, integration times as short as a single line time (64 μsec) are possible. Charge reset is achieved with the same means whereby antiblooming protection is made available.

A second interesting application of charge reset or electronic shuttering is control of the video output of the image sensor via the integration time. If the light input is quite high in terms of W/cm^2, shortening the integration time can reduce the video output level. Optimizing the video signal by control of the integration time involves providing the image sensor with an on-chip iris control. The latter can operate without any effect on the depth of focus of the optical system as a whole.

Chapter 7

ADVANCE IMAGING : LIGHT SENSITIVITY

Continuing research and development effort is proceeding on how to improve the performance of solid-state image sensors. Improving technology, optimizing design, and creating new device architectures, all have the same goal : to increase the signal-to-noise ratio of the sensors for a given application or a given resolution. At the moment, resolution is no longer an issue as far as imaging is concerned for consumer applications, but it returns to the scene as soon as applications such as HDTV are considered. In this chapter attention will be paid to the techniques intended to increase light sensitivity. The next two chapters will deal with decreasing noise levels and different types of device architectures. All items are catalogued under the main heading of "Advanced Imaging", but topics regarded today as being "advanced" may be introduced tomorrow in the consumer sector and described the day after tomorrow as "standard".

This chapter on light sensitivity will give an overview of different methods and techniques for increasing the response of imagers to the incident light. A few of the methods described are only applicable to the interline-transfer type of imagers, others only to frame-transfer devices. But they all have the same objective : to increase the light sensitivity.

The first section considers the influence of the quantum efficiency of solid-state imagers on the signal-to-noise ratio of the device. It will shown that this influence depends on the noise level. The section on the aperture ratio of the pixels is devoted to several optimization methods for enhancing the optical response of interline- and frame-interline transfer CCDs. The introduction of microlenses, photoconversion layers, and optimized designs of the vertical registers increase the aperture ratio of photodiodes considerably. The limitation of the optical sensitivity of frame-transfer CCDs can be traced to the optical characteristics of the multilayered structure. The third section shows how issues to this kind can be handled.

Other possibilities for enhancing the light sensitivity are the illumination from the back-side of the sensor and the incorporation of active pixels with an amplification function. Both items will be dealt with in the last two sections.

7.1. Increasing the light sensitivity

How can an increased light sensitivity influence the signal-to-noise ratio of the imager ? A great deal of effort is being put into the development of solid-state imagers in order to further increase this S/N ratio. This figure of merit that applies here can be simply defined as :

$$\frac{S}{N} \approx \frac{number\ of\ signal\ electrons}{number\ of\ noise\ electrons} \cdot \qquad [7.1]$$

On one hand the number of signal electrons can be specified in terms of the number of photons impinging on the sensor surface per unit area $\Phi_0/(h.v)$, the pixel area A_{pixel}, the quantum efficiency η, and the integration time T_{int} :

$$number\ of\ signal\ electrons = \frac{\Phi_0}{h.v}.A_{pixel}.\eta.T_{int} \cdot \qquad [7.2]$$

On the other hand the number of noise electrons can be divided between three noise sources :
- photon shot noise equal to the square root of the number of signal electrons. The definition of photon shot noise is similar to that of dark-current shot noise given in section 3.4.1;
- noise electrons generated in the CCD channels (e.g. incomplete transfer; dark current, fixed pattern noise, ...), all summated in n_{CCD};
- output amplifier noise : n_{SF}.

Because these various noise contributions are uncorrelated, the number of noise electrons can be written as :

$$number\ of\ noise\ electrons = \sqrt{\frac{\Phi_0}{h.v}.A_{pixel}.\eta.T_{int} + n_{CCD}^2 + n_{SF}^2} \cdot \qquad [7.3]$$

If all these contributions are inserted into [7.1], the definition of S/N, can be written as (White 74) :

$$\frac{S}{N} \approx \frac{\dfrac{\Phi_0}{h.v}.A_{pixel}.\eta.T_{int}}{\sqrt{\dfrac{\Phi_0}{h.v}.A_{pixel}.\eta.T_{int} + n_{CCD}^2 + n_{SF}^2}} \cdot \qquad [7.4]$$

For high levels of light input (Φ_0 high), the shot noise becomes dominant, and the above expression [7.4] can be simply rewritten as :

$$\frac{S}{N} \sim \sqrt{\eta} \ . \qquad\qquad [7.5]$$

For low levels of light input (Φ_0 low) on the other hand, the shot noise becomes negligible compared to the amplifier noise and the noise from the CCD itself. S/N can be reduced in this case to :

$$\frac{S}{N} \sim \eta \ . \qquad\qquad [7.6]$$

In general it can be stated that the signal-to-noise ratio is proportional to at least the square root of the quantum efficiency, but in many situations it is directly proportional to the quantum efficiency. For this reason it is of great importance to increase the quantum efficiency of the devices in order to further optimize the signal-to-noise ratio.

Increasing the quantum efficiency of the devices can be done by optimizing the aperture ratio of the pixels, by increasing the light transmission of the multi-layered structures on top of the pixels, or by illumination the solid-state imagers from their back-side.

WORTH MEMORIZING

For optical input signals with a low light level, the S/N ratio of CCDs is proportional to the square root of the quantum efficiency, and for all other light levels it is actually directly proportional to the quantum efficiency. Increasing the light sensitivity is one of the chief means in optimizing the S/N ratio of imagers.

7.2. Aperture ratio of the pixels

The aperture ratio of a pixel is defined as the ratio of the light-sensitive area of a pixel to the total area of that pixel. From its basic design concept, the aperture ratio of a frame-transfer pixel is relatively high because almost the entire pixel is sensitive to light. In contrast to frame-transfer imagers, the aperture ratio of a (frame-) interline-transfer pixel is quite low. Near each sensitive pixel a CCD cell of the vertical shift register is incorporated which takes up silicon space but is not light-sensitive. Typical values for the aperture ratio are between 25 % and 45 % for IL sensors dedicated to consumer applications. This relatively low figure can be considerably increased by providing the pixels with microlenses, by placing a photoconversion layer on top of the imager or by optimizing the design of the vertical registers to minimize their area. All these options will be described in this section.

7.2.1. MICROLENSES

The result of providing the IL or FIT pixels with microlenses is shown schematically on the cross section of an interline-transfer pixel in Figure 7.1 (Ishihara 83, Sano 90). In the complete two-dimensional matrix of photodiodes each single pixel has its own microlens, consisting of a base-resin layer. This means that part of the

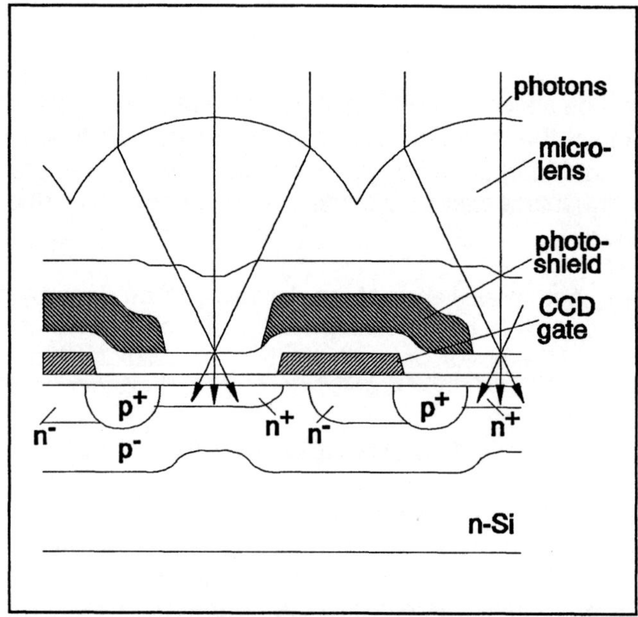

FIGURE 7.1. Cross section of an IL or FIT image cell provided with on-chip microlenses.

light impinging on top of the vertical CCD shift register can also be focused on the sensitive part of the pixel. The sensitivity of sensors provided with microlenses can be easily increased by a factor of two. For smaller pixel dimensions, even as low as 6 μm x 7 μm, the increase in quantum efficiency can in fact be threefold (Furukawa 92). The explanation for this effect is quite simple : the smaller the pixel size, the smaller the aperture ratio of the pixel and the higher the gain in sensitivity can be if the pixels are provided with microlenses.

Mainly because of their benefits with regard to sensitivity of interline- and frame-interline-transfer imagers, the technology of microlenses has been quite extensively studied. Different types of microlenses can be placed on the rectangular photodiodes belonging to these types of image sensors. The various alternatives are displayed in Figure 7.2.

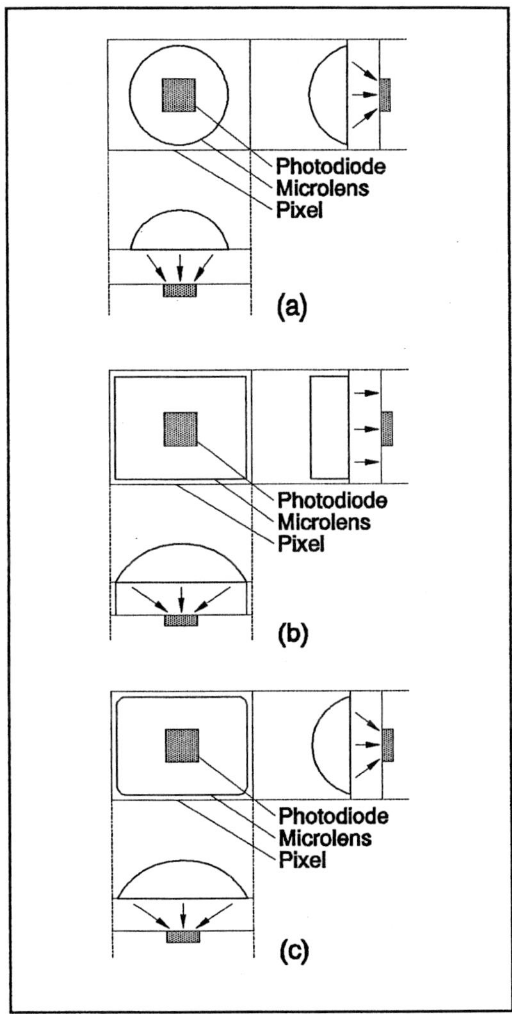

FIGURE 7.2. Top views and cross sections of the different types of microlenses : hemispherical lens (a), semicylindrical lens (b), and rectangular dome lens (c).

From top to bottom : the hemispherical lens (which has a full circular structure), the semicylindrical lens (which is only round in one dimension), and the rectangular dome lens (which is differently shaped along its two orthogonal axes). For example :

the original pixel without a microlens and of size 7.5 μm (V) x 9.6 μm (H) has an aperture ratio which is less than 30 % of the total unit cell area. In the case of this pixel and with a lens spacing of minimum 0.8 μm, the aperture ratio can be increased to 49 % for the hemispherical lens structure, to 67.2 % for the semicylindrical lens and to 81.9 % for the rectangular dome lens (Sano 91). From these figures it can readily be concluded that the rectangular dome lens structure is most suited for incorporation on top of the photodiodes. Special microlens simulation tools were developed to determine the real effective aperture ratio of a rectangular dome lens structure. The effective aperture ratio of the rectangular dome lens is optimized by an appropriate choice of :
- the thickness and width of the lens (radius of lens curvature);
- the refractive indices for the lens resin and (optional) color filter layer;
- the thickness of the color filter layer;
- the area of the photodiode;
- the thickness of the metal photoshield, and;
- the area of the unit cell.

A very important boundary condition encountered during the design of the microlens array is the presence of the main lens of the camera system. The character of the incident rays, which are shaped by the main lens, on the microlenses considerably influences the effective aperture ratio. In the entire optical system containing a microlens array and a main lens, the rays from the exit pupil of the main lens enter the microlens. The rays incident on the microlens are therefore shaped like a cone with a top angle φ, and the principal ray (forming the center of the cone) at the periphery of the sensor has an angle δ between the ray and the optical axis. The two angles φ and δ are graphically defined in Figures 7.4 and 7.6, respectively.

The angle φ varies as function of the size and position of the exit pupil, and consequently also as a function of the F number of the main lens. The exit pupil is by definition the smallest lens aperture seen from the image plane, while the F number is defined as the ratio of the distance of the exit pupil (from the image) and the diameter of the exit pupil. Figure 7.3 illustrates the effect of φ depending on the F number for two situations (Furukawa 92). On the left side a high F number is chosen, the exit pupil is small, the rays falling on the micro lens are nearly parallel to each other, φ is about equal to 0, and the rays are focused almost perfectly into the photodiode of the pixel. On the right side a low F number is chosen, the exit pupil of the main lens is relatively large, the rays falling on the microlens are no longer parallel to each other, φ is quite large, and the rays are focused in the lens material itself instead of in the photodiode. Consequently, not all the rays are absorbed in the photodiode after focusing by the microlens. From this example it can be seen that small F numbers cause a decrease in effective aperture ratio because the rays cannot be converged sufficiently.
To summarize : the microlenses are optimized for a most practical F number. The simple microlens will no longer be optimized for other F numbers. This effect is

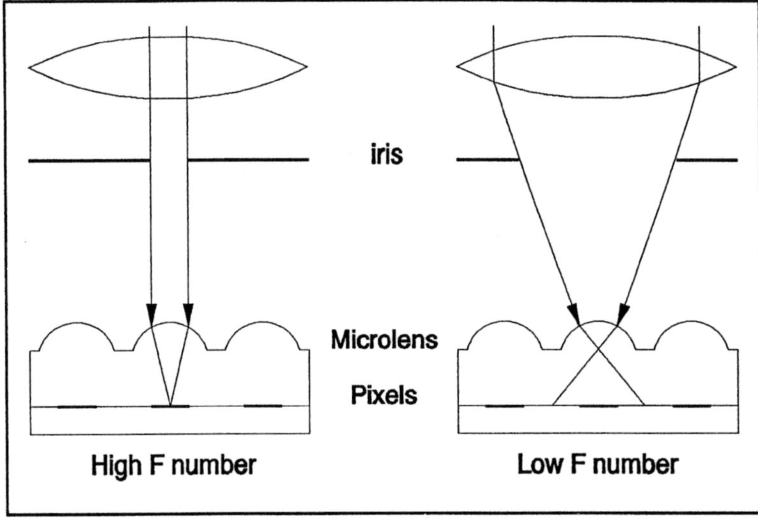

FIGURE 7.3. Effect of the F number of the main lens on the optical light path through a microlens.

also known as optical aberration.

An example of this effect is shown in Figure 7.4 where the ray angle φ and the sensitivity increase (compared to the imager without a microlens array) are shown as a function of the F number (Deguchi 92). The basic theory arising from Figure 7.3 is thus repeated : low F numbers cause high values for φ and consequently a drop-off toward lower values for the sensitivity gain.

As already mentioned, the angle δ varies as a function of the distance to the exit pupil but also depends on the position of the pixel on the sensor surface. Figure 7.5 illustrates this effect. Again two situations are presented : on the left side a pixel at the center of the image section is shown, the principal ray falling on the microlens is parallel to the optical axis, δ is equal to 0, and the rays are focused almost perfectly into the photodiode of the pixel ; at the right a pixel at the periphery of the imager is considered, the principal ray falling on the microlens makes a relatively large angle with the optical axis, δ is quite large, and the rays are no longer focused optimally onto the photodiode.
Consequently, not all the rays are absorbed in the photodiode after focusing by the microlens. From this illustration it can be seen that, when δ is large, the effective aperture ratio at the periphery of the sensor is lower than at the center since the position of the principal ray refracted by the microlens is shifted from the center of the photodiode aperture. This effect is further illustrated by an example in Figure 7.6 (Deguchi 92). The worst-case angle δ (at the pixel most out of center) is shown for an imager with a 5.5 mm image diagonal, as a function of the distance to the

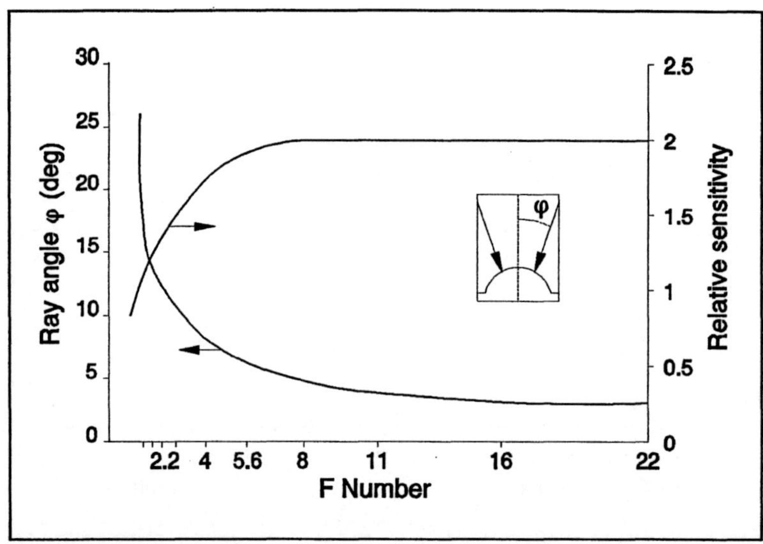

FIGURE 7.4. Quantitative illustration of the effect of the F number on ray angle φ and sensitivity increase.

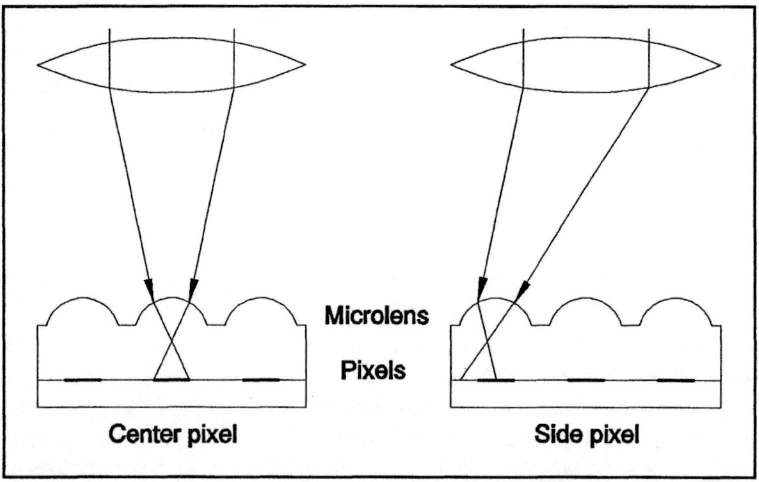

FIGURE 7.5. Effect of the δ angle on the optical light path through a microlens.

exit pupil . If the exit pupil of the main lens is close to the image sensor, the angle δ increases rapidly leading to a nonuniform effective aperture ratio. This is translated into a non-uniform sensitivity profile across the overall image section. (Note that the exit pupil is lens-type dependent : for a short focal length the exit pupil is normally close to the image plane.)

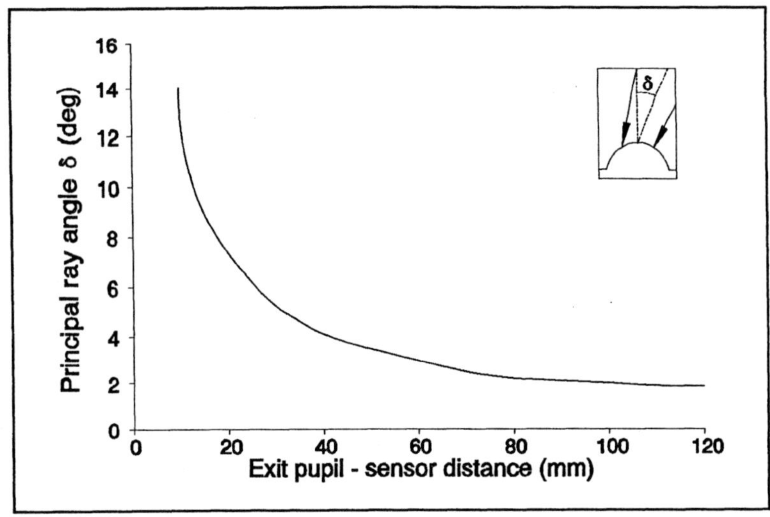

FIGURE 7.6. Quantitative illustration of the effect of the angle δ on the sensitivity increase.

7.2.2. PHOTOCONVERSION TOP LAYER

Depositing a photoconversion layer on top of the imager results in a pixel cross section as shown on Figure 7.7 (Tsukada 81, Chikamura 84, Manabe 88). The small photodiode of the classical interline-transfer CCD is extended vertically with a metal interconnection toward an amorphous-silicon top layer. In the actual device this metal contact is made of molybdenum silicide (Mo polycide) and the amorphous-silicon top layer consists of a sandwich of intrinsic and p-type amorphous hydrogenated silicon, indicated respectively as α-SiC:H(i) and α-SiC:H(p) in Figure 7.7.

Before deposition of the α-Si, the complete structure is planarized by the deposition of the BPSG layer (Boron Phosphorus doped Silicate Glass). The amorphous-silicon photoconversion layer is electrically biased by means of the transparent conducting ITO (Indium Tin Oxide) layer. The photoconversion layer completely covers the image section of the sensor. The initially small aperture ratio of about 25 % is now increased to almost 100 %. This structure exhibits extremely high sensitivity, almost no smear (the Mo polycide is opaque) and, due to its wide vertical CCD channels, has a high charge-handling capability. However, there are of course also disadvantages : in addition to its complicated technology it has a relatively high dark current and the severe drawback of a large image-lag (Nohmi 90). The image lag consists of two different effects : the unpinned storage diode cannot be emptied completely within the limited time, and the amorphous photoconversion layer contains many bulk traps which retain and (later) release photon-generated charges. The first effect is known as capacitive image lag, the second as photoconductive image

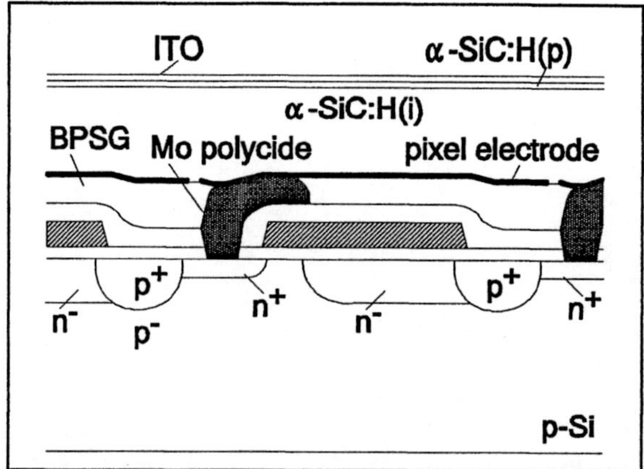

FIGURE 7.7. Cross section of an IL image cell provided with an amorphous-silicon top layer as its photoconversion element.

lag.

A technique used to lower the capacitive image lag drastically can be achieved by resetting the storage diodes to a reference voltage after each field readout. To apply this method, each pixel is provided with an extra storage-diode resetting gate. The construction of the pixels is complicated by this extra resetting gate and the method itself adds some reset noise to all the individual pixels. However, this effort is compensated by a lower image lag : a reduction from 3.5 % to 0.4 % (both numbers are related to the third field) is achievable (Shibata 92).

In short, the technology and the pixel construction of the image sensors overlaid with an amorphous photoconductive top layer are somewhat complicated and the devices show some "tube-related" phenomena, but the imaging characteristics are quite spectacular. Devices with 2M pixels measuring 7.3 μm (H) x 7.6 μm (V), covered with an amorphous-silicon top layer and presenting a sensitivity of 210 nA/lux and a dynamic range of 110 dB, have been reported (Shibata 92).

7.2.3. OPTIMIZING THE VERTICAL CCDS

Instead of increasing the light-sensitive area, it is also possible to decrease the area of the light-insensitive regions, namely the vertical CCD shift register. This is the idea behind the Charge-Sweep Device (CSD), shown in Figure 7.8a (Kimata 85), which is an interline-transfer CCD with an extremely narrow width for the vertical CCD channels. This channel is unable to handle the charge packets delivered by the photodiodes without spilling the charges over across the barrier gates. Spilling-over, however, is allowed if one and only one video line at a time is transferred to the vertical shift registers. The single charge packet per vertical CCD spreads

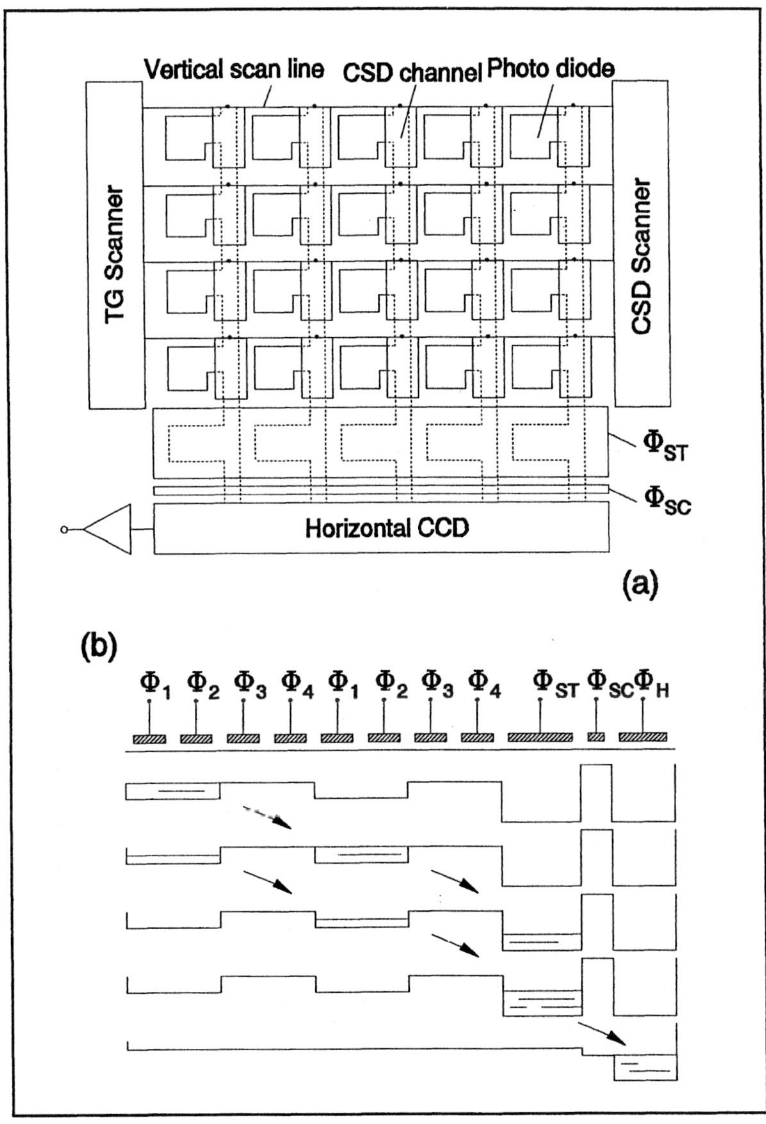

FIGURE 7.8. The Charge-Sweep Device : its device architecture (a) and its working principle (b).

across the excessively narrow CCD channel over several CCD cells. This process is shown schematically in Figure 7.8b. The charges in the various CCD cells of one vertical CCD are swept into a storage site close to the horizontal register, where all subpackets are again summated to restore the original charge packets as delivered by the photodiodes.

The CSD imager is characterized by :
> - slightly more complicated clocking (the vertical shift registers have to empty each line period);
> - the need for some extra provisions to select a single line to transfer its charges from the photodiodes into the vertical registers;
> - operation with somewhat low clock voltages (e.g. 1 V peak-to-peak for the vertical CCDs);
> - high charge-handling capability;
> - high vertical transport efficiency.

This same idea of making the width of the vertical CCDs on the silicon smaller than that of the standard interline-transfer devices is applied in the trench CCD (Yamada 89). Its principle is shown in Figures 7.9a and 7.9b, which are respectively, a top view and a cross section of the trench CCD.

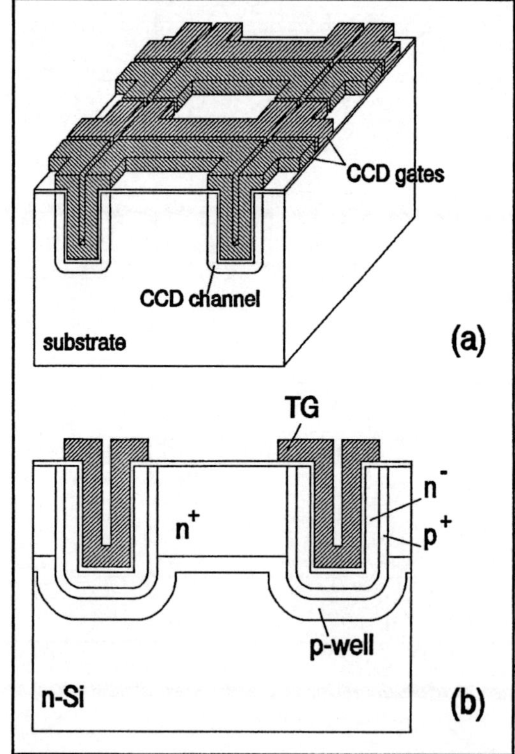

FIGURE 7.9. The Trench CCD : a generalized perspective view (a) and a cross section perpendicular to a U folded CCD channel on an n-type substrate (b).

The vertical CCDs are folded into a U form and inserted into grooves etched in the silicon. In this way the silicon area taken up, as seen in a top view, is relatively small and the aperture ratio of the device can be made large. However, the Si-SiO_2 interface of this trench CCD coincides with the groove. Careful processing is needed because this can give rise to a high density of defects and interface states, resulting in high levels of dark current and dark-current nonuniformities.

WORTH MEMORIZING

Pixels of interline- and frame-interline-transfer CCDs are characterized by low aperture ratios. Several new developments to compensate for this drawback have been reported :
- **covering each photodiode of the imaging array with an individual microlens. The gain in sensitivity is relatively high but depends on the F number of the main lens of the camera system and on the position of the pixels on the array;**
- **covering the imager with a photoconverting top layer which increases not only the light sensitivity but unfortunately also the image lag;**
- **"clever design of the vertical shift register by introducing a charge-sweep device or a trench CCD.**

7.3. Light transmission of multilayered structure

The aperture-ratio optimization is a typical issue for (frame-) interline-transfer devices because, as already mentioned, the aperture ratio for a frame-transfer device is relatively high. A typical issue involving the frame-transfer principle is the limited light transmission of the CCD gates on top of the pixels. Almost every (frame-) interline-transfer sensor is provided with photodiode pixels, which are only covered by fully transparent silicon oxide and/or silicon nitride. This means that the pixels of these types of imagers can be more light sensitive in the blue and near-UV spectrum, compared to the frame-transfer imager. However, much effort is also being devoted to increasing the sensitivity of frame-transfer devices. The most common means of enhancing the light sensitivity have to do with the minimization of the light absorption in the CCD gates : adapting the optical thicknesses of the multilayered structure, and minimizing the number of gates or incorporating transparent conductive gates.

7.3.1. ADAPTING THE OPTICAL THICKNESSES

The easiest way to minimize the absorption in the multilayered structure (of gate oxide, gate, silicon oxide, and/or silicon nitride) is to adapt their relative thicknesses to keep the absorption as low as possible, their transmission as high as possible, and their reflection as low as possible. Another optical effect plays an important role in the multilayer structures : interference peaks. These interference peaks are created by the interaction of different thicknesses and refractive indices of the layers of the multilayered structure. Interference peaks can hardly be avoided but on the other hand it may sometimes be useful to position the interference peaks at wavelengths of interest. Two different examples are shown in Figure 7.10a (Anagnostopoulos 80) and Figure 7.10b (Weijtens 85), in which, respectively, the polysilicon thickness of the gate is adapted and a gate dielectric sandwich of SiO_2

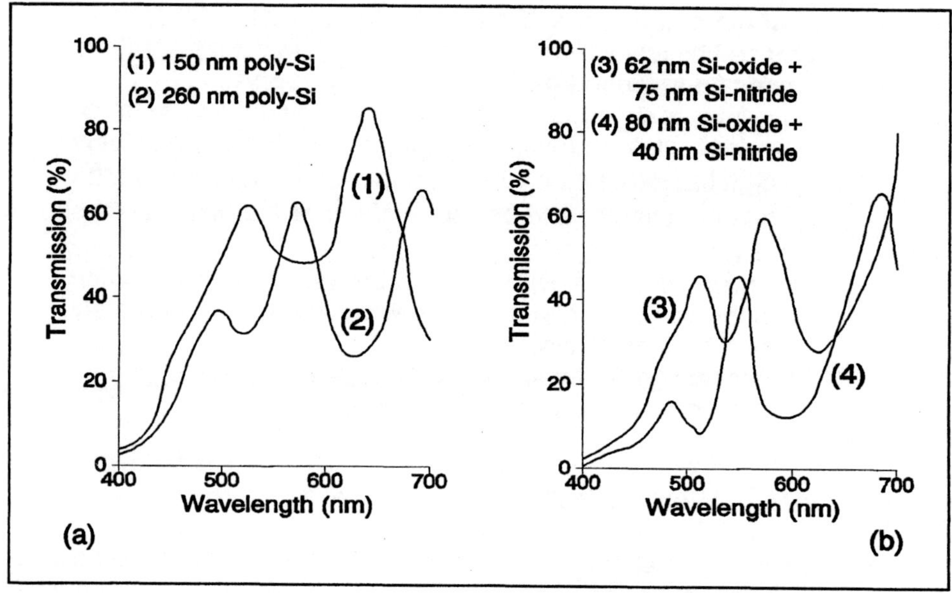

FIGURE 7.10. The influence of the layer stack on the light transmission of a multilayered structure is shown : (a) for various gate thicknesses and (b) for different gate dielectrics.

and Si_3N_4 is optimized underneath the poly-Si gate material of 270 nm thickness. Note in Figure 7.10a the overall increase of the light sensitivity by changing the polysilicon thickness from 260 nm to 150 nm. The same effect is shown in Figure 7.10b, where the overall transmission of the multilayered structure increases without any changes in the thickness of the absorbing layer. But the gate dielectric sandwich is optimized from 40 nm of Si_3N_4 on top of 80 nm of SiO_2 to 75 nm of Si_3N_4 on top of 62 nm of SiO_2. All curves are clearly dominated by a series of peaks and valleys. These are due to the multiple reflections in the multilayered structure on top of the CCD substrate.

7.3.2. MINIMIZING THE NUMBER OF GATES

The introduction of the virtual-phase CCD is based on minimizing the number of gates on top of the charge-coupled device (Hynecek 81). With the virtual-phase concept (see also section 2.1.5) shown in Figure 7.11a, a two-phase clocking device is constructed in which one polysilicon gate is fully replaced by an appropriate doping profile in the silicon substrate. Because only half of the pixel is covered with polysilicon, the optical response of these imager types is quite high. A typical example is shown in Figure 7.11b. Possible drawbacks of virtual-phase CCDs are their limited transport speed, their limited charge-handling capability, and the quite high electric fields in the semiconducting substrate, induced by the externally applied gate voltages.

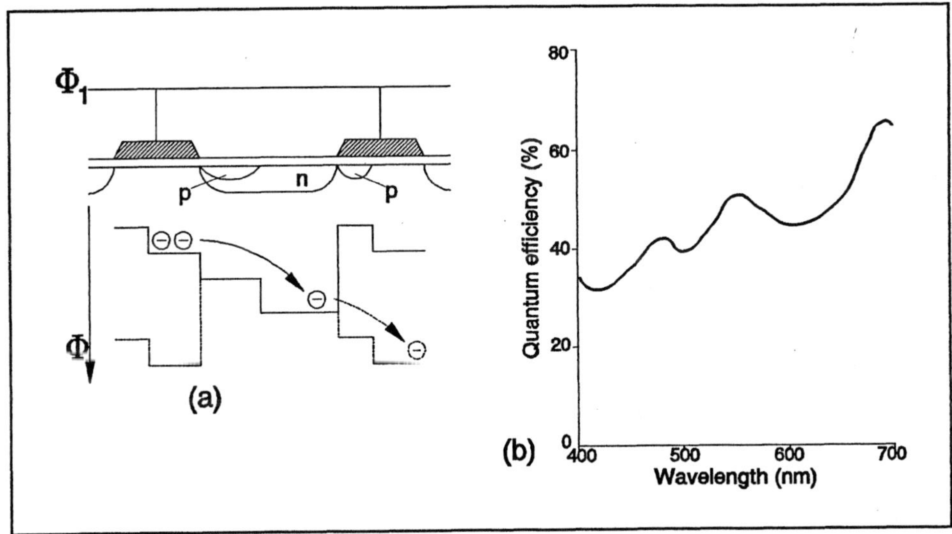

FIGURE 7.11. A cross section of a virtual-phase CCD (a) and its photoresponse (b).

Instead of omitting half the gates, it is also possible to cut light windows in the gates themselves (Hagiwara 78). Figure 7.12 illustrates an example of a four-phase CCD in which two gates are provided with light windows to enhance the device's blue sensitivity. The number of gates is unchanged, but the area of absorbing polysilicon gates is reduced (van de Steeg 85). This can be seen by comparing the cross sections BB' and CC' in Figure 7.12. The former cuts through the light windows. The blue sensitivity contribution is mainly due to the windows, but green and red light also generates electrons underneath the gate areas themselves.

The so-called cross-gate imager (Mitani 84) uses the same principle to enhance the blue response of the CCD. Basically, it is also a frame-transfer imager. Its

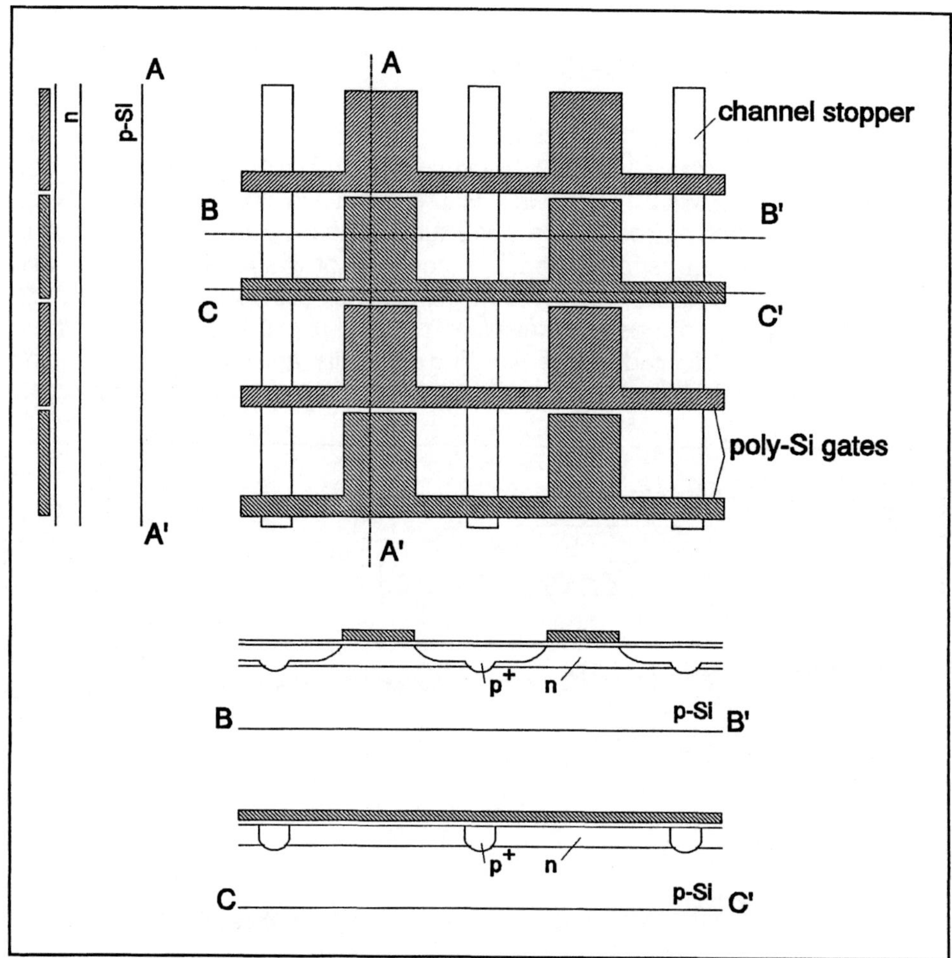

FIGURE 7.12. *Blue response enhancement by light windows, cut in the CCD gates.*

cell structure is shown in Figure 7.13.

Two polysilicon layers define the four phases of the transport channels. By placing them alternately vertically and horizontally, the transport path of the charge packets is defined as indicated by the dotted line.

7.3.3. TRANSPARENT CONDUCTIVE GATES

The ultimate solution, of course, is not to have any polysilicon at all on top of the imager, but a fully transparent and conductive gate material instead. The replacement

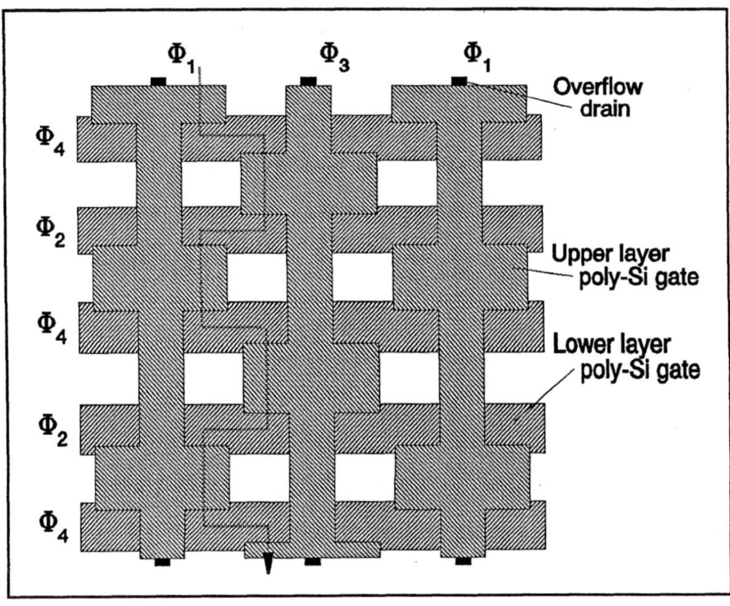

FIGURE 7.13. The cell structure of the cross-gate CCD with a lateral overflow drain. The dotted line indicates the path of charge transfer.

of poly-Si by these kinds of materials (Indium Tin Oxide or ITO, Tin Oxide), has been studied at different research institutes (Brown 76, Schroder 78, Harada 80, Theuwissen 84a, Kosman 90). The response of a linear image sensor in which some of the pixels are photodiodes, some of them MOS capacitors with a poly-Si gate and, the remainder MOS capacitors with an ITO gate is shown in Figure 7.14 (Van der Spiegel 84). The already familiar drawback of using poly-Si gates (covering the pixels completely), namely a somewhat low blue response is again shown in this figure. This can be remedied by using ITO as the gate material. ITO absorbs almost no light and, moreover, its refractive index matches very well with SiO_2 and Si_3N_4. However, as can also be learned from Figure 7.14, the sensitivity of bare silicon in the case of the photodiode is also quite high. The sensitivity results of MOS capacitors with ITO as gate material are very attractive, but the introduction of these "strange" materials in standard silicon technology creates problems of surface and interface damage and incompatibility with existing silicon technology. In particular, the introduction of a double ITO layer process has rarely been reported because of problems with inter-ITO isolation. For these reasons, the introduction of ITO or SnO_2 is always kept on a low laboratory level. More recent work on this subject proposed using a sandwich of a very thin poly-Si layer (25 nm) and an ITO gate of normal thickness : 150 nm (Weijtens 92). The numbers of interface states and oxide charges reported come close to those which are standard in IC technology, but the issue of inter-ITO isolation is still an unsolved problem.

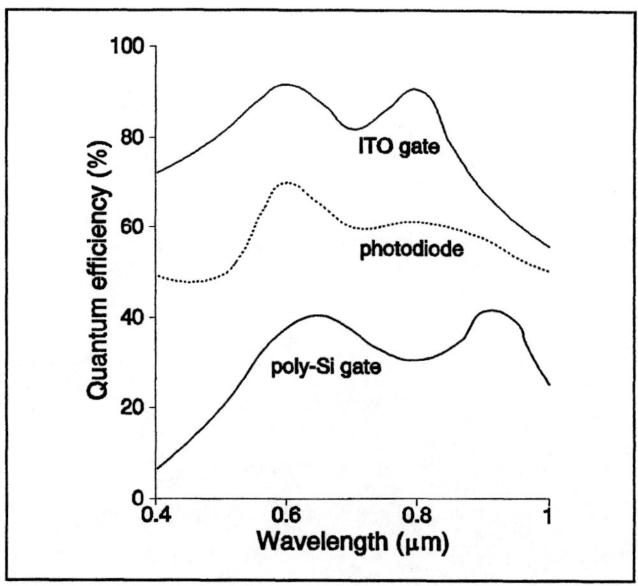

FIGURE 7.14. Response of a linear CCD imager with photodiodes, MOS capacitors with a poly-Si gate and MOS capacitors with an ITO gate.

WORTH MEMORIZING

Minimizing the absorption of incoming light in the polysilicon gate material enhances the quantum efficiency of frame-transfer devices, particular for blue light. The simplest way to optimize transmission through the multilayered structure is to adapt the optical thicknesses of the different layers to each other in such a way that reflection in the overall structure is minimized.

Another interesting option is to minimize the number of gate material on top of the pixels. This can be done by means of the virtual-phase CCD, less than 50 % of whose cell area is covered only by the absorbing polysilicon. Extra light windows in the gate material or the cross-gate configuration have the same background : less absorbing material on top of the cell. With respect to this issue the ultimate solution will be to introduce fully transparent and conducting gate material, e.g. indium-tin oxide. Although the CCD becomes very sensitive with this solution, ITO technology is not very readily compatible with the existing IC technology.

7.4. Back-side illumination

To obviate all problems of light absorption and reflection on top of a CCD, it is also possible to illuminate the device from the back (Blouke 78, Levine 84). On this side of the chip, bare silicon is available. The problem remaining is the "transport" of the generated electrons toward the front surface of the devices. This has to take place by diffusion, which severely degrades the diffusion MTF. To overcome this problem, the devices are thinned in such a way that the rear surface of the devices comes to about 10 μm or less from the front side of the CCDs. Figure 7.15a shows a cross section of an imager thinned from the rear to a final thickness of about 10 μm (Woody 91). Figure 7.15b shows the quantum efficiency for a full-frame device illuminated from the front and also after thinning, from the back. The back-side illumination results in a very high quantum efficiency with a very smooth and flat decay.
The following also applies to these devices : interesting figures for sensitivity, but quite complicated technology needed to thin the devices, passivate them and handle them afterwards. More information about back side illumination can be found in the section 10.3.3 on X-ray imaging.

WORTH MEMORIZING

Back-side illumination of charge-coupled devices solves all problems of quantum efficiency because absorbing gate material is completely absent and the aperture ratio is 100 %. New and even harder problems emerge : thinning, passivating and handling the very vulnerable image sensors.

7.5. Pixels with an amplification function

In theory, one of the most ideal ways to increase the sensitivity of the imager is to amplify the photoconverted signal locally in each pixel. If, however, each individual pixel has to be provided with its own amplifier, construction of these pixels will become rather complicated. Nevertheless, research effort has resulted in a few different pixel organizations and configurations including an amplification function. All imagers produced by this technique use the MOS-XY organization and have in addition to the simple combination of the photodiode and the switching transistor, the amplifier.

A few types of pixel amplifiers will described briefly in this section : the Static-Induction Transistor (SIT), the Charge-Modulation Device (CMD), the Bulk Charge-Modulation Device (BCMD), the Amplified MOS Intelligent Imager (AMI), and the Base-Stored Image Sensor (BASIS).

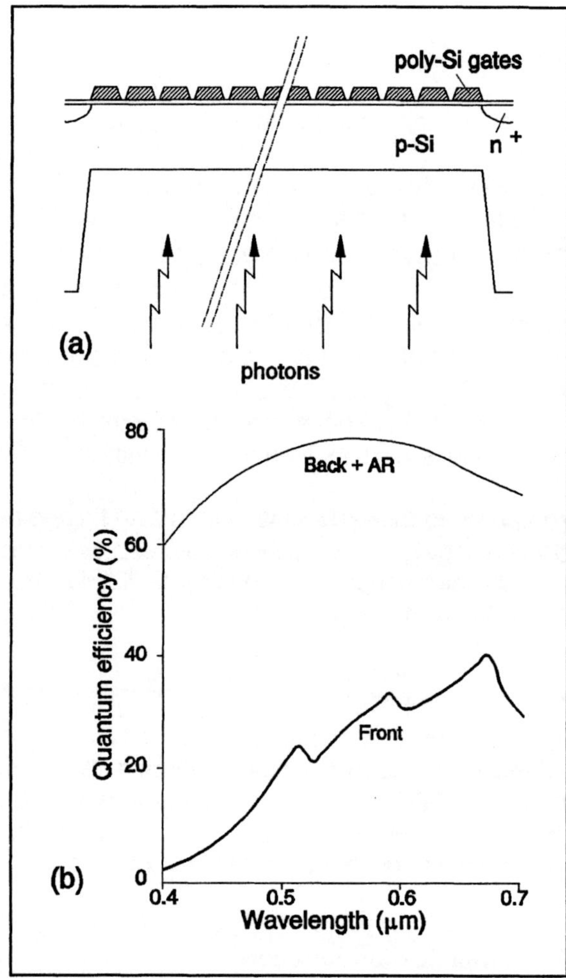

FIGURE 7.15. Cross section of a back-side thinned CCD (a) and its enhanced response to light (b).

7.5.1. STATIC-INDUCTION TRANSISTOR

The equivalent circuit of a pixel of the SIT imager is shown in Figure 7.16 (Yusa 85). A light-sensitive MOS capacitor is placed on top of a static-induction transistor. Charges generated by the incoming light are accumulated on the bottom plate of the MOS capacitor. During the cell-readout stage the gate of the structure is pulsed, the accumulated charges are dumped in the base of the SIT device and the amplified output current delivered at the emitter side of the transistor is measured. Typical values for the current amplification are of the order of 1000. The SIT image

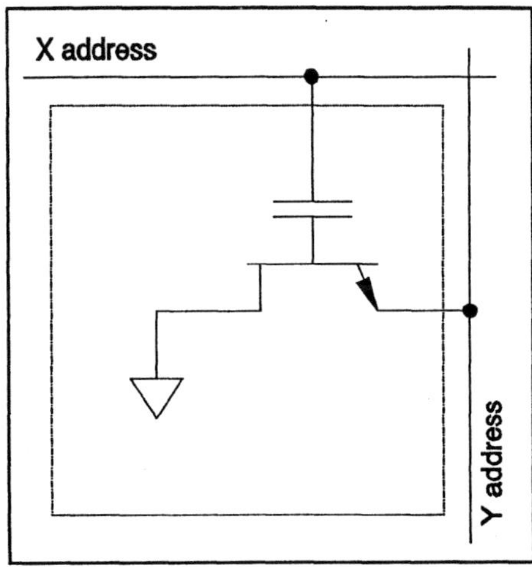

*FIGURE 7.16. The equivalent circuit for a pixel
of the static-induction transistor (SIT) imager.*

sensor readout is nondestructive.

7.5.2. CHARGE-MODULATION DEVICE

Figure 7.17 shows the basic setup of a pixel from a charge modulation imaging
device (Nakamura 86). The MOS transistor is built around an n^+ source and drain,
the channel is of the n-buried channel type. Charge carriers are generated in the
potential well under the gate electrode. The holes of these are stored in the Si-SiO_2
interface under the gate, while the electrons flow to the source or drain regions.
The holes stored increase the surface potential, thus lowering the current barrier
height for the electrons. The electron-current flow consequently varies with the
number of stored holes, and an amplified light-dependent current is obtained. In
addition, the spatial separation of holes and electrons enables the device to operate
in a nondestructive readout mode.
One of the advantages of the CMD pixel is its compact structure. Pixel sizes of
5.0 μm x 5.2 μm are reported (Nomoto 93).

7.5.3. BULK CHARGE-MODULATED DEVICE

The bulk charge-modulated device is related to the CMD, not only by its name
but also by its operation. While the unit cell of the CMD is constructed around
an MOS transistor, the pixel of the BCMD is formed with the aid of a buried-channel
device. It is shown schematically Figure 7.18 (Hynecek 88). The BCMD consists
of a buried p-channel readout transistor back-gated by signal electrons in an n-type
confinement region below the readout transistor. Optically generated electrons

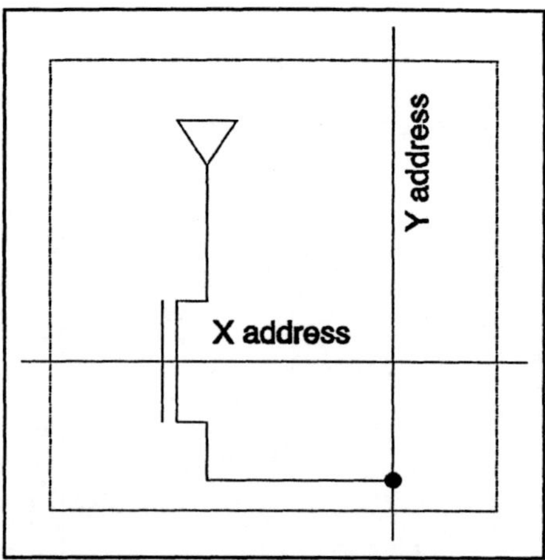

FIGURE 7.17. The basic setup of a pixel from the charge-modulation imaging device (CMD).

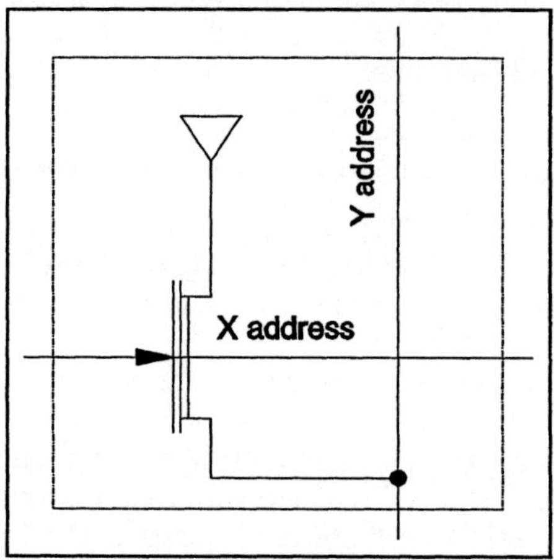

FIGURE 7.18. The basic setup of a pixel out of the bulk charge-modulation device (BCMD).

are collected and integrated into the confinement region. These charges modulate the bulk current from electrons flowing from the drain to the source contact. A reset operation dumps the signal electrons into the substrate.

7.5.4. AMPLIFIED MOS INTELLIGENT IMAGER

As far as its construction is concerned, this type of imager is the most complex of those described here. Its unit cell, shown in Figure 7.19 (Andoh 90), has functions

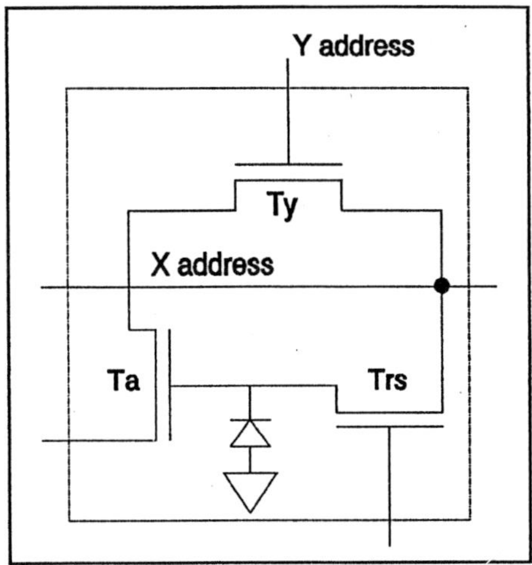

FIGURE 7.19. Unit cell of the amplified MOS intelligent (AMI) imager.

divided over different transistors, while in the SIT, the CMD and the BCMD, for instance, a single transistor is responsible for photoelectric conversion, amplification, read-out, and reset.

In the AMI, the following operations take place :
- the photodiode converts the incoming photons into an electrical signal;
- these photon-generated charges are stored on the capacitance of the photodiode in parallel to the input capacitance of T_a;
- amplification is performed by T_a;
- the amplified output signal is fed to the outside world through the selection switch T_y;
- T_{rs} resets the pixel.

Using this type of amplified MOS intelligent configurations, amplification factors of about 100 have been reported (Andoh 90).

7.5.5. BASE-STORED IMAGE SENSOR

The vertical bipolar transistor-based device consists of a p-type base sandwiched between an n^+ emitter and an n^+ collector (Nakamura 91). This can be seen on Figure 7.20. The base is reset to a reference voltage using a clamp transistor, followed by capacitive coupling of the base-emitter junction into forward bias for a short period of time. Optically generated holes flow to the p-type base where they are integrated. During readout, the base voltage is sampled using the bipolar transistor in an emitter-follower mode.

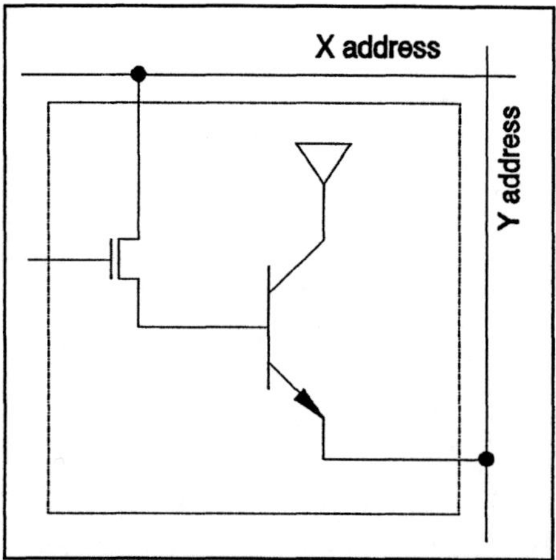

*FIGURE 7.20. The pixel configuration of a unit
cell of the base-stored image sensor (BASIS).*

Progress in the development of image sensors with amplification at each pixel site
is quite marked. Several improvements on the various architectures have optimized
the devices as far as sensitivity, lag, smear, and dynamic range are concerned.
However, most of the structures suffer from the same problem : nonuniformity in
spectral response from pixel to pixel in combination with nonuniformity in dark current
or fixed-pattern noise (FPN). Nevertheless, research efforts are continuing to reduce
these artifacts also. Perhaps in the coming digital era, when field memories become
very cheap, pixel correction for nonuniformities may be made possible and MOS
type image sensors with an amplification function within each pixel will enter the
scene of the solid-state imaging field.

WORTH MEMORIZING

**Pixels with an amplification function and organized in an MOS-XY
addressable matrix may emerge as challengers to CCDs in the
coming decade. Their optical-sensitivity characteristics are
outstanding (due to their internal amplification) and they can be
fabricated using more or less standard IC technologies. Most of
them, however, deal with one or other problem such as : nonuniform
spectral response (from pixel to pixel), nonuniform dark current,
fixed-pattern noise, or reset noise.**

7.6. Conclusions

Efforts made to increase the light sensitivity or the quantum efficiency of solid-state image sensors are immediately rewarded with an increased signal-to-noise ratio of the imager. For low-light-level optical input signals, the S/N ratio of CCDs is proportional to the square root of the quantum efficiency, for all other light levels it is in fact directly proportional to the quantum efficiency. Increasing the light sensitivity of CCDs is one of the chief means of optimizing the overall performance of imagers.

On the basis of the various CCD architectures, completely different technologies have been developed to optimize the light response of the imagers. For interline-transfer type devices, efforts are being made to increase the aperture ratio of the pixels; in the case of frame-transfer type imagers, more attention is paid to decrease the absorption of incoming photons in the layers on top of the active area of the image sensors.

Increasing the aperture ratio of IL or FIT pixels can be done in several ways, of which the most popular today is the introduction of microlenses. Every single pixel is provided with its own microlens to concentrate the incoming light as fully as possible on the light-sensitive photodiode and no longer on the insensitive vertical CCD register as well. Microlenses are quite effective : sensitivity increase up to a factor of three is no longer an exception. But microlenses have also some drawbacks : the response of the imager is no longer linear with the F number of the main lens and depends on the exact location of the pixel on the matrix. Whether nonlinearity is a problem or not is very much application-dependent. This dependence on location on the sensor area can be compensated for by the design of the microlens array.

Another means increasing the aperture ratio almost up to 100 % is the incorporation of a photoconversion (amorphous silicon) layer on top of the image sensor. The sensitivity increase is very high but some "old" drawbacks of imaging tubes also show up again, namely : image lag and burn-in effects. New techniques and technologies to counteract these artifacts are in the process of development.

Increasing the aperture ratio can also be achieved by decreasing the silicon area occupied by the vertical shift registers. Examples of this method are the charge-sweep device and the trench CCD. With the former a very small CCD channel is constructed and provided with a very clever clocking system to handle single packets over the complete length of the vertical register. In the latter case, the CCD channel of the vertical register is U-folded and placed in a groove in the silicon substrate.

So far as frame-transfer image sensors are concerned, their sensitivity can be further optimized by paying extra attention to the absorption and reflection of the incoming photons in the multilayered structure on top of the active silicon.   Just straightforwardly, reflections can be minimized by proper choice of the different optical thicknesses of the multilayered structure. If it is possible to reduce the active area covered by the absorbing gate material, the quantum efficiency is also increased. Introduction of the virtual-phase CCD makes use of this effect. More than 50 % of the active area is free of any gate material, and the absorption (in the gate material itself) of incoming photons is kept as low as possible. As a result, virtual-phase devices have a very high optical response over the complete visible spectrum. The same idea of leaving out as much as possible of the absorbing gate material can be found in the cross-gate CCD or in the frame-transfer devices which have "blue-sensitive" windows cut in their gates.

As far as frame-transfer devices are concerned the ultimate solution is to get completely rid of the polysilicon gates. This can be achieved by the introduction of fully transparent and conductive gate material or by illumination from the back-side of the devices. The first solution produces extremely high values for the quantum efficiency, but deals with compatibility problems by introducing "foreign" material into the classical IC technology. The latter solution looks pretty attractive but introduces new handling, thinning, and passivating issues into the CCD technology.

Theoretically, the most ideal way to increase the sensitivity of solid-state imagers is to amplify the photoconverted signal locally in each pixel. Solid-state imagers according to this principle are all configured in an MOS-XY addressable device. Examples discussed are arrays with pixels based on the static-induction transistor (SIT), the charge-modulation device (CMD, the bulk charge-modulation device (BCMD), the amplified MOS intelligent imager (AMI), and the base-stored image sensor (BASIS). Although the first attempts to fabricate these kinds of devices look promising, there still is a long (developing) road to go before the optical and electrical characteristics of these new devices can be brought to the level at which CCDs are presently operating. Perhaps in the coming digital era when field memories become very cheap and processing is done fully digitally, correction for non-uniform response of the pixels, nonuniform dark current and fixed-pattern noise can be implemented. The combination of such new digital technology with the amplified pixel sensors will be able to compete with CCDs. But today CCDs are dominant and will remain so for years to come.

Chapter 8

ADVANCED IMAGING : NOISE AND SMEAR

As learned from the preceding chapter, where the signal-to-noise ratio was stated to be one of the key performance factors, the noise behavior of a solid-state imager is a very important parameter. It determines the lower limit of the imager's operational range. But the noise of a charge-coupled device is more than just the denominator of the S/N ratio. The overall noise figure is a combination of different noise sources which all have their own specific frequency spectrum and temperature dependence. This means that, depending on the frequency range or temperature range of interest, one or other of these noise sources will predominate. Because of this behavior, optimization of charge-coupled devices so far as noise performance is concerned becomes a difficult task. An imager optimally performing at room temperature may have a noisy output at elevated temperatures. And to design a device which has a low noise level at all temperatures, several developments dedicated to decreasing the noise level of the sensors have to be applied.

In this chapter the most important insight into noise performance of charge-coupled devices will be described. Starting with the technology-related noise sources, the origin of white and black point defects, column defects, transfer noise, striations, pixel non-uniformities and dark-current shot-noise will be explained. Some specific measures taken to lower these artifacts will also be included in the section dealing with technology-related noise sources. Extra noise electrons are added to the CCD output signal at the output node and in the output amplifier. This addition is on account of the thermal noise of the MOS transistors and the $1/f$ noise of the devices but mainly on account of the reset operation of the reset transistor. Optimized design and balanced technology can lower these noise effects at the output stage. The reset noise in particular, can be drastically reduced by dedicated signal processing. A couple of video-processing techniques are also discussed in this chapter, together with a few completely new designs which tackle the problem of output-amplifier noise.

An interesting phenomenon encountered in solid-state imagers is smear. Although this effect is not a typical noise source, it is dealt with in this chapter because it shows up as an artefact on the monitor, especially when the image sensors are used in situations where highlights are included in the scene. Smear depends very much on technology, on design, on driving method and even on the device's architectures. All these connections are also covered in this chapter.

8.1. Decreasing noise levels

One way of increasing the signal-to-noise ratio is to optimize the sensitivity of the devices as discussed in chapter 7. At extremely low-light input, the detection limit of any signal is determined by the remaining noise floor. This is very clearly illustrated by Figure 8.1, which gives an overview of the various noise levels showing up in a typical CCD imager. The noise figures are given in relation to the level of the input signal. The amplifier noise and the noise generated in the CCD itself are independent of the amount of incident light. But the photon shot noise is proportional to the square root of the number of electrons generated, which depends on the quantum efficiency of the device.
Also included in Figure 8.1 is the definition of the dynamic range of the imager, namely the ratio of the saturation level or full-well content to the minimum signal

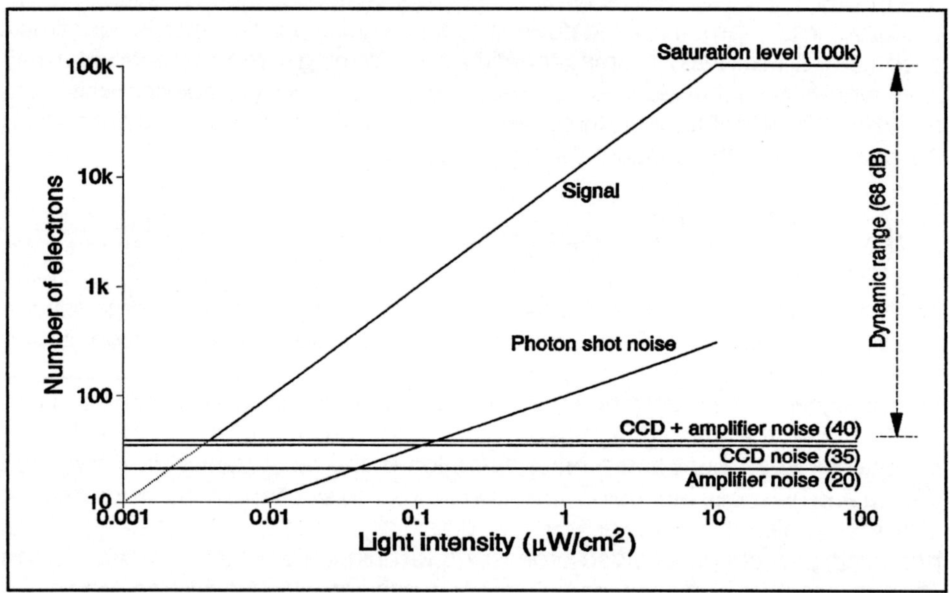

FIGURE 8.1. Illustration of the different noise sources of a CCD imager, in terms of number of electrons and as a function of the incident light intensity.

level at the output stage, i.e. the noise floor of the CCD itself and its output amplifier. Reducing the various noise sources is equally as important as increasing the light sensitivity of the devices. Because most kinds of noise sources also have different origins, it is not possible to tackle them all at once. Moreover, several mechanisms are technology-related while others have to do with the design of the device, and some can be minimized by the video-processing circuitry or technique, others not.

WORTH MEMORIZING

CCD and output amplifier noise are independent of the intensity of light impinging on the imager. The photon shot noise is proportional to the square root of the quantity of electrons generated by the optical input.

8.2. Technology-related noise

The noise sources which have to do with technology-related aspects are probably the most difficult to deal with. In some cases the generation sites of all types of noise are already present in the starting material on which the imagers are fabricated (Jastrzebski 87). In other situations, the noise sources are introduced during the processing of the CCDs, e.g. by dust particles. In all cases it is very difficult to overcome these problems, but, in general, ultraclean processing is of vital importance. Everything in the clean room, the apparatus, the gases and the liquids, has to be as pure as possible, and the processing has to be carried out by a very disciplined operating team. Only with these extremely exacting parameters it is possible to fabricate high-quality image sensors with low noise levels.

In this section not all but only a few of the technology-related noise sources will be discussed. For most of those mentioned, a possible means of minimizing the particular noise source will also be presented. In most cases it is impossible to totally eliminate them.

8.2.1. POINT DEFECTS

White point defects or spots visible on the display or monitor are the result of a local (electron-) generation site which in all cases causes additional electrons in the charge packet integrated near it. The origin of the white spots can be :
    - a local crystal defect in the silicon substrate (dislocation, O-precipitate, stacking fault) (Ogino 83);
    - mechanical damage;
    - a local gate dielectric problem;
    - the inclusion of a metal atom, e.g. Fe, Cu, Au (Jastrzebski 90).

Extremely clean processing techniques can keep the number of defects as low as possible. A special processing method, known as gettering can attract the (charge-) generation sites to the back of the wafer. Their influence will then not be "seen" at the front of the wafer. Gettering can be done by means of heavy phosphorus doping on the back of the wafer or by means of a "damaged" back-side already present on the wafers when they enter the clean room. Another well-known gettering technique is internal gettering. Special temperature treatments of the wafers in the early stages of processing attract the oxygen atoms to the center of the wafers and create a defect-free toplayer (denoted zone).

Black point defects are associated in most cases with dust on the sensor surface itself, or even incorporated in one of the layers on top of the silicon wafer. The incoming light information is disturbed and cannot reach the pixels in the silicon. Cleaning the sensor surface (if still possible after packaging) may help to overcome the problem of black pixel defects.

8.2.2. COLUMN DEFECTS

If the generation of excess electrons at a point defect is so great that the potential wells are filled with dark current, even during the transport phase of the CCD, the point defects will result in a column defect.
Column defects can, however, also be generated by local transport problems : channel narrowing, gate shorts, and gate interruptions, can all give rise to a local obstruction in the CCD transport channel and can be seen on the monitor as a column defect. To keep artifacts of this kind as insignificant as possible, clean chemicals during the lithographical and etching processes are of vital importance.

8.2.3. TRANSFER NOISE

Any inhomogeneity in transport efficiency can be seen on the monitor as a fixed disturbance pattern. Although a locally changing transport efficiency can also result in a column defect, not all do. This type of noise source can be caused by irregularities in the gate definition or in the definition of the CCD channel itself.

8.2.4. STRIATIONS

The saturation level or the onset of antiblooming may not be equal for the pixels across a sensor. A possible source can be inhomogeneities in the substrate doping. These effects are known as striations (Senda 85). The homogeneity of the substrate doping can be sufficiently increased to overcome the striation effect by means of an epitaxial toplayer on the starting material, e.g. an n-type layer on top of the n^+-type substrate.

8.2.5. PIXEL NONUNIFORMITIES

Any nonuniformity in geometry, layer thickness, doping level or doping profile, on gate definition, can change the response from pixel to pixel and can introduce nonuniformities in sensitivity, saturation level, and dark-current generation. These noise sources cannot be reduced by changing or optimizing layers, implantations, or doping levels. Again, it is entirely a matter of strict processing discipline : cleanness in all stages and of all materials involved in the fabrication of the CCDs.

8.2.6. DARK-CURRENT SHOT NOISE

The generation of dark current is, like the generation of electrons by photons, a stochastic process. Variation on this generation process is the source of dark-current shot noise. The level is equal to the square root of the total number of electrons involved in the dark-current charge packet. Dark-current shot noise can only be reduced by decreasing the dark current itself. Again, this parameter is greatly influenced by the processing conditions in the clean room. Discipline by everyone and cleanness of everything involved can limit the amount of dark current.

Dark-current generation sites are interface states located at the Si-SiO$_2$ interface or bulk states in the silicon substrate. Generation of dark current by interface states is greater than the generation by bulk states by at least a factor of ten. As already pointed out in section 3.3, it is not the dark current itself, but the temperature-dependence and the nonuniformities of the dark current that pose difficulties in handling dark-current effects.

A few interesting techniques are introduced for shielding the electrons in the CCD channels from these interface states. With these techniques, generation from the most important source of dark current can be drastically reduced and the amount of dark current remaining is mainly due to the presence of bulk states. Figure 8.2 illustrates how a shallow p$^+$ layer shields the photodiode of a (frame-) interline-transfer

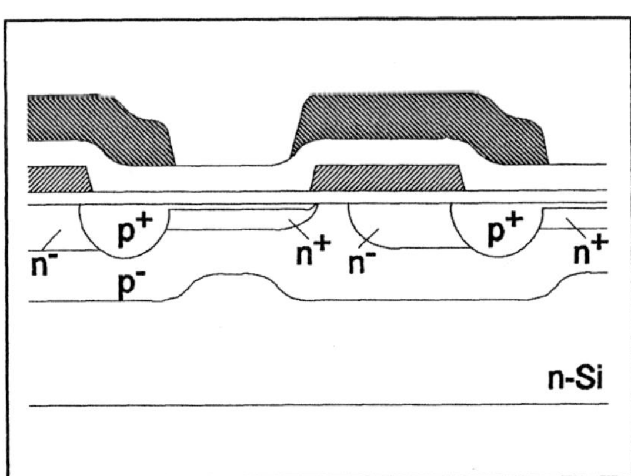

FIGURE 8.2. Shielding interface states in an interline-transfer or frame-interline-transfer image cell by means of a shallow p$^+$ layer.

device from the interface states (Teranishi 82). The latter are all filled by holes in the inverted top layer of the silicon. This simple construction of a p$^+$-n$^+$-p$^-$ photodiode

has a second attractive advantage : it increases the storage capacity of the photodiode.
(This method of shielding the interface by means of a very thin p^+-type layer is known by the name of Hole-Accumulation Device or HAD.)

The above-mentioned technique is not possible in a frame-transfer device, because a p^+ layer shields the CCD channel not only from the influence of the interface states, but also from that of the gate voltages. An alternative based on gate pinning can be applied to frame-transfer imagers and also to the vertical registers of a (frame-) interline-transfer device to fill all interface states with holes and make them more or less inactive with respect to their interaction with the CCD channels. These gate-pinning suppressing methods are more extensively described in the section on scientific imaging (10.1.5).

WORTH MEMORIZING

Technology-related noise sources are manifested by white and black point defects, column defects, transfer noise, striations, pixel nonuniformities, dark-current shot noise, and dark-current nonuniformities. Technology-related noise sources are very hard to tackle. Only optimized processing steps, first-class starting material, very pure chemicals, clean gases and a very disciplined operating team can minimize the effects of this type of noise sources.

8.3. Output amplifier noise

The other important noise source, besides the noise sources in the CCD itself, is the output amplifier. The small analog circuitry used to convert the electrons into a voltage and to buffer the output voltage toward the outside world, adds some uncertainty to the signals in the form of noise electrons. The overall noise from the output amplifier can be split up into various elements depending on the mechanisms by which they are generated.

To enable these items to be studied separately, Figure 8.3 schematically shows the most commonly used output amplifier : a double or triple source-follower stage together with a reset transistor (the operating principle of the output amplifier with reset has been explained in section 2.4.1). In the example of Figure 8.3, all transistors are of the depletion type. This is not necessary, neither is the on-chip load of the last stage. In almost all practical applications, the load is placed off-chip to decrease the power dissipation on-chip. The noise characteristics of these kinds of output stages can be separated into thermal noise, reset noise, and $1/f$ noise.

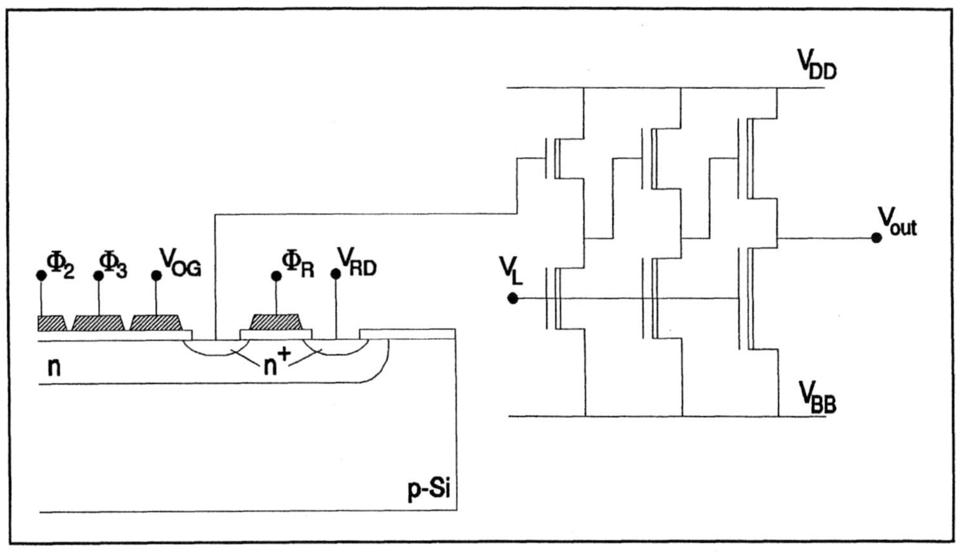

FIGURE 8.3. *A commonly used output stage for CCD imagers : a double or triple source-follower connected to the floating diffusion with a reset transistor.*

8.3.1. THERMAL NOISE

The thermal noise of the complete system is mainly due to noise generation in the inversion channels of the MOS transistors. An optimized design and a well-considered choice of the transistor areas and their ratio of channel width W and channel length L, both for the drivers and for the loads of the output stage (Heidtmann 87), can keep the noise figure of the output amplifier as low as possible. In general, the power spectrum of the thermal noise will be inversely proportional to the W/L ratio (Ozaki 91). To illustrate this behavior, a common measure to express the noise performance of a CCD is defined : the Noise Electron Density (NED).
This figure of merit is determined by the square root of the product of a noise level (spectral density) and an equivalent noise bandwidth, and is simply given by (Centen 91) :

$$NED(f) = \left[\frac{e_n(f).C_t}{q} \right]^2 \qquad [8.1]$$

where :
- $e_n(f)$ is the total equivalent noise voltage present at the detection node of the output stage (= floating diffusion); this parameter depends to a great degree on the transistor type and its geometry;

   - C_t is the total (physical) capacitance present at the detection node of the
          output stage. It includes the floating diffusion capacitance, the gate
          capacitance of the first driver transistor, and all parasitics.
(The NED is expressed in electron^2/Hz.)

Figure 8.4 shows the dependence of NED as a function of the width W of the first

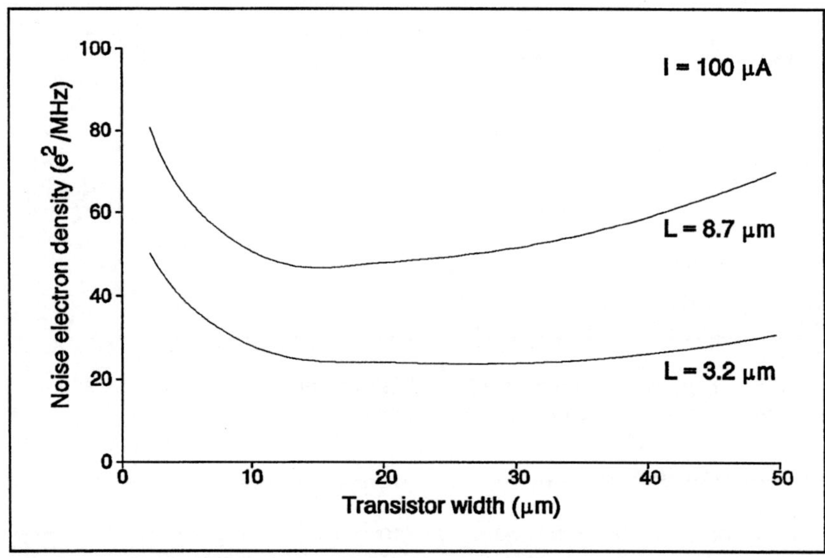

FIGURE 8.4. Influence of the width W and the length L of the first driver transistor on the noise equivalent density of a CCD output stage.

driver transistor for two values of its length L (Centen 91). Changing the dimensions
of the transistor influences both NED parameters in [8.1] : increasing W while keeping
L constant also increases the total capacitance C_t but lowers the total noise voltage
e_n. This last effect is dominant for values of W smaller than about 15 μm. For greater
values of the width, NED is almost constant for a short channel, e.g. L = 3.2 μm,
but increases again for greater values of L, e.g. L = 8.7 μm. In Figure 8.4 the current
through the first source-follower stage is set to 100 μA, but the bias current itself
also influences the NED. This is shown in Figure 8.5, keeping W and L respectively
to 47 μm and 3.2 μm. Increasing the bias current lowers the noise equivalent density,
via the e_n parameter.
Realistic values for L, W, and the bias current can be deduced from these two figures.
Bearing conversion, bandwidth and power dissipation restrictions in mind, a value
for the bias current (through the first stage of the source-follower configurated
amplifier) might be 100 μA. The gain in noise performance by increasing this bias
current is marginal and higher values only increase the power dissipation. As far

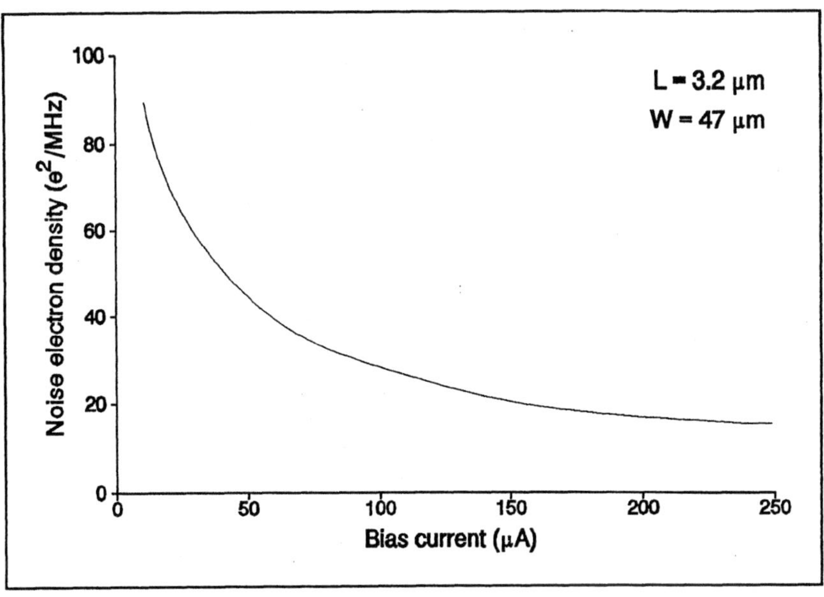

FIGURE 8.5. Influence of the bias current through the first source-follower stage on the noise equivalent density of a CCD output stage.

as noise is concerned, a low value for L is favorable, e.g. $L = 3.2 \ \mu m$. This value for L makes the choice of W more or less noise-performance independent as long as $W > 15 \ \mu m$. A design Including too wide a transistor lowers the conversion factor (see section 8.4), so keeping W close to the value of $15 \ \mu m$ seems optimal.

8.3.2. 1/f NOISE

The 1/f noise is mainly generated in the driver of the first follower stage of the output amplifier. Interaction of the interface states, located at the $Si\text{-}SiO_2$ interface of the MOS transistor, with the electrons in the inversion channel causes fluctuations in the voltage at the output of the first follower stage. The interactions between the interface states and the charges can be kept to a low level by using a depletion type MOS transistor because the inversion channel in this type of device, is pushed somewhat deeper in the silicon. In general, the power spectrum of the 1/f noise will be inversely proportional to the channel area of the transistor or to the product W.L (Ozaki 91).

8.3.3. RESET NOISE

Inherent to the reset action of the floating-diffusion capacitance is an uncertainty about the voltage on the capacitor C_{FD}. The uncertainty or reset noise can be quantified and is equal to kTC_{FD} (k represents Boltzmann's constant, T the temperature). It is very difficult to minimize this reset noise, and even impossible to get rid of it by adapted design or by technological optimizations.

An additional component in the reset noise is the "partitioning noise" : at the moment the reset transistor is switched off, the charges from the inversion layer underneath the gate of the transistor have to be directed to either the source or the drain side of the device in order to diminish the inversion channel. Which charges will move to the drain and which to the source or floating diffusion ? This uncertainty causes some extra voltage fluctuation from one reset action to the other. The partitioning noise can be minimized by an appropriate design of the reset transistor (Watanabe 84). For instance, its channel can be funnel-shaped with the "funnel output" as the floating diffusion. If in this situation, during the switching-off of the reset transistor, the channel width is narrow enough, the narrow-channel effect will push the electrons out of the inversion channel in the direction of the widest side of the channel : the reset-drain side. Partitioning will be minimized and the partitioning noise on the floating diffusion will be as small as possible.

8.3.4. ELIMINATION OF THE RESET NOISE

The only way to deal with the reset noise is to "measure" its value and compensate (electronically) for it afterwards. This is done with a method known as Correlated-Double Sampling (CDS) and shown in Figure 8.6 (White 74). The output signal

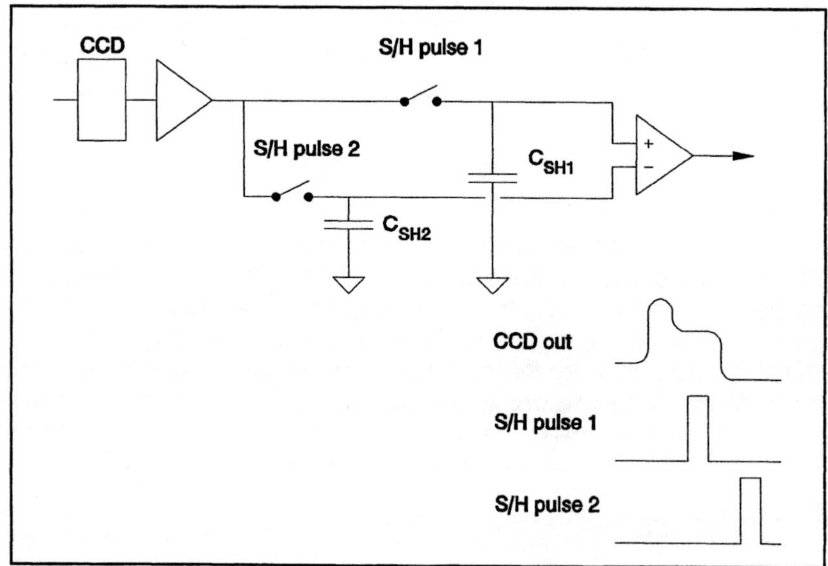

FIGURE 8.6. Reset-noise reduction by means of correlated-double sampling.

of the CCD is sampled twice : once for its reset-hold level and once for its actual video output (for the explanation of the waveforms, see section 2.4.1). The first sample, stored on C_{SH1}, is used to measure the reset noise because the reset-hold level is equal to a preset DC voltage with the reset noise kTC_{FD} added to it. The second sample, stored on C_{SH2}, naturally measures the video signal which is superimposed on the reset noise. If one of these samples is subtracted from the other, the video signal remains, the reset noise being cancelled out.

The circuit shown in Figure 8.7 performs as described above in the CDS situation,

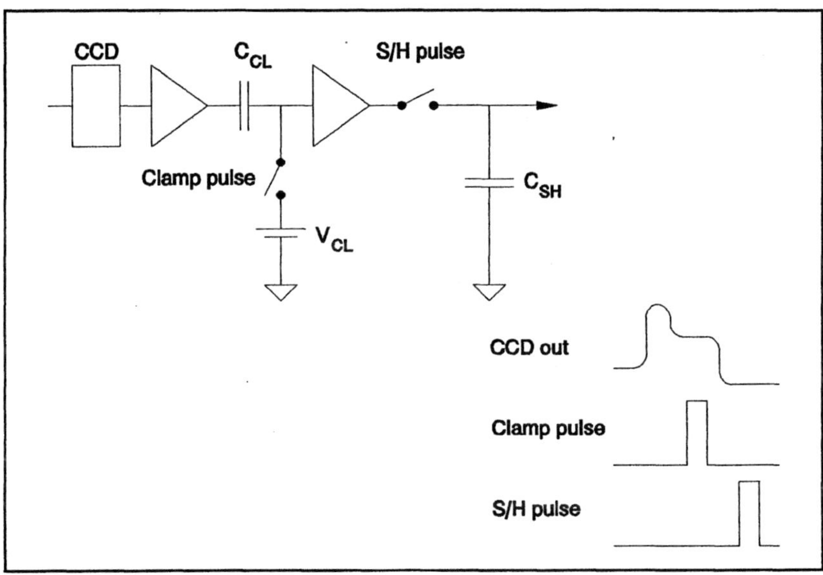

FIGURE 8.7. Reset-noise reduction using a clamping circuit.

but electronically in a slightly different way : during the reset-hold period, the CCD output signal is clamped to a fixed clamping voltage V_{CL}, and then the video signal is sampled on C_{SH} by means of the sample-and-hold circuit. The net result is the same as with correlated-double sampling : a video-preprocessing circuit compensates electronically for the reset noise.
The CDS circuit can be integrated on-chip (Hynecek 86), which makes processing by the end-user fairly straightforward.

However, due to the sampling-and-holding process, noise components with frequencies exceeding the Nyquist limit fall into the baseband. Furthermore, as a result of high clock-rate sampling, the clamping characteristic becomes flawed. Problems of this kind can be avoided with a delay-line processing. The main idea of delay-line processing (DDS) is illustrated in Figure 8.8. With this method the feedthrough period for the CCD output signal, delayed by the delay line, is adjusted to the signal period for a nondelayed signal. The difference between the two levels is detected by the operational amplifier and the differential signal is gated. The degree of aliasing is much less than for the CDS method because the signal is held in the hold capacitor.

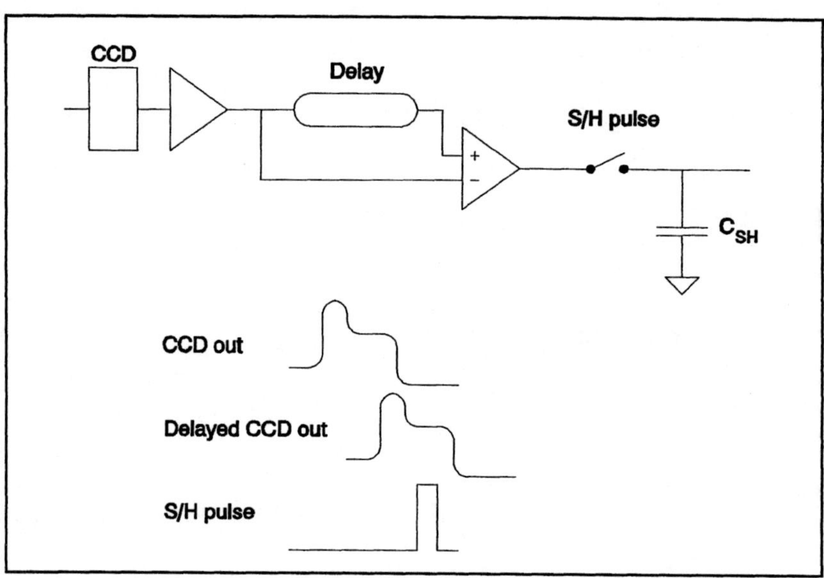

FIGURE 8.8. Reset-noise reduction using video-preprocessing based on a delay line.

A comparison of the two video-preprocessing techniques is shown in Figure 8.9 (Ohbo 88) : the noise content in the CCD output signal as a function of frequency is compared with the noise content in the correlated double-sampled signal and the signal obtained after delay-line processing. The increase in S/N ratio resulting from both video-preprocessing techniques is due to the lowering of the amplifier and reset noise. The shot noise and the fixed-pattern noise remain.

Including a CDS circuit or a delay-line processing operation in the preprocessing electronics also removes part of the 1/f noise, especially in the low-frequency range. For high-frequency noise signals, a CDS circuit is not quite satisfactory. A solution can be found by replacing the MOS transistor of the first follower stage by a junction FET. It is well known that a JFET has better 1/f-noise characteristics than a MOS device, due to less interaction between the interface states and the charges from the transistor channel.

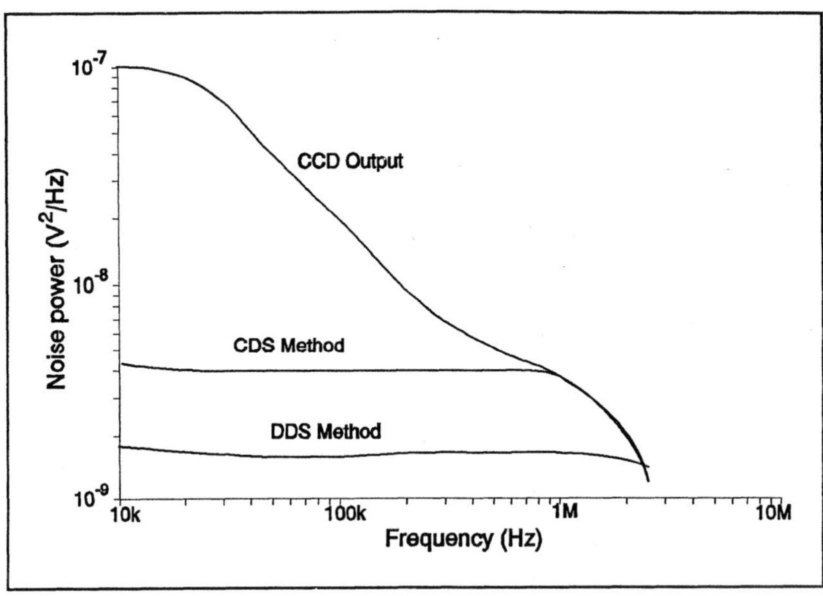

FIGURE 8.9. Comparison between video-preprocessing with a CDS and a delay-line circuit.

8.3.5. NEW OUTPUT-AMPLIFIER ARCHITECTURES

A new output architecture was launched in which the driver MOS transistor of the first source-follower stage is replaced by a JFET. In previous structures the MOS transistor was placed close to but next to the CCD. In the new structure, the JFET is physically located in the output diffusion itself (Mutoh 89). This is shown in Figures 8.10a and 8.10b, in which respectively the top-view and the cross-section through the JFET are shown. The structure looks very similar to a classical output structure with a reset transistor. But characteristic of the architecture implementing the JFET is the fact that the drain of the p-channel JFET (D in Figure 8.10a) is combined with the stopper implantation around the floating diffusion, that the gate of the JFET is the floating diffusion itself, and finally that the source of the JFET (S in Figure 8.10a) is a small p^+ diffusion placed in the floating diffusion. Note that only the source of the JFET is added to the classical design. All other transistor parts were already available in the original floating-diffusion configuration. The structure is extremely compact, minimizing parasitic capacitances and resistances. The electrons are transferred toward the floating diffusion by a classical CCD transport system. At the floating diffusion they are shifted on the gate of the JFET and are able to modulate the hole current through the first source-follower stage. The resetting of the floating diffusion is done in the classical way : by means of a simple reset transistor.

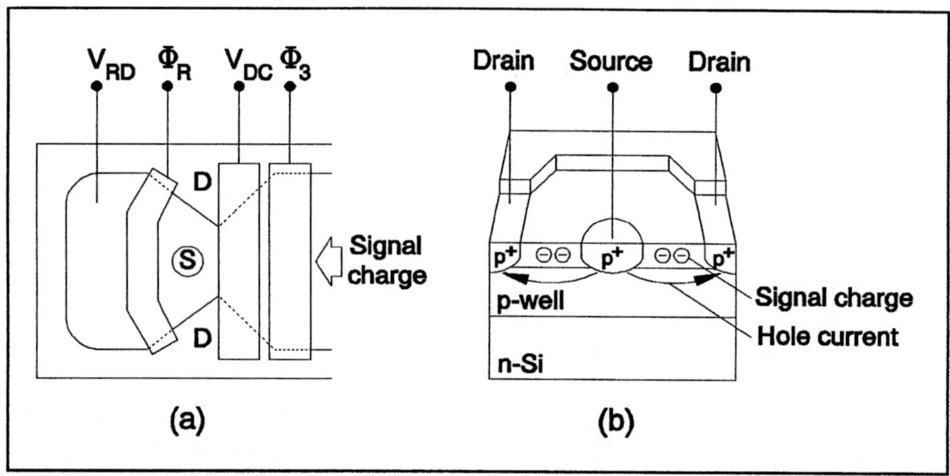

FIGURE 8.10. Top view (a) and cross section (b) of the output stage including a JFET as first follower, constructed as part of the floating-diffusion output node.

An alternative to the JFET transistor designed into the floating diffusion can be found in the design of a MOS transistor across the floating diffusion (Brewer 80, Matsunaga 91, Roks 92). The basic configuration is illustrated in Figure 8.11. As in the foregoing JFET idea, the hole current through the MOSFET is modulated with the electrons put by the CCD on the floating diffusion located underneath the gate of the MOSFET. The electrons are stored in the bulk of the MOSFET and the hole current through the transistor can be a surface current or a bulk current (as indicated on Figure 8.11b), depending on the bias of the sensing gate. The MOSFET operates as the driver of a first source-follower stage. The transport of the electrons from the channel of the driver transistor can be performed by a simple reset operation.

Further optimizations of this type of output-node configuration are possible, namely :
- the sensing gate can be covered by a second gate, which is negatively fed-back to the output signal. This construction leads to extremely low noise levels of 1 electron equivalent-noise signal and consequently a very high dynamic range of, for instance, more than 90 dB (Matsunaga 91);
- a bipolar transistor can be incorporated of which the base current is equal to the hole current supplied by the MOSFET across the output diffusion. This construction ensures very low noise levels in combination with extreme values for the conversion factor, e.g. 25 μV/electron (Roks 92).

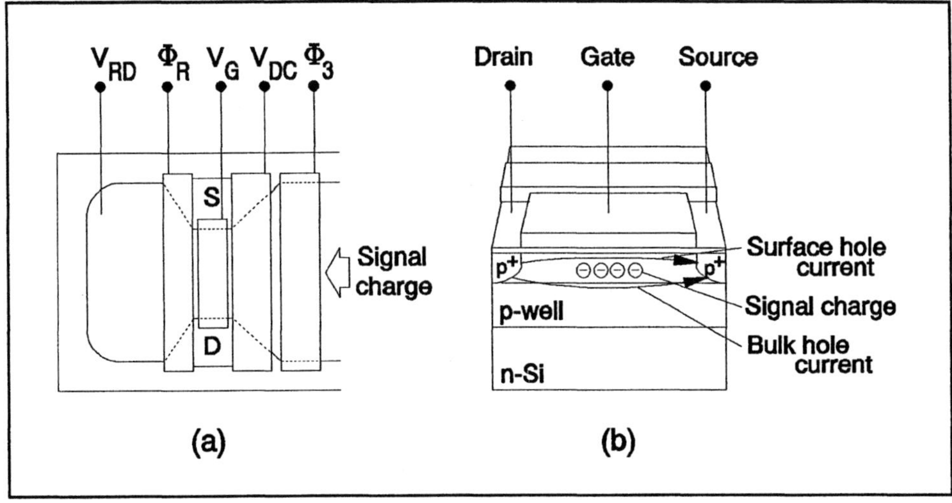

FIGURE 8.11. Top view (a) and cross section (b) of the output stage including a MOSFET as first follower, constructed as part of the floating-diffusion output node.

WORTH MEMORIZING

Output-amplifier noise can be divided into different components : thermal noise, 1/f noise, and reset noise. Thermal noise cannot be eliminated, but can be considerably reduced by appropriate design and layout of the driver of the first source-follower stage. The same holds for 1/f noise : suitable design of a buried-channel transistor can keep the 1/f noise component to an acceptable limit. On the other hand, it is very hard to get rid of the reset noise at the CCD level because this is a fundamental process. Fortunately, the reset-noise level can be compensated for by the video processing circuits. Techniques such as correlated-double sampling or delay-line processing can drastically lower the influence of the reset noise.

New output-amplifier architectures are being developed in order to increase the overall noise figure of CCDs. An example is the incorporation of a JFET or a MOS transistor at the site of the floating-diffusion mode to minimize the thermal noise and 1/f noise components.

8.4. Output-amplifier sensitivity

Up till now, the discussion of S/N optimization has been limited to the S/N performance of the image sensor, including some preprocessing to reduce the noise components. But, in addition to the imager itself, the electronic circuit following the CCD can also determine the signal-to-noise characteristics. Apart from the noise of the output amplifier, the conversion factor or sensitivity (expressed in μV/electron) is a very important parameter.

A large conversion factor prevents extra S/N reduction caused by peripheral signal-processing circuits. The conversion of charges into voltages is done by dumping the electrons on a floating-diffusion capacitor and subsequently sensing them by means of the source followers. There are several methods to increase the sensitivity of the output amplifier :

- boosting the gain of the source-follower stages by decreasing the channel conductance of the transistors and suppressing the back-gate effect of the driver transistor;
- decreasing the input capacitance of the first source-follower stage by optimizing the gate length L and gate width W of the first-stage driver transistor, bearing in mind short-channel effects and thermal-noise performance when determining and designing L and W respectively;
- minimizing the parasitic capacitance of the output node of the horizontal output register by careful layout and optimized design. An example of an optimized design is shown in Figure 8.12, from left to right :
    + a classical output stage with floating-diffusion capacitance C_{FD}, a reset transistor controlled by its reset pulse Φ_R, and a single output stage with a current source I. The parasitic capacitance is denoted by C_p;
    + an output amplifier with a reset transistor provided with a screening gate. The screening gate shields the reset transistor from the floating diffusion and drastically reduces the clock feedthrough from Φ_R on C_{FD}.
    The DC bias V_{DC} has to be chosen such that part of the reset transistor, underneath the screening gate, is always turned on. As a consequence, part of the channel of the reset transistor is added to C_{FD} and, instead of increasing the sensitivity of the output stage, the conversion factor is decreased because the capacitance on which the charges were dumped has become higher;
    + to avoid the aforementioned effect, the screening gate is positively fed-back using the voltage available at the source node of the first source follower, and together with this the parasitic capacitance is highly reduced too (Theuwissen 88, Akimoto 91).
    The gain of the first source-follower stage A_{SF} is close to unity, making the added capacitance of the transistor channel negligible because

the voltage in the channel itself is almost equal to the voltage at the screening gate.

The same applies to the parasitic capacitance : if the interconnect from the source of the first source follower to the screening gate is physically located underneath the interconnect from the floating diffusion to the gate of the first source follower, the remaining parasitic C'_p will be given by :

$$C'_p = \frac{C_p}{1 - A_{SF}} .$$

[8.2]

With this simple construction, reasonable high conversion factors can be attained : 16.2 μV/electron with a total input capacitance of 6.7 fF (Akimoto 91).

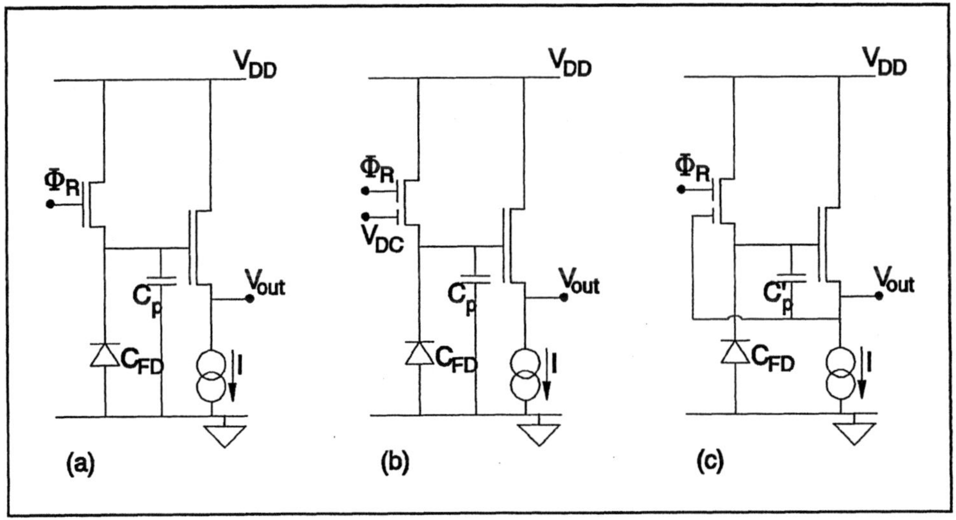

FIGURE 8.12. Three different output-stage constructions : the classical configuration (a), with a screening gate (b), and with a positively fed-back screening gate (c).

WORTH MEMORIZING

The sensitivity of the output amplifier or conversion factor is determined by the value of the floating-diffusion capacitance and the parasitic capacitance in parallel with the former. An increase of the conversion factor can be effected by optimizing the total capacitance at the input of the first stage of the output amplifier.

8.5. Smear

Smear is a spurious signal seen more or less as a bright vertical column on the monitor. The artifact runs from top to bottom on the monitor (if no charge reset is applied) and precisely through the highlight. Although the smear signal is generated in different ways in different type of imagers, it takes the same form in all type of devices. Because of its various sources, however, it cannot be counteracted in the same way for all of them. This section describes the origin of smear and how to eliminate or minimize its effect.

8.5.1. SMEAR IN FRAME-TRANSFER CCDS

Considering the basic operating principle of the frame-transfer CCD, smear is inevitable. It is generated at the moment when the charge packets are transferred from the image section to the storage section. The imager is not shielded from the incoming light during this transfer, charge-carrier generation continues and spurious signals are added to the charge packets in transport.

Figure 8.13 shows the situation where a white rectangle on a black background is captured by the frame-transfer device. The height of the rectangle is 10 % of

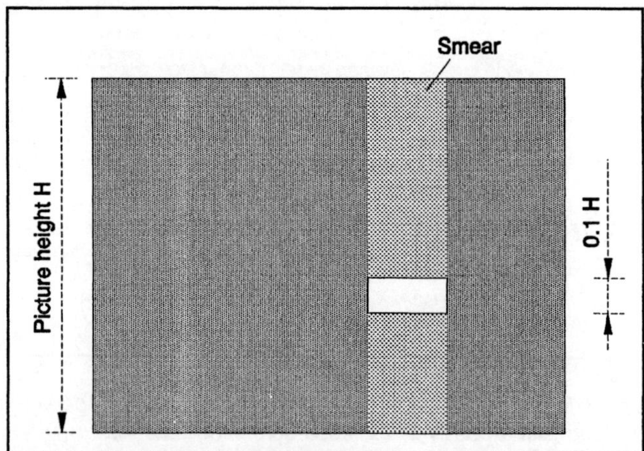

FIGURE 8.13. Definition of smear measurement with a white rectangle
(10 % picture height) on a black background.

the total picture height. If T_{int} is the integration period, and T_{tr} the transport time, the smear level Sm (in %) in the black portions of the scene is :

$$Sm = 10 * (T_{tr} / T_{int}) .$$ [8.3]

To minimize the smear signal, transport from the image section to the storage section should be done as fast as possible, or in the complete absence of any further light input. Fast transport can be ensured by high-frequency vertical clocks, but in most cases the large RC values of the polycrystalline-silicon gates impose an upper limit on the vertical transport frequency. New techniques making use of double-metallization technologies, reduce the R values of the gates considerably by means of a strapping method. This issue will be discussed in the section on high-speed clocks for the vertical registers (9.4).

A second way to reduce smear is to provide the camera, which uses the image sensor, with a mechanical or electro-optical light shutter which shields the sensor completely from light during the frame shift. With this method operation of the frame-transfer CCD can be made 100 % smear-free.

8.5.2. SMEAR IN (FRAME-) INTERLINE-TRANSFER CCDS

Although the origin of smear in the case of an interline-transfer CCD is completely different from its generation in a frame-transfer device, it is visible in the same form for both and, surprisingly their amplitudes are also about the same.

In the interline-transfer-imager situation, smear is caused by either :
- stray electrons generated underneath the photodiode area and diffused into the vertical CCD shift registers, or;
- stray photons which arrive in the vertical CCD shift registers via, for instance, light pipes (multiple reflections at the Si-SiO$_2$ interface and at the lower surface of the light shield) and generate electron-hole pairs locally.

These two mechanisms are illustrated in Figure 8.14, from which it can be concluded that smear is generated in the vertical shift registers (directly or indirectly) by photons originally "intended" to be converted into neighboring pixels.

Highlights can cause spurious smear electrons in the vertical CCD shift-register cells so that each charge packet passing the highlighted photodiode receives some extra electrons. Note that, during the readout of the video information, transport in the vertical shift registers takes 20 msec or 16.7 msec, and each charge packet belonging to such a CCD register can dwell quite a long time (= 1 active line time) near a single highlighted photodiode.

There are three ways by which the smear in interline-transfer devices can be reduced :
- reducing the diffusion of stray electrons by the introduction of an optimized n-p-n structure in the photodiode (Kuroda 86) and by adding an extra diffusion barrier underneath the vertical CCD shift register (Sakakibara 91, Negishi 91). The latter is shown in Figure 8.15. Comparison of this cross section with the original one in Figure 8.14 will show that a p$^+$-implanted buried-well structure has been added underneath the n-type

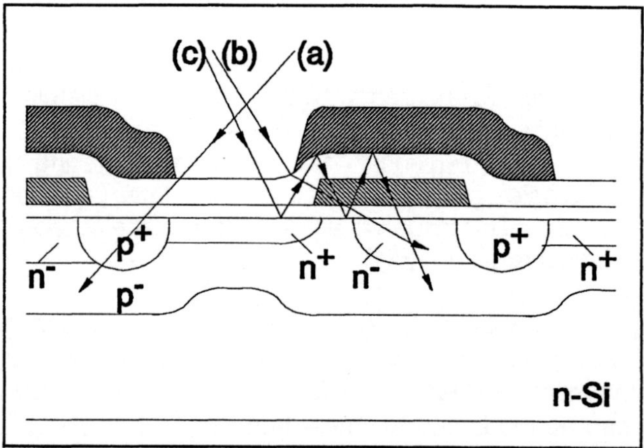

FIGURE 8.14. Diagram showing the origin of smear in a (frame-) interline-transfer device. The artifact is caused by stray electrons (a), scattered photons (b), and light piping (c).

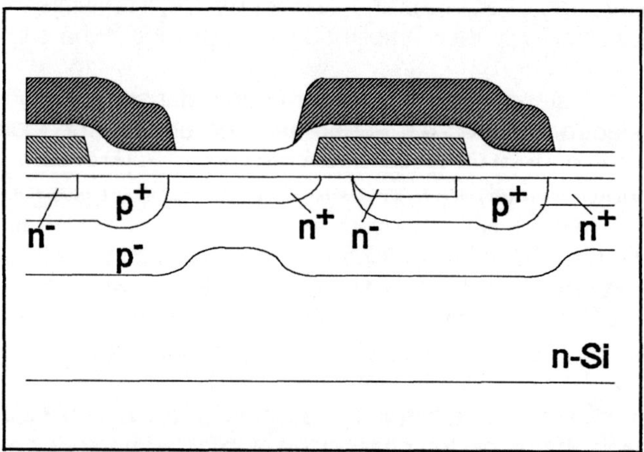

FIGURE 8.15. New techniques to reduce smear in IL and FIT imagers : an extra p⁺ well under the CCD channel, and a thinner dielectric underneath the light shield.

buried channel. This additional layer increases the charge-handling capability but more important it increases the electric field which tends to prevent the "indiffusion" of stray electrons;
- reducing the light piping by optimizing the light shield (Teranishi 87) and locating it, for instance, as close as possible to the silicon surface, making use of new tungsten technologies (Losee 89). This technique is also

schematically shown in Figure 8.15. Comparison of this illustration with
Figure 8.14 illustrates that the dielectric layer between the poly-Si gates
and the light shield has been reduced. The result of this technique is
a strong reduction in light piping and consequently an increase in smear
performance;
  - reducing the time during which smear electrons can be generated and added
    to the charge packets by shortening the time the charge packets remain
    in the vertical shift registers. This technique has also led to the introduction
    of the frame-interline-transfer CCD (Horii 81, Horii 84). With this FIT
    configuration the vertical transport of the charge packets is much faster
    than in the classical interline-transfer device.

Although the smear level in interline-transfer devices can be reduced, its complete
elimination is never possible. The reduction of the smear by means of the frame-
interline-transfer CCD can be as high as 110 dB. This is an acceptable value for
consumer applications, but in professional and broadcast TV cameras, 110 dB
reduction is still not enough. Again, only a mechanical shutter in combination with
the FIT can make this device fully smear-free.

8.5.3. SMEAR COMPENSATION TECHNIQUES

Note that, for both frame-transfer and (frame-) interline-transfer devices, the total
amount of smear is generated in two parts :
  - the first part before the start of the integration, in fact during the last clocking
    cycles of the frame shift of the previous field;
  - the second part after the end of the integration, during the frame shift of
    the existing field, when the smear is added to the video signal.
The smear signal can thus be split into Sm_b, smear before the integration, and Sm_a,
smear after the integration. The Sm_b component is generated during the frame
shift of the previous field. When the video information of this field has been shifted
to the storage section, new empty image lines roll from the top of the image section
into the image section immediately after the video lines containing the information
of the previous picture. These empty image lines will capture smear signals during
the period when they are shifted downward to the integration sites which they will
occupy in the next field integration.
This typical smear-generation mechanism is responsible for a smear signal on the
monitor both below and above the highlight.

It is quite simple to get rid of the Sm_b signal by using the charge-reset technique.
With this method all integrated charges up to a certain moment in the integration
cycle are drained, including all smear signals generated before the actual video.
On the other hand charge-resetting does not affect the absolute value of the Sm_a
content, but may increase its relative value compared to the video amplitude. For

extremely short integration times, the amount of smear can be of the same order as the amount of video.

Much effort is put into eliminating the Sm_a signal to get rid of the complete smear component after Sm_b is dumped by means of charge resetting. The only electronic way reported up to now is to compensate the video output of the CCD for the Sm_a signal. This can be done by "measuring" the Sm_a content of the video signal and compensating it electronically. This measuring operation can be performed in two extremely different ways :
- a single line in the two-dimensional CCD is shifted through the complete image section, taking a sample of the overall and average value of the smear across the complete device. During the normal active integration time this "smear line" is shielded from incoming light to make sure it contains only smear signals. When this CCD line containing the Sm_a sample is moved out, it is stored in a line memory and each video line of the CCD is compensated for it. The advantage of this system is its compactness; its disadvantage is the "low temporal resolution" of the sample : the smear correction is very sensitive to moving objects, and if these are present in the picture, the corrected signal can look still worse than the signal without smear compensation;
- the CCD takes a smear sample in the form of a CCD line after each video line (Esser 88). This system is quite complex with regard to clocking and it has to store twice as many CCD lines as the system without smear compensation, but it has the great advantage of being practically insensitive to moving objects.

The optimum compensating technique will be situated somewhere between the two mentioned : one complete sample of the smear across a complete CCD line for a few video lines. In all cases, however, the electronic smear-compensating methods are severely limited when extreme overexposure occurs. For instance : what happens if there are highlights in a scene with an intensity so high that even the smear signal on its own has an amplitude comparable to a full well ?

8.5.4. SMEAR IN MOS-XY AND CID IMAGERS

In an MOS-XY imager, smear is generated in about the same way as it happens in an interline-transfer device : stray electrons and stray photons also influence the amount of electrons which are available on the capacitance of the sensing lines. Figure 8.16 illustrates this phenomenon. To decrease the smear level in MOS-XY devices, the same remedies are valid as have just been described for the interline-transfer case : vertical n-p-n structures and optimized light shielding.

A breakthrough for MOS-XY devices was the introduction of the Transversal Signal Line (TSL) architecture (Noda 86), which minimizes the amount of smear drastically.

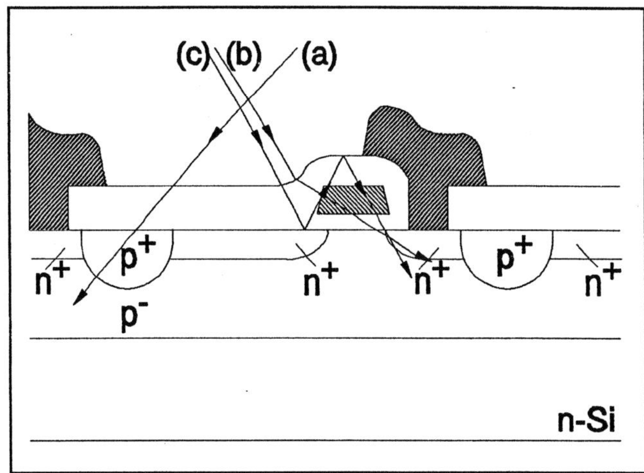

FIGURE 8.16. Illustrating the origin of smear in an MOS-XY : stray electrons (a) and scattered photons (b),(c) cause the artifact.

The working principle of the MOS-TSL device is described with the aid of Figures 8.17a and 8.17b, in which a block of two-by-two pixels is shown for the classical MOS-XY imager and the TSL device, respectively. In the conventional MOS imager, the signal charge is transferred during the readout first from the photodiodes to the vertical signal lines by the vertical scanner. All charge packets from one video line are transferred at a time. The smear charge is then accumulated during a complete line time of 64 μsec. In the TSL imager every pixel has a horizontal switch and signal readout takes place directly in the horizontal direction. In this configuration the horizontal signal line accumulates smear only during one pixel time, e.g. 200 nsec. The smear is therefore reduced drastically compared to the smear level of conventional imagers.

In addition to this interesting smear feature, the TSL has some extra advantages over the classical MOS-XY imager : the horizontal switch of the TSL switches a capacitance which is much smaller than the capacitance switched by the horizontal scanner of the conventional imager, and consequently the kTC noise is much lower. Also, the fixed-pattern noise of the new device is different from the original : in the conventional imager fixed-pattern noise was caused by nonuniform crosstalk and leakage through the horizontal switches. The shape of the fixed-pattern noise was a set of vertical lines because every pixel in a same column was read out through the same switch (Noda 86).

CID imagers are fully smear-free. This ideal characteristic is due to the fact that the output signal is a displacement current generated in a gate located on top of the silicon and fully isolated from the silicon substrate. Stray electrons and stray photons causing the trouble in the other type of imagers remain inside the substrate

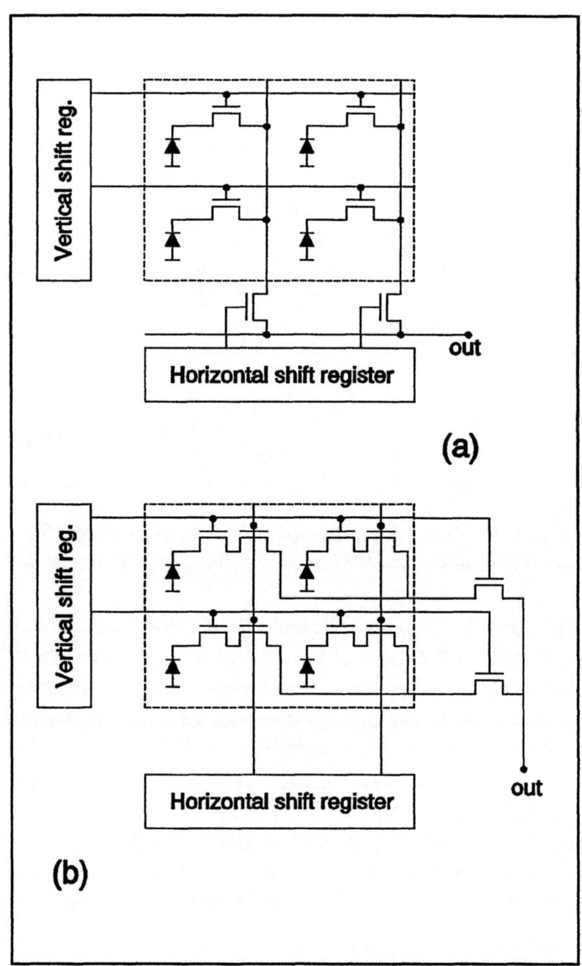

FIGURE 8.17. A block of two-by-two pixels for the classical
MOS-XY imager (a) and the TSL device (b), respectively.

and have no influence on the displacement current in the read-out lines/gates.

8.5.5. STATE OF THE ART IN SMEAR SUPPRESSION

The smear signal is spurious and degrades the video signal in all applications.
The fact that almost all solid-state imagers suffer from smear is a drawback compared
to the old imaging tubes. Tubes are also smear-free, although their performance
was vitiated by other highlight problems such as image lag and burn-in effects.
On the other hand, CIDs are also fully smear-free.

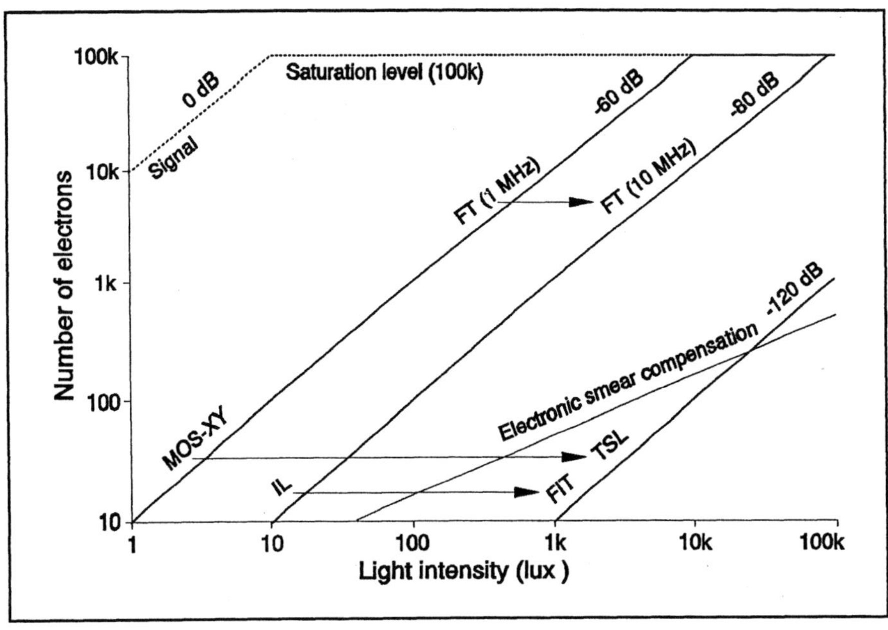

FIGURE 8.18. Comparison of the various solid-state image sensor types so far as smear
is concerned (dotted line : video, solid lines : smear).

Much effort is being put into the development of different techniques to suppress
smear in charge-coupled devices when used as imagers. Smear levels nowadays
are quite low, but there is not a single solution at device level to fully eliminate smear
in FT, IL, FIT or MOS-XY image sensors. The "state of the art" so far as smear
is concerned is summarized in Figure 8.18. It shows the output voltage of a
hypothetical image sensor with a sensitivity of 10,000 electrons per lux of incoming
light energy and a saturation level of 100,000 electrons. In the case of a frame-
transfer imager with a frame shift at 1 MHz, the smear level can be calculated by
means with formula [8.3] to be - 60 dB. About the same value is valid for the MOS-XY
device. The smear performances of these two sensor types can be increased by
a faster frame shift and a new device concept, respectively. The architecture of
the MOS-XY imager can be redesigned to give a TSL device with a smear level
as low as -120 dB. In the case of the frame-transfer device, 20 dB can be gained
by increasing the vertical clocking of the device by a factor of 10. With the relatively
high frame shift the frame-transfer CCD is comparable in smear performance to
the standard interline-transfer imager. Lowering of the smear signal in an interline-
transfer imager can be achieved using the alternative device architecture, i.e. the
frame-interline-transfer CCD, in which the vertical clocking frequency is drastically
increased. The FIT performs as well as the TSL imager.
In the case of electronic smear suppression with a frame-transfer imager, the end
result is expected to be fully smear free. But what is shown in Figure 8.18 is not

the resulting smear level, but the extra noise introduced due to the compensation technique. This noise component is proportional to $(2.Q_n)^{0.5}$: the original smear signal as well as the smear copy have their photon shot noise proportional to $Q_n^{0.5}$. Due to the uncorrelated nature between both, the previous relationship is valid (Esser 88).

WORTH MEMORIZING

Smear is a spurious signal running from top to bottom through a highlight in the picture. The origin and also the countermeasures vary according to the type of device :
> **- in frame-transfer CCDs, smear is generated by impinging photons during the frame shift. A faster frame shift or shielding the device from light by means of a shutter during the frame shift lowers and may even obviate the smear signal entirely;**
> **- in interline-transfer CCDs, smear is generated by scattered photons entering the vertical CCD shift registers or by stray electrons collected in the vertical shift registers instead of collected in the photodiodes. Countermeasures include optimized photodiode design, optimized doping profile of the vertical registers, and the introduction of the frame-interline-transfer device;**
> **- in MOS-XY sensors, smear is generated in the same way as in interline-transfer CCDs. The same countermeasures are also used, or an adapted architecture may be introduced : the transversal signal line;**
> **- of all the solid-state imagers, only CID sensors are completely smear-free.**

8.6. Conclusions

In this chapter attention has been paid to noise sources present in the CCD itself or in the output amplifier and to smear. To complete the overall discussion of noise, the conversion factor or sensitivity of the output amplifier has been also described.

The noise characteristic of a solid-state imager is one of the latter's most important parameters because it defines the lower limit of light input at which the device can be operated. When working with a camera in very poor light conditions, noise should be kept as low as possible. It is not surprising that a good deal of R&D effort is put into lowering the noise level of the CCD imagers. The overall noise figure of a charge-coupled device is composed of several components. Some noise-generation mechanisms are typically CCD-bound while others are associated with the output amplifier. CCD noise and output-amplifier noise are independent of the light intensity impinging on the imager, but another noise component, the photon shot noise, is proportional to the square root of the amount of electrons generated by the optical input.

The CCD noise sources themselves are very much technology-related. They show up as, for instance, white and black point defects, column defects, transfer noise, striations, pixel nonuniformities, dark-current shot noise and dark-current nonuniformities. Technology-related noise sources are very hard to tackle. Their origin may be situated in the starting material used to process the devices. But even silicon substrates of the highest quality only yield defect-free imagers if the wafers are processed in an ultraclean room and if processing is done by a very disciplined team. Other possible methods of tackling defects in the wafer are gettering techniques : locating defects at appropriate places in the wafer (at the back or in the center) so as to attract all the impurities.

Output-amplifier noise can also be divided into different components : thermal noise, 1/f noise, and reset noise. Thermal noise, which has a flat frequency spectrum, cannot be eliminated but it can be considerably reduced by a suitable design and layout of the driver of the first source-follower stage. The same is true of 1/f noise, with its typical dependence on the inverse of the frequency. With proper design and the incorporation of a buried-channel transistor, the 1/f-noise component can be kept to an acceptable limit.

On the other hand, however, it is very hard to get rid of the reset noise at CCD level because it is a fundamental process. Charging and discharging a capacitor through a resistor always add some uncertainty to the voltage level across the capacitor. Fortunately, the reset noise level can be compensated for by the video-preprocessing circuits. Techniques such as correlated-double sampling or delay-line processing drastically lower the effect of reset noise.

As technology-related noise sources diminish, the noise level of the output amplifier becomes more and more predominant. To deal with this effect, new output-amplifier architectures are being developed to increase the overall noise figure of CCDs. In some cases incorporation of a JFET or a MOS transistor at the site of the floating diffusion further decreases the thermal noise and 1/f-noise components.

The sensitivity of the output amplifier or conversion factor is determined by the value of the floating diffusion capacitance and the parasitic capacitance in parallel with the former. Increasing the conversion factor can be done by optimizing the total capacitance at the input of the first stage of the output amplifier. Although the conversion factor or the sensitivity of the output amplifier has no direct relation to the noise performance of the latter, it can influence the noise performance of the video-processing circuitry placed behind the imager. If a high conversion factor can give rise to a low gain in the processing electronics, the overall S/N ratio of solid-state imager plus processing board can be further optimized.

Smear is a spurious signal running from top to bottom through a highlight in the image on the display. Depending on the device type smear may be from totally different origins. In frame-transfer CCDs, smear is generated by photons impinging during the frame shift. A faster frame shift or shielding the device from light during the frame shift by means of a shutter lowers and may even eliminate the smear signal completely. Although a frame-transfer device can never be made smear-free, it is the only charge-coupled device which can be operated in a completely smear-free camera by the incorporation of a mechanical or electro-optical shutter.
In interline-transfer CCDs, smear is generated by scattered photons entering the vertical CCD shift registers or by stray electrons collected in the vertical shift registers and not in the photodiodes. Countermeasures are optimized photodiode design, optimized doping profile of the vertical registers, and the introduction of the frame-interline-transfer device. As regards smear, MOS-XY sensors look very similar to interline-transfer CCDs because smear is generated in the same way in both cases. Methods to minimize the smear are also identical. Even an adapted architecture has been developed : the transversal signal line.
Charge-injection devices are fully smear-free. They are operated via measurement of an induced gate current. The gates of the device are located on top of the silicon and are fully separated from the silicon bulk where the photons generate the electron-hole pairs.

Chapter 9

ADVANCED IMAGING : DEVICE ARCHITECTURES

In the two preceding chapters attention was focused on solid-state imaging developments concerned with measures to improve the performance of the image sensors. Improving technology and optimizing design have the same objective : to upgrade performance for a given application or a given resolution. Further optimization of the overall performance of a device can also be achieved by adapting or redesigning the architecture of the device. Examples of this option can already be found in the descriptions of new types of output amplifiers, but there the idea is only adapted to a small part of the total imaging chip. In this chapter the accent will be placed on the device architecture in general. So far as device architectures are concerned, it is impossible to present a comprehensive study of all types of device concept reported. So, without having the intention of being complete, the most important developments are reported here.

This chapter starts in its first section with a technique to increase the horizontal pixel density. This method was introduced in the early eighties in order to adapt the pitch of the horizontal output register of a two-dimensional imager to the smaller horizontal pixel pitch in the image section. Since then, however, technology has improved to such an extent that small dimensions are permissible thus rendering implementation of that particular technique unnecessary. Surprisingly, the method is still in use but the arguments in its favor have changed, namely from layout issues to speed issues. The solution adopted also allows the frequency of the horizontal output register to be lowered.

Although the technique under study seems very simple as far as its concept is concerned, the reality is that narrow-channel effects in the CCD channels have to be dealt with. Solutions to circumvent or even to solve these types of problems are described : a design adaptation, gate tapering, extra channel implantations, extra channels, or a combination of two or more of these methods.

Attention will be paid first to optimization of the horizontal structure, then to several designs of the vertical shift registers. A very-high-resolution vertical shift register, called accordion CCD and a very small vertical shift register for IL type imagers, called a charge-sweep device, will be described. Most attention will be paid to speeding up of the vertical clocks of the FT and FIT sensors. Additional measures in device architecture and in device technology are needed to provide the horizontally running CCD gates with an extra metal strap, running vertically between the photosensitive areas of the sensor. Driving the image section of a frame-transfer in various transporting mode, with a four-phase or a three-phase clocking system, results in an image section with a tunable number of video lines. This technique can be applied to make a CCD imager with a switchable aspect ratio.

Electronic still pictures have to be captured progressively but have also to be displayed or stored in an interlaced scanning mode. Combining the capturing process and readout of the imager in a different mode can be handled by ingenious design of the architecture of the device. The same applies to time-delay and integrating devices. The architecture of the imager is adapted in such a way that it combines the very high resolution of the linear array (one-dimensional) with the sensitivity of the two-dimensional imagers.

9.1. Increasing horizontal pixel density

As already stated, frame-transfer image sensors are very suitable for high-resolution applications. The unit-cell structure is very simple and in the horizontal direction, the pixel is built up with only an n-type implantation to define the CCD channel and a p^+ implantation to separate the channels with the aid of a channel stopper. This compact structure permits very high pixel densities but, combined with the very tight horizontal pixel pitch, it also imposes some very exacting requirements on the construction of the horizontal register. The problem is illustrated in Figure 9.1 : how to fit one CCD cell of the horizontal output register into the horizontal pitch of the imager ? In its simplest configuration, the horizontal register is a three-phase CCD, consisting of three poly-silicon layers.

Its minimum dimension (in the direction of the serial charge transport, which is from right to left in Figure 9.1) is set by the layout rules for the polysilicon layers because these layers are used to define the CCD shift register (in the charge transport direction). The pitch of the horizontal output register, and consequently also of the image cell, is completely defined by a combination of the minimum poly-silicon to poly-silicon overlap, the minimum polysilicon spacing and a minimum polysilicon width.

If the application of the imager requires higher pixel densities in the image section than can be managed from the point of view of the horizontal register, the only way to overcome this problem is charge multiplexing (Beck 82, Oda 85, Orihara 86). This is a technique in which the horizontal output register is divided into several parallel horizontal output registers, as in the case of quadrilinear imagers.

A potential solution of charge multiplexing by dividing the charge packets of a single horizontal video line over three horizontal (parallel) CCD registers is shown in Figure 9.2 : hatched areas are channel-stopper implantations, polysilicon transfer gates (TG_1, TG_2 and TG_3) run horizontally underneath the three phases Φ_{H1}, Φ_{H2} and Φ_{H3} (Collet 85). The areas defined by the channel-stopper implantations are forbidden

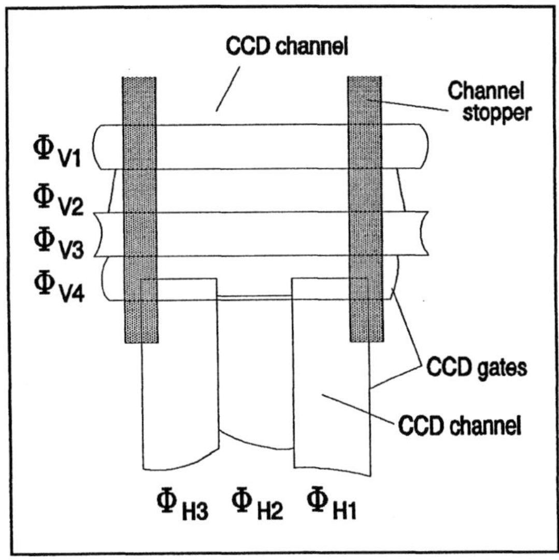

*FIGURE 9.1. Illustration of the limiting factor in defining
the horizontal resolution of a frame-transfer imager.*

areas for the charge packets. The construction of the transfer gates is such that they are made from the first polysilicon layer. This means that the transfer gates define the channel potential in the silicon underneath them, whether they are covered with extra polysilicon gates (from the second and third layer) or not. This structure allows control of the transfer regions between two horizontal registers by the transfer gates TG_1, TG_2 and TG_3, and control of the horizontal CCD registers itself by Φ_{H1}, Φ_{H2} and Φ_{H3}. The three charge packets indicated by A, B, and C, can be separated and transferred to sites A', B', and C' by appropriate clocking of Φ_{H1}, Φ_{H2}, and Φ_{H3}, in combination with three transfer gates TG_1, TG_2, and TG_3 and with the gates of the storage region.

More detailed principles together with appropriate clocking diagrams are shown respectively in Figures 9.3a and 9.3b :
- the starting condition is presented in the first illustration in Figure 9.3a, at time t_0. The structure is identical to that in Figure 9.2. The charge packets, represented by the black rectangle, the black circle, and the black diamond, are stored underneath the last gate of the storage region Φ_{V4} and are separated from the horizontal output register by a low-biased TG_1;
- at time t_1 the biasing conditions for the various gates are such that the charge packets are all pushed toward TG_1 and two of the three are stored under this gate because all clocks of the horizontal registers, except Φ_{H1}, are blocked. The black rectangle can pass into the first horizontal register across TG_1 toward Φ_{H1};

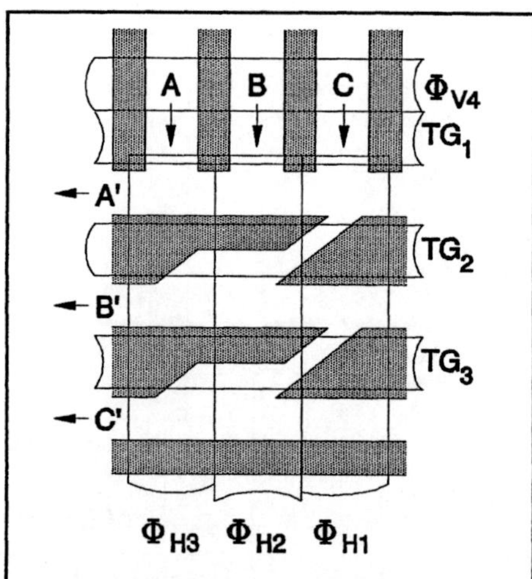

FIGURE 9.2. Charge-multiplexing structure to "hustle" the parallel video information (A, B, C) into three serial output registers arranged in parallel (A',B', C').

- at time t_2 the transfer gate TG_1 goes low again, and all charge packets are pushed away from TG_1. The black rectangle moves completely underneath Φ_{H1}, but the black circle and the black diamond are pushed back toward Φ_{V4}. At this moment the first charge packet is separated from the others, which still remain in the storage part of the imager. The black rectangle is stored in the top one of the three parallel registers;
- at point t_3 the transfer from the top register toward the middle one is initiated. The transfer gate TG_2 is biased at a high potential and the low-going Φ_{H1} forces the black rectangle underneath this transfer gate;
- at time t_4 the next separation step is started : while Φ_{H2} is biased high, the last gate of the storage area is clocked once again and the black circle and the black diamond are pushed underneath transfer gate TG_1. The circle can move further toward Φ_{H2}, but the diamond is stored underneath transfer gate TG_1 because the horizontal clock Φ_{H3} is changed to a low potential;
- at t_5 the black diamond is allowed to flow back into the storage area by a high voltage on Φ_{V4} and in the meantime the black rectangle is forced completely under Φ_{H2};
- at time t_6 the black diamond is also forced back into the storage area because TG_1 also goes low again then. The next separation is completed : the black rectangle is stored in the middle register and the black circle in the top registers while the black diamond is still stored in the storage area of the imager;

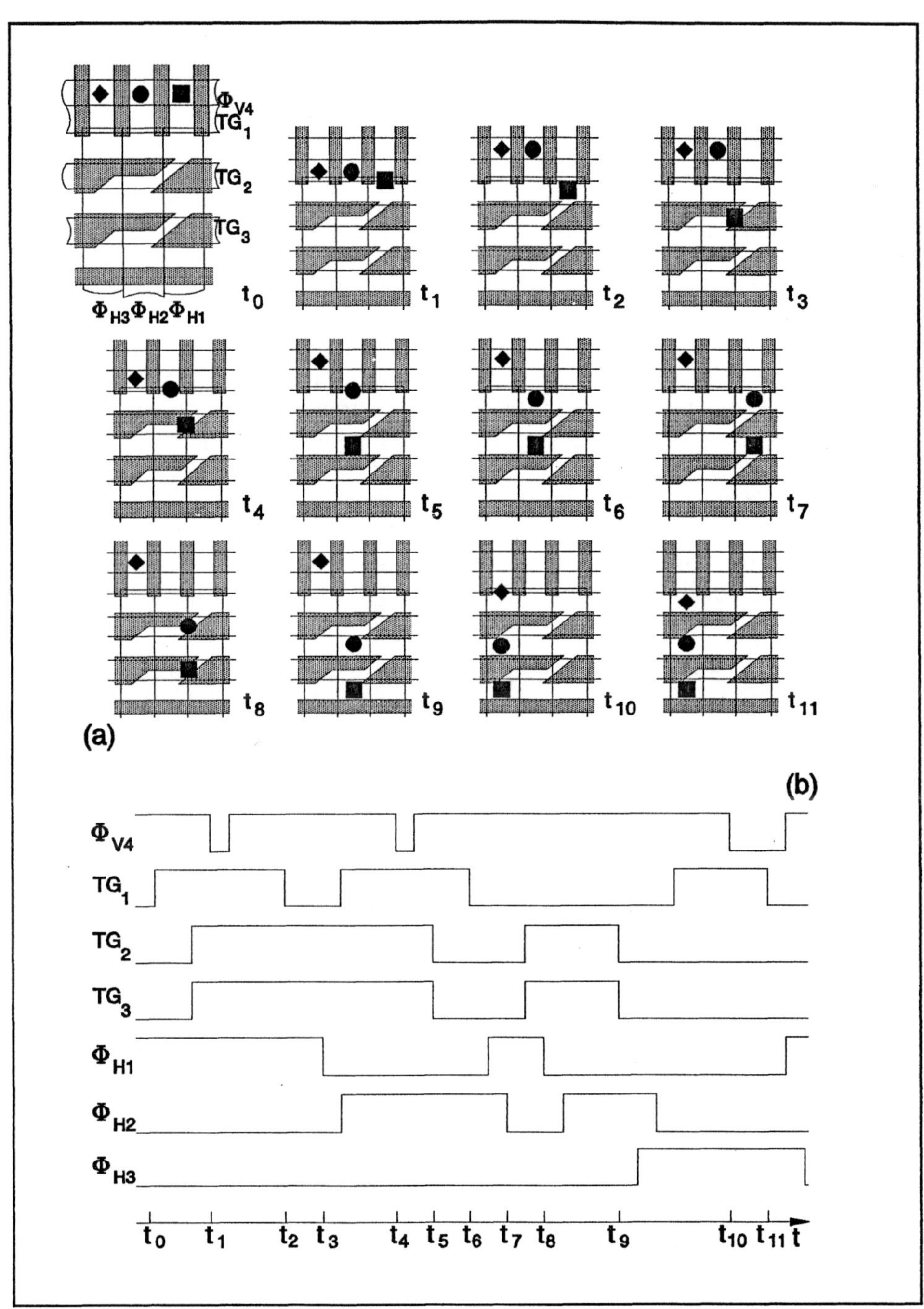

FIGURE 9.3. Timing diagram needed to perform the "hustling" depicted in Figure 9.2.

- at time t_7 transfer of the black rectangle and the black circle toward the horizontal register one level lower starts. Both charge packets are forced backward from Φ_{H2} to Φ_{H1} into the horizontal registers. The main reason for this operation is to bring the charge packets toward the opening in the stopper implantation underneath the transfer gates TG_2 and TG_3;
- at time t_8, all clocks of the horizontal registers are biased low, only TG_2 and TG_3 being biased high. Consequently, the black rectangle is forced underneath TG_3 and the black circle underneath TG_2, both in the direction of the horizontal register one level lower;
- at time t_9 the next transfer is induced, namely away from the transfer gates and directly into the horizontal register below them : the black rectangle into the bottom channel and the black circle into the middle channel;
- at time t_{10} both charge packets are moved forward to Φ_3 and, in the meantime the one still in the memory region is also transferred into the top register : the black diamond is pushed out of the storage area by a low biased Φ_{V4};
- at time t_{11} the "hustling" action is completed : all charge packets from the last line of the storage area are divided over the three different horizontal output channels and they are all stored temporarily underneath the Φ_{H3}, ready for the general transport through the output channel toward the output amplifiers.

Figures 9.2 and 9.3 illustrate the charge multiplexing of the packets into three horizontal registers, each with a three-phase CCD transport mechanism (each CCD gate of the horizontal register system corresponds to one pixel pitch), but alternatives multiplexing the packets into two horizontal registers, each with a two-phase or a four-phase CCD transport mechanism, are also possible, as well as other combinations.

As a consequence of the multiplexing, the clock rate of the horizontal transport system is decreased by a factor equal to the number of horizontal parallel registers. Lowering the clock rate has a favorable effect on on-chip and off-chip power dissipation. On the other hand the design of a "high-speed" CCD shift register is more complicated than that of a "low-speed" CCD register. To comply with the specification as far as clocking frequency is concerned, extra channel implantations could be needed in the case of a high-frequency clocking (Azuma 91).
A lower pixel rate per horizontal register is also attractive with regard to the bandwidth of the output stage if each horizontal register is provided with its own output stage. Another advantage of the multiplexing can be found in color sensors; for instance, with a three-color stripe filter and a multiplexing system extending over three horizontal channels, each channel contains the video information of one and only one color.

Besides the above mentioned advantages, charge multiplexing also has a few disadvantages : each channel needs its own output amplifier, but the individual output stages have to be exactly identical to each other in all respects. Additionally, their mutual timing can be critical or has to be at least fully synchronous.

A second possible source of trouble may be the bottleneck in the charge transport channel if the charge packets have to be transferred from a horizontal register toward the transfer region under the transfer gates. This bottleneck can cause narrow-channel effects resulting in local potential barriers and incomplete charge transport. The net result on the display can be a fixed-pattern noise of black and white vertical columns. These narrow-channel effects can be reduced and/or the fringing fields at these bottlenecks can be increased by several means, e.g. an adapted design, tapered gate configurations, extra channel implantations, a compound channel, or a combination of several techniques.

9.1.1. DESIGN ADAPTATION

The effect of an optimized design is shown in Figure 9.4 (Bisschop 88). A small portion of the horizontal register architecture at the "entrance" to the transfer region is sketched in Figure 9.4a. As can be seen, the entrance to the transfer region, which is the opening in the stopper definition, is rather small compared to the gate Φ_{H2}, which drives the charge packet. The reverse funnel effect causes a narrow-channel problem. This narrow-channel effect can be decreased by incorporating a small extension of the transfer gate into the horizontal CCD channel, because as, seen from the horizontal register, the charge packets are not only pushed by Φ_{H2}, but also pulled by TG_2.

The potential diagrams for three different voltages on TG_2 are shown in Figure 9.4b. (In this cross section along AA', the crossing of Φ_{H2} over TG_2 is not shown for sake of simplicity.) The potential barrier at the entrance to the transfer region is shown for a low voltage (V_1) on TG_2. It will keep some charges behind in the horizontal register. The barrier becomes smaller with increasing voltage (V_2) on TG_2 and will finally disappear when TG_2 reaches its highest value (V_3). In the design phase of the imager, electrostatic potential simulations have to determine W', W, L', and L in such a way that smoothness of transport from the horizontal register toward the transfer gate can be guaranteed.

9.1.2. GATE TAPERING

The technique of gate tapering makes use of properties opposite to the narrow-channel effect : broadening the channels can increase the fringing fields and can even support any charge transport (Theuwissen 88, Yonemoto 90). The impact of tapered gates on the design is shown in Figures 9.5a and 9.5b. In the former, a classical structure is shown with two horizontal registers, driven for instance by a four-phase horizontal transporting system (2 gates per horizontal pixel pitch). The gate tapering itself is shown in Figure 9.5b : in the four-phase transport system, divided between two parallel horizontal channels, one of the four gates is designed as a trapezoid in order to increase the fringing fields toward the transfer gates.

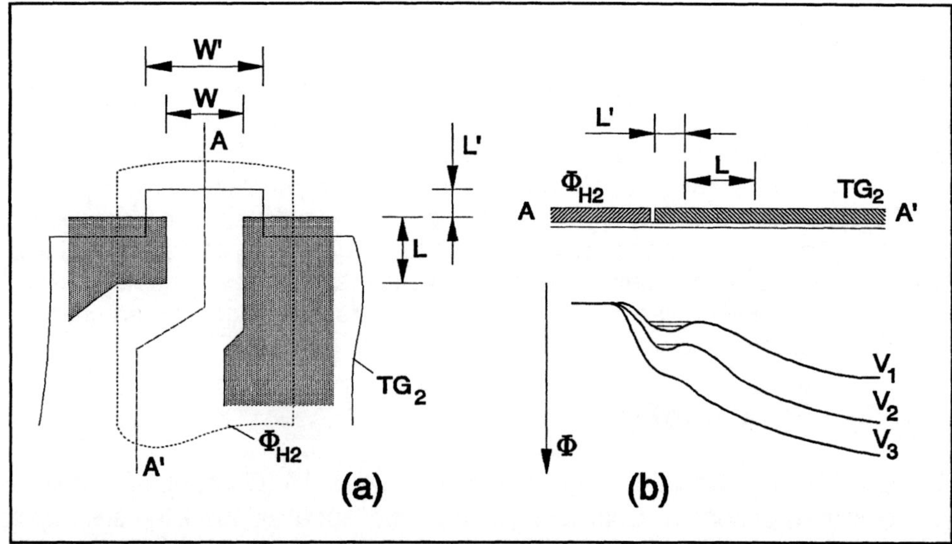

FIGURE 9.4. A layout concept (a) of the transfer region between two parallel CCD registers with its potential diagram (b).

Besides its effect of increasing fringing fields, also narrow-channel effects are lowered by the gate tapering technique. Narrow-channel effects are more or less defined by the ratio between the width of the "entrance" in the transfer region and the length of the driving horizontal clocking gate. In the case of tapered-gate design this ratio is closer to unity, and the narrow-channel effect itself is reduced.

9.1.3. EXTRA CHANNEL IMPLANTATIONS

The same idea of increasing the fringing fields locally to ensure complete charge transport can be recognized again in the design shown in Figure 9.6a (Oda 89) : an extra lightly doped p region is included in the top regions of the horizontal registers and parallel to the charge transport in these registers. The resulting charge storage region underneath the CCD gate can again be designed in a tapered form, for the same reason as above. The extra implantation provides an extra "downward" driving force on the electrons.

The extra implants in the two-phase system in Figure 9.6b are not uniformly distributed across the horizontal register as a whole : they form an L-shape together with the original implantations in the two-phase system (Stevens 89, Lee 90), but the net result is the same : additional electric fields push the charge packets "downward". Note the absence of transfer gates in this system, but the inclusion of a second Φ_{H1} clock. The two-phase system is operated with three clocks : Φ_{H1A}, Φ_{H1B} and Φ_{H2}.

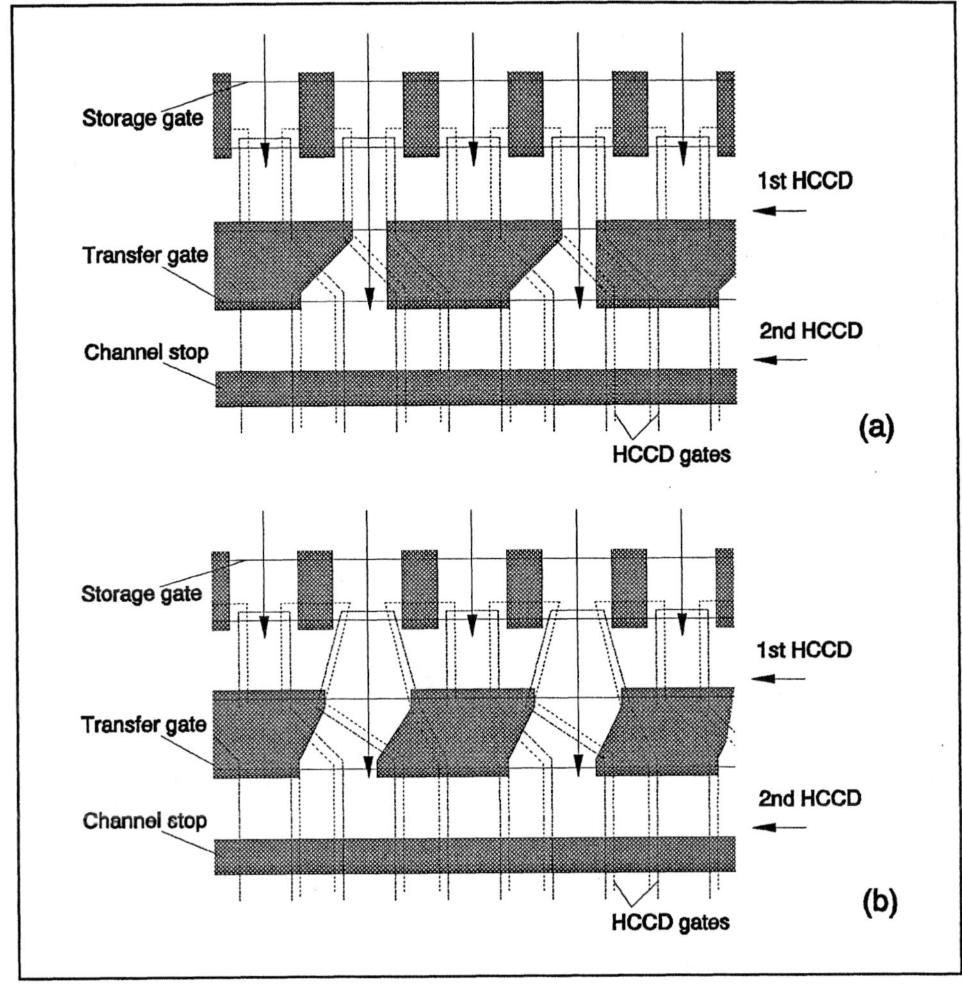

FIGURE 9.5. Design optimization by gate tapering (b), compared with the classical rectangular-gate structure (a).

The last example, shown in Figure 9.6c, has an additional uniform n implantation in the bottom channel of the multiplexing system (Iesaka 88). This higher-doped n channel pulls the charges to the bottom channel. Although the horizontal registers have different channel dopings, care has to be taken to ensure that the channel doping near the output gate at the output node is exactly the same for both registers. This requirement can be fulfilled by doping the last gate before the output gate of the lightest-doped channel (the top channel in Figure 9.6c) with the same implantation used to dope the highest-doped channel.

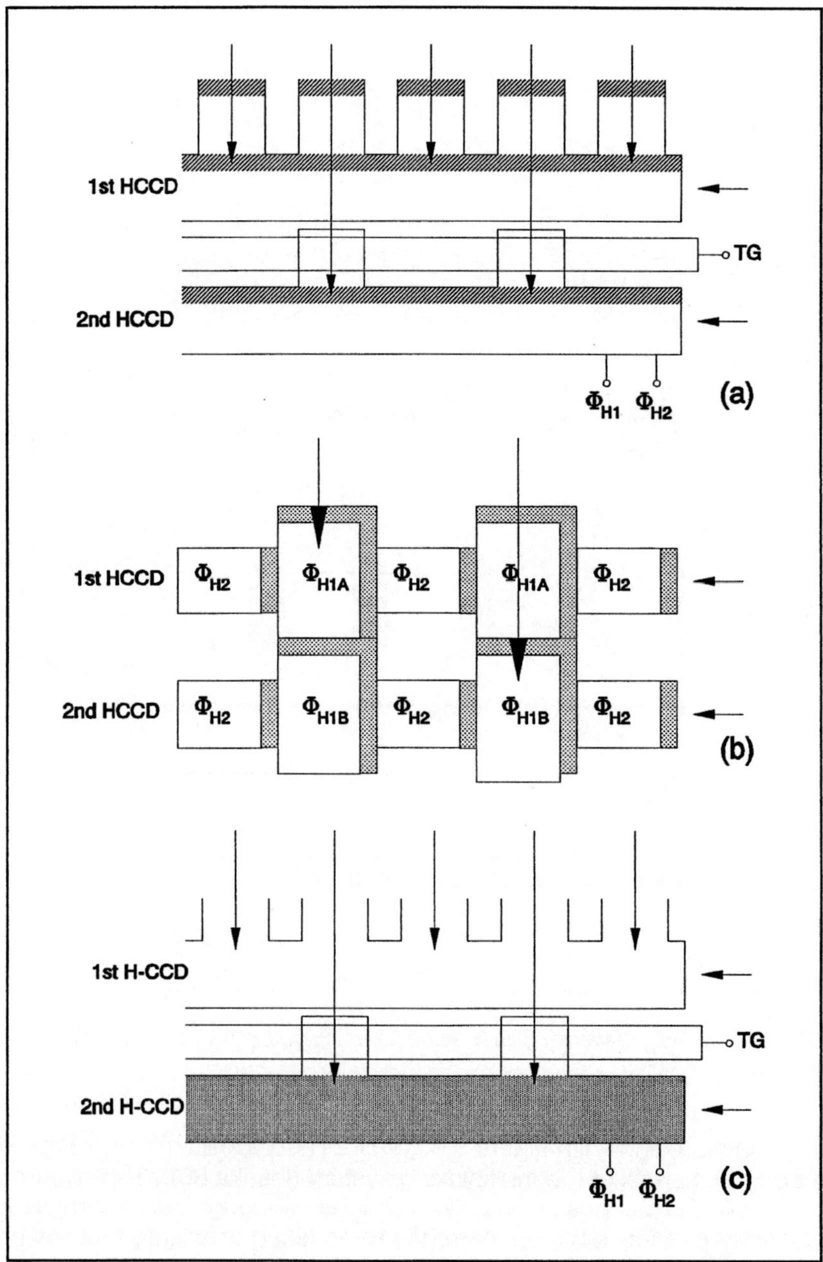

FIGURE 9.6. *Extra implants in the CCDs : parallel to the readout registers (a), an adapted two-phase structure (b), different channel dopings in the different horizontal registers (c).*

9.1.4. COMPOUND CHANNELS

During the multiplexing cycle the transfer of the charge packets through the relatively wide top channel to the bottom one (for instance, in a structure similar to those shown in Figure 9.6) can be a time-limiting factor. In the case of wide horizontal channels with a short gate length, the perpendicular transfer of the charges during multiplexing involves a small channel with a fairly large gate length. The width and length of the CCD gates are interchanged depending on the direction of transport of the charge packets. To avoid speed limitation or incomplete charge transfer, this problem can be obviated by splitting the wide top channel into two smaller parallel channels. This solution is illustrated in Figure 9.7 (Kobayashi 93). The charges which have to move to the bottom channel are passed through both smaller top channels (1a HCCD and 1b HCCD), while the bottom channel (2 HCCD) retains its original width. At the output stage of the top channels, the two charge packets are mixed together to restore the original charge packet. In other words, the charge packets from 1a HCCD and 1b HCCD are added together before they are transferred to the floating diffusion.

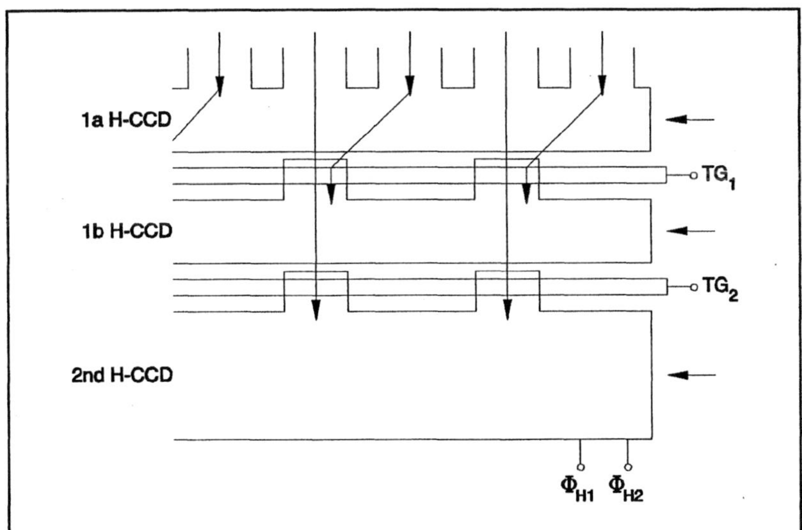

FIGURE 9.7. Basic idea of the compound-channel construction using to avoid excessive transfer length in the top channel.

9.1.5. COMBINED TECHNIQUES

A combination of two of the foregoing techniques, e.g. a tapered-gate structure combined with different channel dopings (Manabe 91) or a tapered-gate structure combined with an extra implantation between the horizontal output registers (Harada 92), can be made.

Originally, the charge-multiplexing technique was used with frame-transfer devices only, to increase the resolution of the devices. In the case of (frame-) interline-transfer devices, there is no need to incorporate a multiplexing system. In most applications the horizontal pitch of these devices is larger than frame-transfer pitches due to the unit cell geometry of (frame-) interline-transfer imagers : a CCD transport cell near each photodiode. Nowadays, however, (frame-) interline-transfer devices also employ the multiplexing technique, especially if high pixel-rate imagers are involved. In these situations, charge multiplexing offers appreciable advantages : it decreases the pixel rate per horizontal channel, and can keep the clocking rate lower, while an exaggerated bandwidth for the output stage is unnecessary.

WORTH MEMORIZING

Increasing the horizontal pixel density has resulted in the creation of a multiplexing technique applied to the horizontal output register of two-dimensional image sensors. The "old" technique is still used nowadays to limit the speed of the horizontal registers, resulting in lower on-chip power dissipation, simpler driving electronics and a lower bandwidth requirement for the output amplifier.
The multiplexing technique can cope with transport problems, which can be avoided with :
- optimized design of the transfer gates;
- gate tapering;
- extra channel implantations;
- compound channel construction;
- a method, combining two (or more) of the above.

9.2. New clocking schemes for vertical registers

So far as consumer applications of solid-state imagers are concerned, the only device parameter kept constant is the number of horizontal lines on the sensor. This number depends on the TV standard but otherwise fixed and, this boundary condition being known, it is easy to explain why not very much effort is being devoted to further development of the vertical pixel structure. Requirements for higher vertical densities or for faster vertical transport speeds are only set for deviating image formats. This section discusses four solutions in which the vertical imager structure is adapted to achieve a higher resolution requirement (the accordion imager), or a denser horizontal pixel pitch (the charge-sweep device), or to make a considerably higher vertical transport speed possible, or to change the number of video lines in the image section (dynamic pixel management).

9.2.1. ACCORDION CCD

High resolution in combination with a small image format also necessitates a compact architecture in the vertical direction. Solid-state imagers for endoscopy are probably the most exacting application for high resolution or high pixel densities in both directions. How the horizontal resolution can be optimized by charge multiplexing has been explained in the preceding section. The resolution can be doubled vertically by introducing the accordion transport mechanism for vertical CCDs (Theuwissen 84, Theuwissen 88). Although the principle is applicable to all types of CCD imagers, only the combination of an accordion with a frame-transfer imager will be described.

Figure 9.8 explains the idea behind the accordion transport system. It shows a cross section of part of the image area of a frame-transfer imager. Instead of integrating pixels with a length (or height) of four gates, integration is limited to pixels with only half of that length (or height) : two gates. One of these two gates is the blocking electrode, the other the integrating electrode. In this compact, two-phase system it is impossible to transport the charge packets using conventional clocking methods. That is why the accordion transport mechanism is applied. First only the right-most pixel is enlarged from two to four gates; if the first packet is moving in a fully four-phase mode, the same is done with the second packet, enlarging its length, and it, too, can be hooked on to the four-phase shifting system. This conversion from a two-phase integrating system to a four-phase transporting system is done pixel after pixel till all packets are moving in the four-phase mode or till the accordion is completely stretched.

If the imager chip is designed to have a compact vertical structure in the image section, the same can be done for its storage section. Alternatively the densest structure for the frame-transfer imager can be obtained with an accordion structure in both the image area and the storage area. These two accordions have to be matched to each other in such a way that a complete frame shift from image to storage section passes through the following stages :
- opening of the image accordion toward the storage section;
- once the first charge packets arrive at the bottom part of the storage section, the storage accordion is folded up;
- once the storage accordion is completely folded, the image accordion can be similarly folded so as to initiate the next integration in its two-phase mode;

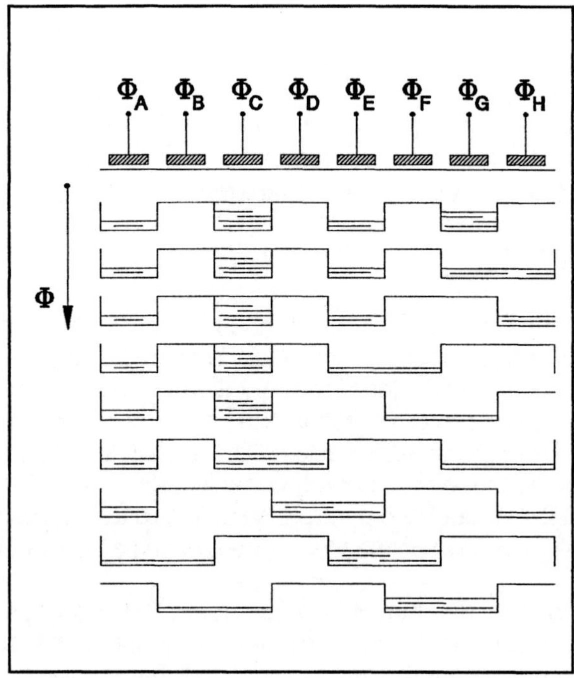

FIGURE 9.8. Basic working principle of the accordion CCD.

- the storage accordion opens up again, line by line, during the readout of
  the video information.

Although the fully synchronous clocking of the two accordions seems cumbersome,
the opposite is true if the clocking of the image and the storage section is done
on-chip, by means of two digital shift registers. This is illustrated in Figures 9.9a
and 9.9b. In a CCD technology using the vertical n-p-n structure these digital shift
registers can be built in (dynamic) CMOS logic (Theuwissen 84).
All individual gates of both sections of the image sensor are connected to two digital
shift registers. Instead of clocking the image sensor, the correct voltages are offered
to the gates by these two digital shift registers. A few clock pulses to operate the
digital shift registers are shown in Figure 9.9b : Φ_1 and Φ_2 are common to the two
sections, IM and ST are the two inputs of the shift registers. The "content" of the
unit cells of the digital shift registers is designed such that : if IM and/or ST are/is
clocked synchronously with respect to Φ_2 but with only half its frequency, the CCD
sections are driven in the four-phase transport mode, keeping IM and/or ST constant.
This closes the appropriate accordion and can, for instance, change the image
section to its integration mode. Proceeding from a constant value on IM and/or
ST to a clocking mode opens the accordions. Note that a logical "0" or a logical

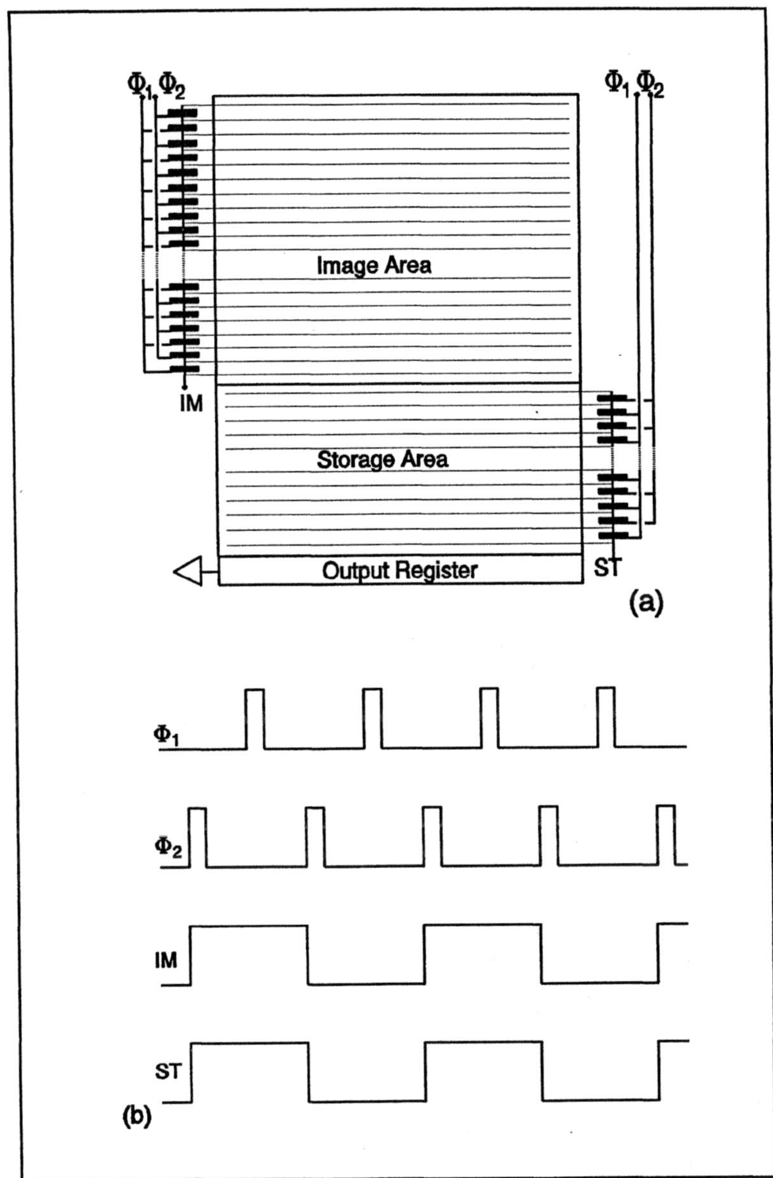

FIGURE 9.9. Driving the accordion by means of two digital shift registers (a) built in CMOS technology, together with is clocking diagram (b).

"1" on IM at the moment the image accordion is closed, selects the integration field of the interlaced system.

As well as a compact structure in the vertical direction, the accordion CCD has quite a simplified clocking system. An additional advantage is its potentially higher sensitivity to light because less polycrystalline-silicon gates cover the surface. One of the main drawbacks during the device testing and/or the product evaluation and engineering stage is the lack of a direct access to the CCD gates. These are all connected to an intermediate output node of the digital shift registers and not directly to a bonding pad.

9.2.2. CHARGE-SWEEP DEVICE (CSD)

The concept of the charge-sweep device has been developed to tackle the issue of the small vertical CCD registers in the IL and FIT imagers (Kimata 85). The vertical shift registers are becoming smaller and smaller as the horizontal pixel pitch shrinks but also in the course of the optimization of the aperture ratio of the photodiodes. To overcome this problem, the vertical shift registers are designed very small but, instead of emptying the contents of a complete field in the vertical registers at one go, only one horizontal video line is emptied. The vertical registers are designed so small that their individual CCD cells cannot handle the charge from one photodiode but, due to the fact that only the information from one photodiode is available per vertical register, spilling-over of the charges into other CCD cells of the vertical registers is no problem. As long as the total information, from one photodiode, divided over several CCD cells, is gathered together again in a kind of a storage cell, the problem of having too small a vertical channel is solved.

A more extensive description of the charge-sweep device was included in section 7.2.3.

9.2.3. HIGH-SPEED CLOCKS FOR VERTICAL CCDS

In applications requiring large-area imagers at "TV speed", such as HDTV sensors and a few scientific imagers, the RC constant of the polysilicon gates can become so large that they dictate the upper limit of vertical transport and perhaps making it impossible to drive the imagers at the classical "TV rate". To overcome this problem, the vertical transport can be speeded up by strapping the high-ohmic polysilicon gates by means of low-ohmic metal bars. Fortunately, these metal bars are already available in the (frame-) interline-transfer devices, namely in the form of the metal light shields above the vertical shift registers (Nobusada 89). This is illustrated in Figure 9.10a where the horizontally running polysilicon gates are strapped by means of the individual light shields, alternately as Φ_{V1}, Φ_{V2}, Φ_{V3} and Φ_{V4}. This pattern is repeated horizontally for every fourth pixel and vertically for every unit cell of the vertical CCD register.

Although relatively simple, this strapping technique can cause some technological problems due to the contact opening above the active CCD channel. Alloying of

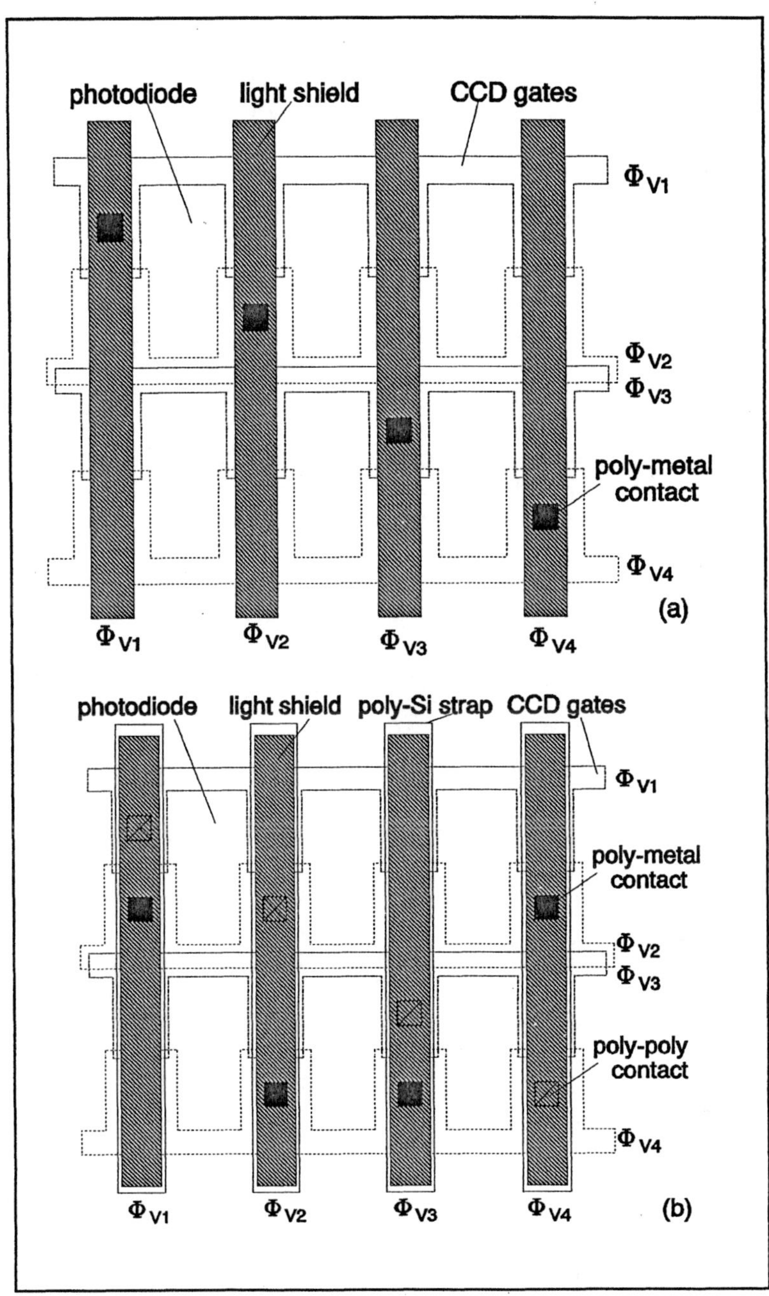

FIGURE 9.10. Speeding up the shift registers of an IL or FIT imager by strapping the poly-Si gates to the aluminum light shields : without (a) and with (b) an additional poly-Si layer.

the metal light shields by the polysilicon gates can change the threshold voltage above the CCD channel locally and disturb the CCD transport. However, alloying can be prevented by introducing refractory-metal light shields (Morimoto 92) or adding an extra intermediate polysilicon layer (Yonemoto 90). The metal contacts this new polysilicon layer, which in turn contacts the polysilicon CCD gates in such a way that the poly-poly contacts and the poly-metal contacts do not overlap each other (see Figure 9.10b). Any further alloying problems are always shielded by the VCCD gates and are limited to the newly introduced polysilicon layer.

In the frame-transfer concept, the said metal bars are not available and, if strapping is necessary, additional technological steps have to be taken in the production process to incorporate them. An example is shown in Figure 9.11 (Theuwissen 91) : very small vertical metal lines strap the horizontally running CCD gates as was done with the (frame-) interline-transfer devices. If the width of the bars is chosen small enough, and if they are located on top of the (almost light insensitive) channel stops, introduction of the strapping technique causes no loss of light sensitivity in the frame-transfer imagers, either.

In all cases the metal strips decrease the RC constants of the polysilicon gates, speed up considerably the vertical transport capabilities of the imagers, but necessitate the addition of a new polysilicon layer for the (frame-) interline technology or a metal layer for the frame-transfer devices. On the other hand, the strapping technique seems to be indispensable for HDTV imagers.

9.2.4. DYNAMIC PIXEL MANAGEMENT

Dynamic pixel management is a technique which allows a switchable aspect ratio of a frame-transfer CCD (Centen 94). The imager considered is driven by a four-phase clocking system in its so-called "full size" architecture. In the case that the same image section is provided with a three-phase clocking system, the number of video lines available on the imager is increased by a factor 1.33 (= 4/3). This effect is illustrated in Figure 9.12 : at the left-hand side of the image section of the frame-transfer device, the clocking system has four phases ($\Phi_{4,1}$ ... $\Phi_{4,4}$) and every pixel has a length of four gates. At the right-hand side the clocking system is reduced to a three-phase CCD ($\Phi_{3,1}$ ... $\Phi_{3,3}$) and the length of the pixels is only three gates. Dumping 16.5 % of the video lines at the beginning of each field and dumping the same amount of video lines at the end of each field reduces the overall number of video lines back to the original value. However, in the meantime the height of the image section of the "new" imager is reduced by 25 % compared with the original full size device.

The dynamic pixel management method is successfully applied on a frame-transfer CCD which makes its possible to switch its aspect ratio back and forth between 4:3 and 16:9. The full-size FT imager contains 492 four-phase video lines in its

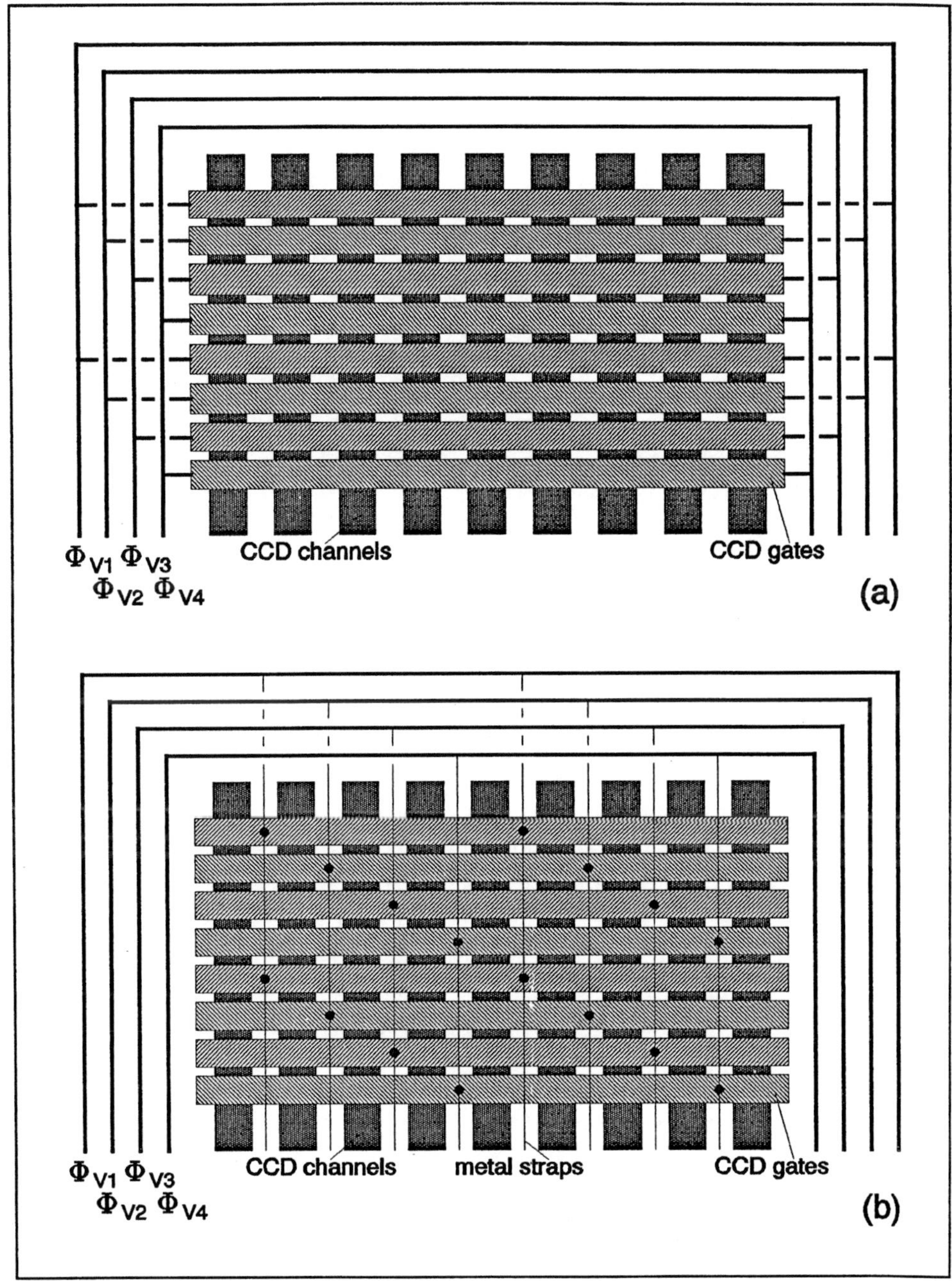

FIGURE 9.11. Speeding up the vertical frame shift of a classical frame-transfer imager (a) by strapping the polysilicon gates to a second metal layer (b).

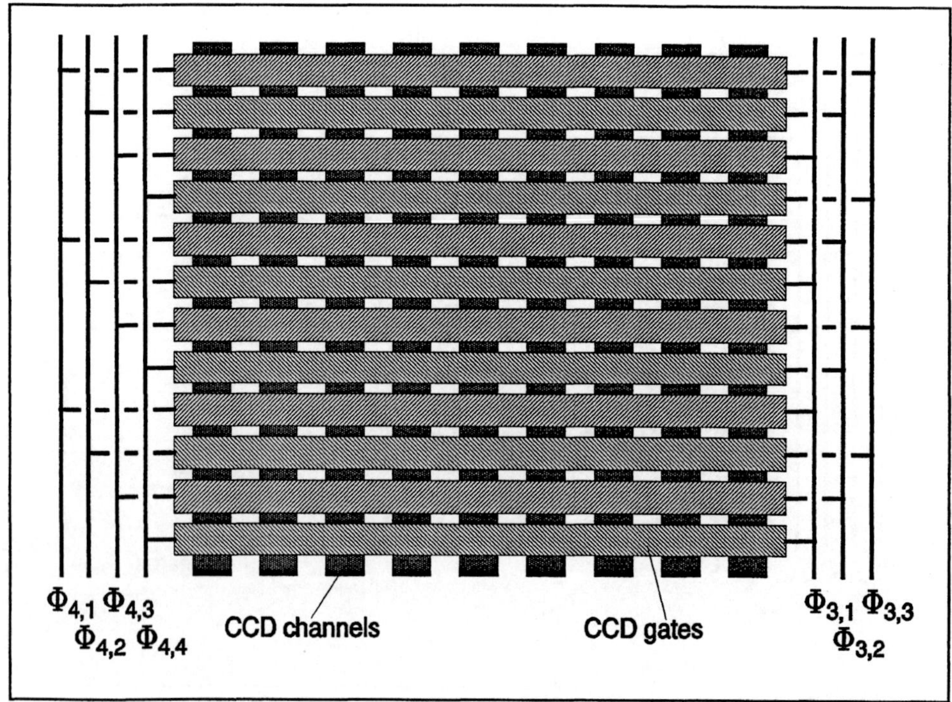

FIGURE 9.12. Working principle of "Dynamic Pixel Management" : the image section of a FT CCD is switched back and forth between a four-phase and a three-phase transporting system.

4:3 mode. Switching the image section to a three-phase transporting system results overall in 656 video lines. Dumping two times 82 lines reduces the image section back to 492 active video lines and to a height of 75 % of the original 4:3 mode. Keeping the original width of the image section, combined with only 75 % of the height gives an overall aspect ratio of 16:9 in the switched mode.

WORTH MEMORIZING

Adapted clocking schemes or optimized device architectures to increase the performance of the vertical CCD shift registers are reported. Examples of ways to meet various requirements are :
- **accordion clocking to increase the vertical resolution of a two-dimensional image sensor;**
- **charge-sweep device to optimize the aperture ratio of the photodiode at the cost of width of the vertical shift registers;**

- metal strapped polysilicon CCD gates to lower the RC constant of the CCD gates and speed up the driving capability of the vertical CCD registers in (frame-) interline-transfer and frame-transfer devices;
- dynamic pixel management to provide a frame-transfer CCD with a switchable number of video lines or a switchable aspect ratio.

9.3. Electronic still picture

In the Electronic Still Picture (ESP) application a solid-state imager and a floppy disk replace the classical photographic film in a photocamera. Users can display their still pictures from the floppy disk on a normal TV receiver, select their favorites and make hard copies of them on a hard-copy unit. Once the picture information is available electronically, picture enhancement, picture restoration, and picture manipulation are also possible. This option of video processing of pictures stored on a floppy disc is attractive particularly for professional users. With regard to this application, ESP looks like being a extension of the video application in which the 8 mm film has been replaced by a solid-state imager and a magnetic tape. Replacing the classic silver-halide film imposes some severe requirements on the sensor : high resolution, good light sensitivity, high response uniformity, high dynamic range, high color fidelity, low noise, low dark current, high protection against blooming, and extremely low smear.

Using solid-state image sensors in a still video system means some extra requirements on the imager because most devices are suitable for capturing interlaced pictures compatible with the interlaced-scan mechanism of the monitor. However, in between the two interlaced pictures there is a time lag of 20 msec or 16.6 msec for the European and the US standard, respectively. If moving objects are present in the image, time lag can introduce artifacts by displaying continuously the two interlaced fields taken with a "single" shot image sensor. "Single" shot here means a single one-frame shot comprising one odd field and one even field. In other words, a full-frame picture (odd and even field) in electronic still-picture mode has to be taken at one single shot, but it has to be displayed or stored in the conventional interlaced mode in use with standard TV receivers and to be compatible with still-video floppy standards.

Interline-transfer, frame-interline-transfer devices and XY-addressed sensors look attractive for the ESP function because in a single shot they capture twice as many video lines as are necessary for a TV monitor. But the aforementioned solid-state imagers produce their interlaced fields by summating the video lines two-by-two. If this summation is avoided, the device has the required number of video lines available. By reading out first the charge packets belonging to the odd field - keeping the charge packets from the even field at their locations - and storing or displaying them, and afterwards reading out the remaining charge packets of the even field, the full-frame picture on the CCD is read out in an interlaced mode. A drawback of these architectures in the case of the ESP application is the relatively low resolution, sensitivity, dynamic range, and difference in dark current between the two interlaced fields. These effects can be minimized by the application of a multiple-frame-interline-transfer CCD, or M-FIT.

Increased resolution and sensitivity can be obtained by providing the electronic still camera with two imagers, e.g. two MOS-XY sensors as shown in Figure 9.13 (Aizawa 85). Converting full frames into interlaced fields should be no problem

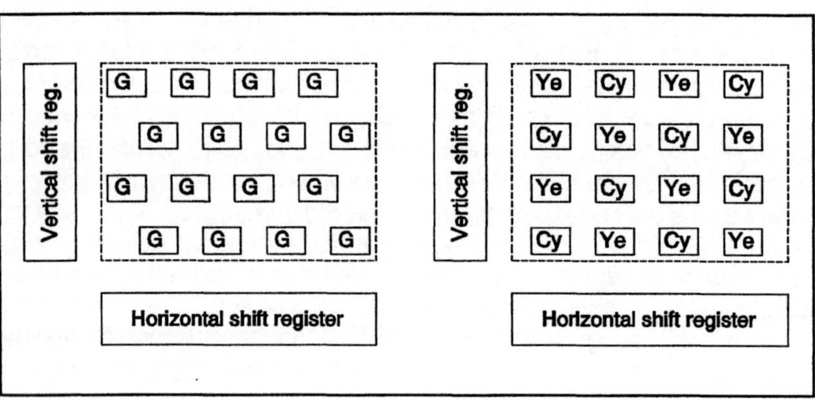

FIGURE 9.13. Two MOS image sensors are used in a still-picture camera to enhance its resolution and sensitivity.

with these sensors, but aligning two imagers in the same camera does not make the product cheaper and will not make it attractive to the consumer market. Note the different functions of the two imagers in Figure 9.13 : one sensor supplies the "green" information, the other the "red" and "blue". The green sensor can, of course, have an off-chip filter, but the color information has to come from an imager with a mosaic or stripe filter on-chip.
Note also the staggered organization of the pixels on the "green" MOS imager; this is to increase its horizontal resolution, one of the key characteristics required for ESP (Hanma 83).

The higher resolution, sensitivity and dynamic range make the frame-transfer imager more attractive for ESP, but here the problem of two interlaced fields derived from a single full frame, single shot, remains. Two frame-transfer devices applicable to electronic still pictures and tackling the said problem were announced in the late 80's :

- The more compact structure (Bosiers 88) makes use of a full-frame image section (484 video lines for the US standard), but the number of storage lines is only half the number of image lines. By providing the frame-transfer device with an extra horizontal readout register in between the image and storage regions, the interlaced readout mode can be generated as illustrated in Figure 9.14. The situation at the end of integration is shown in Figure 9.14a : all video lines, odd and even, are available in the image section. After integration, the even lines (field 1) are read out via the intermediate output register, while the odd lines are shifted down into the storage section (see Figures 9.14b and 9.14c). The odd lines are then read out through the lower output register (see Figure 9.14d). After the separation of the two interlaced fields, storing or displaying them in the interlaced mode should no longer be a problem.

An interesting additional feature of the device described above is its optional use in movie mode as well. In movie mode, the full frame in the image section is compacted to an odd field by adding the image lines two-by-two : 1+2, 3+4, 5+6, ... or to an even field by adding 2+3, 4+5, 6+7, ... In the case of movie mode, the second, intermediate horizontal readout register is not used. Thanks to the construction of the intermediate horizontal register, the device can be made very compact. However, the design of this CCD register is anything but simple : it has to make complete charge transport possible in two directions : "vertically" through the register from the image section to the storage section at a low speed and "horizontally" from right to left at a high speed during charge readout. These requirements result in quite a complicated layout of the intermediate CCD register if it has to be constructed entirely according to the same technology as the standard devices (Bosiers 91).

- The second solution for frame-transfer devices in the ESP mode does not use an intermediate register, so it has the disadvantage of not being very compact (Hynecek 89). The device concept is shown in Figure 9.15. It has a full-frame image section and its storage area can contain the entire information of both fields. It operates as follows : a line of signal charge is loaded into a multiplexer located between the image and storage section. The horizontally arranged pixels are divided into groups of three and rearranged vertically. They are then loaded into the memory in groups of three as a vertical stack. When the second signal line is processed by the multiplexer, it is possible to direct the resulting stack of pixels into memory columns located in between the memory columns of the first signal line. The interdigital structure of the two-field memory makes it possible to transfer the charge from either memory into the horizontal output register. This makes it possible for the device to read out interlaced fields.

In the same way as described above, the device shown in Figure 9.15 can be driven in a movie mode : adding the video lines two-by-two.

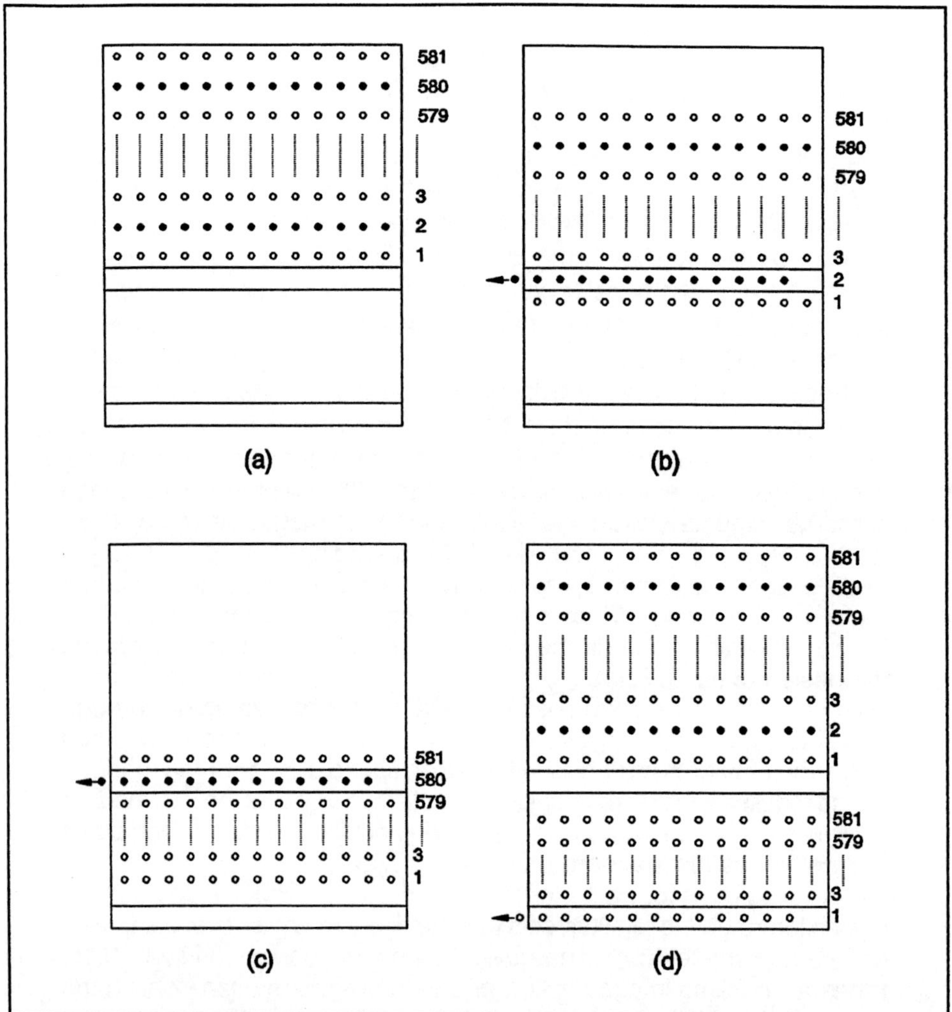

FIGURE 9.14. *The device architecture of a frame-transfer imager fully dedicated to the ESP application, and which still has the facility to operate in accordance with the video standard.*

As already mentioned earlier, one of the typical features of interline- and frame-interline-transfer devices is the presence of the two-field photodiodes. The two fields are captured simultaneously with a single shot exposure. The consecutive readout can be effected field by field and, when stored in an external memory, the frame built around the two fields can be displayed interlaced.

The literature records a further adaptation or development of the interline-transfer device for a typical still-picture application and of the frame-interline-transfer CCD

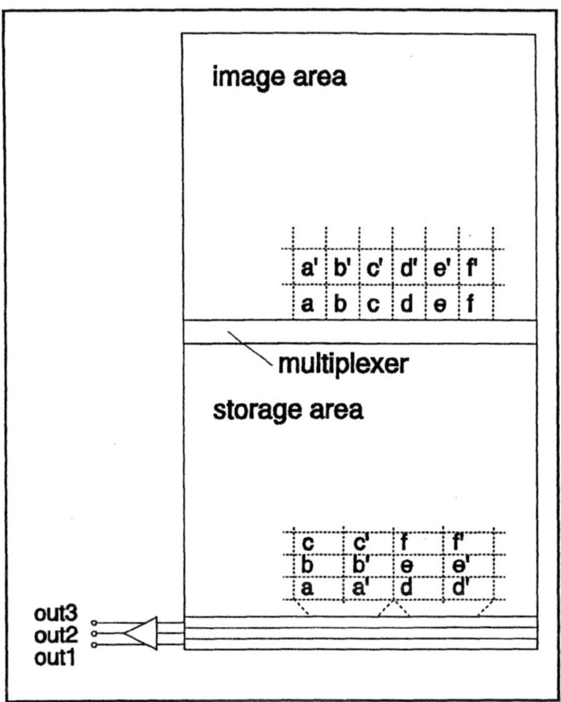

FIGURE 9.15. The device architecture of a frame-transfer imager fully dedicated to the ESP application, and which still has the facility to operate in accordance the video standard.

for progressive-scan capturing and interlaced readout of the image in general :
- A new architecture for the frame-interline-transfer device is introduced with a doubled storage area. Instead of storing only a single field in the (original) FIT device, the new one can store two fields or a complete frame in its memory part. The structure of this so-called Multiple-Frame-Interline Transfer CCD or M-FIT (Itakura 93, Toyoda 94) is shown in Figure 9.16 and resembles very much the architecture of the FT device depicted in Figure 9.15. In the M-FIT the two fields, integrated in the (photodiodes located on) odd and even rows, after shifted one after another in the storage area and are stored in odd and even columns. This action is indicated in Figure 9.16 by the pixels a, b, c, ... and a', b', c', ...
The readout of the device is done in an interlaced mode : during the first field the odd columns are emptied via the horizontal output register, in the second field the even columns are transported to the output node. It should be clear that the reshuffling of the charge packets from the image section into the storage section and the separate transfer of odd and even columns from the memory zone requires a complicated clocking

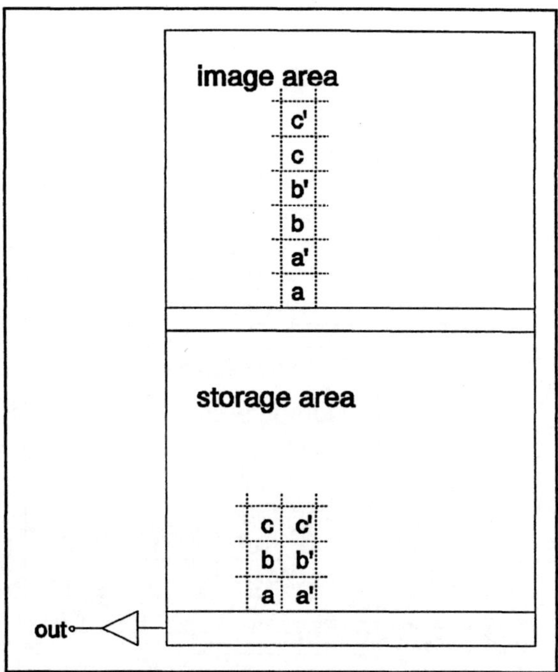

FIGURE 9.16. The device architecture of a multiple-frame-interline-transfer imager fully dedicated to progressive-scan capture and interlaced readout of the image.

scheme. The image section and the storage section of the M-FIT, respectively, have 8 and 6 clocking lines.

- With a new design concept for the IL both the resolution and the sensitivity of the conventional video-format imager are increased (Tabei 91). This new CCD is similar in concept to the interline-transfer CCD, but without the light shield on the vertical charge-transfer lines. Additionally, the color-filter arrangement has also been modified. Blue and green are sensed by the classical interline photodiodes and red by the vertical charge-transfer lines. With this architecture, shown in Figure 9.17, both the interline photodiodes and the vertical charge-coupled shift registers are utilized to capture an image signal. As can be seen from the schematic drawing of the device, the red filter is arranged in a series of vertical stripes on the vertical charge-transfer lines and the green/blue color filters are alternated vertically on the interline photodiodes. This setup is made possible by the fact that the red photons are scarcely at all absorbed in the polysilicon gates on top of the vertical shift registers. On the other hand, the area of a CCD cell of the vertical shift register is about twice the area of a single photodiode. The photons which are absorbed most in these gates are detected in the photodiodes. As known from the basic working principles of IL devices, these areas have good short-wavelength quantum efficiency.

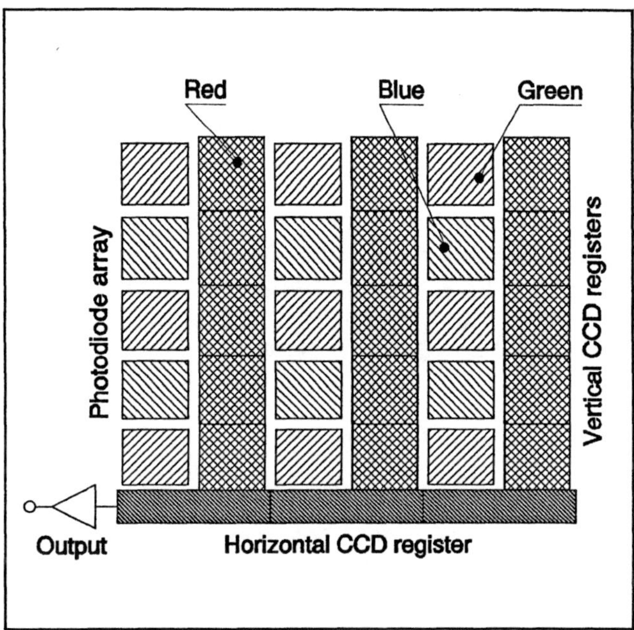

FIGURE 9.17. Schematic drawing of the interline-transfer CCD with adapted vertical shift registers and color filters to be suited for ESP applications.

The sequence of operation of this device can be described as follows :
- the whole sequence starts with a clear mode : all electrons in the photodiodes are dumped into the vertical shift registers and shifted out;
- after the clear mode, integration starts with a shutter trigger. The shutter can be a mechanical one or a strobe light;
- readout mode : the signals of the red field are read out first because they occupy the vertical CCD transfer lines; they are followed by the green field and the blue field. All signals are stored in external frame memories and are reshuffled to reconstruct the video signal on the display.

The advantages of this new device concept are straightforward so far as sensitivity and resolution are concerned. Nevertheless, doubts arise about color separation between the vertical registers and the photodiodes (classical IL devices have optimized color separation from the photodiodes toward the shift registers, but not the other way around), and antiblooming (standard IL imagers have only anti-blooming means in the photodiodes). Another area of potential trouble is the time the charge packets remain in the device. To avoid smearlike effects, the electronic still-picture camera has to be provided with a mechanical shutter. But dark-current nonuniformities can also show up due to the different locations and different storage times of the charge packets.

In all still video or electronic still-picture applications, a single-shot full frame is taken after clearing the sensor of smear, dark current or other spurious signals. After the integration, however, the sensor has to be shielded from further incoming light to prevent smear. This can be done quite simply by introducing a mechanical shutter which covers the sensor during its readout action. For standard operation of the devices, the complete readout cycle takes two fields of, respectively, 40 msec and 33 msec, depending on the TV standard used. These are quite long times and the amount of smear generated during them can be considerable. For this reason, a mechanical shutter is almost indispensable for solid-state imagers in the ESP mode.

WORTH MEMORIZING

Electronic Still-Picture applications are concerned with capturing a progressively scanned image with the image sensor and displaying it in an interlaced pattern. Special architectures for frame-transfer CCDs are designed to fulfill the requirement of the ESP option. Although interline-transfer type sensors always have the capability to capture images progressively, they have to store the second field in the photodiodes during the readout of the first field. Nevertheless, a shutter in the camera is needed for ESP and the electronic still-picture application involves nonuniform dark-current effects owing to the difference of storage time between the two types of fields.

9.4. Time-Delay and Integrating CCD

Time-delay and integrating imagers (TDI) are in fact two-dimensional arrays used basically for imaging applications in one dimension. Their main application is in very high-resolution scanning (Schlig 86) or in low-light-level imaging (Farrier 80) because they combine two characteristics : a high pixel number in one dimension and a high light sensitivity due to its two-dimensional construction.

The way they operate can be described as follows : their organization looks very similar to that of a full-frame or frame-transfer device with a very high number of pixels along a horizontal line and a relatively low number of horizontal lines. Their application has to do with observations or image pickup in which the scanning is done by the imager in one dimension (along the high-resolution horizontal lines of the imager) and the scanning in the second dimension is done mechanically. However, the CCD behaves continuously in a vertical-transport mode such that the vertical transport of charge packets in the short vertical columns is fully synchronized with the mechanical scanning speed applied to the second dimension. The situation is shown schematically in Figure 9.18.

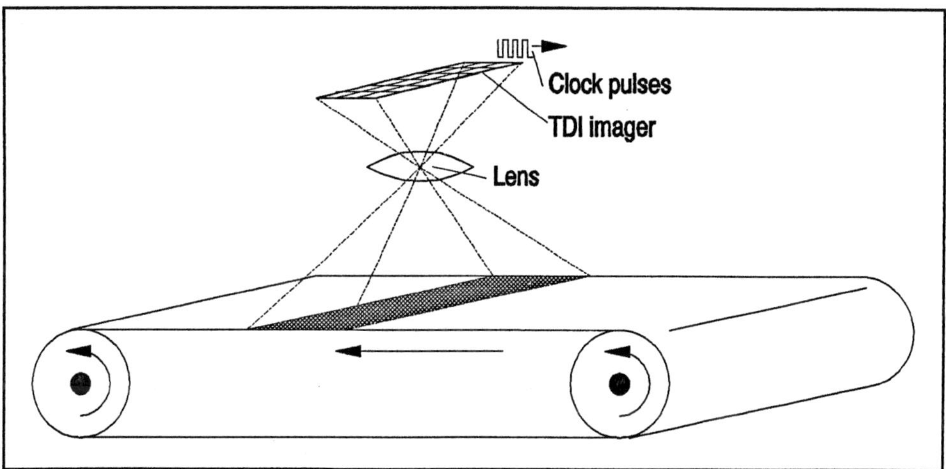

FIGURE 9.18. *Working principle of a TDI : the scanning is done electronically in one direction (perpendicular to the drawing), and mechanically in the other.*

Although the charge packets move continuously in the CCD, each is focused on the same area of the object to be imaged. Because the (moving) charge packet has a much longer integration time compared to its full linear equivalent, the sensitivity of the TDI device can be quite large. The sensitivity of the TDI imager with n lines is therefore n times that of the same imager with only a single stage and using the same line-scan time.

After the movement of the charge packets along the vertical CCD channels, they are transferred line by line to the horizontal output register and shifted to the output.

The main advantages of TDI image sensors include the following :
- enhanced photosensitivity without sacrificing the output data rate. This translates into a high scan speed and/or the ability to image at low light levels;
- increased signal-to-noise ratio . This allows very low-contrast imaging and provides greater detail in the low-light areas of an image;
- improved sensitivity uniformity of the pixels as a result of averaging the sensitivity of many pixels in the TDI column;

> - improved dark-current uniformity of the pixels as a result of averaging the dark current of many pixels in the TDI column.

Concerning TDI imagers, an interesting subject is the vertical resolution of the devices. Since the TDI CCD is not a staring imager, the relative motion between the imaging aperture and the image of the focal plane results in an increase in the effective aperture. Additional to this effect is the continuously changing aperture of the pixel itself due to and depending on the clocking mechanism (Wong 92). Besides this complicated interaction, several important sources of degradation in the modulation transfer function that apply specifically to the TDI mode of operation are :
> - mismatch between the velocity of the scanned image and that of the signal charge being clocked in the TDI direction;
> - discrete charge motion in the TDI direction;
> - angular misalignment of the columns of the imaging device in relation to the direction of mechanical scan.

WORTH MEMORIZING

Time-delay and integrating image sensors are based on the linear array concept but, instead of a single line, they have several lines in parallel to each other. Their construction resembles very much the full-frame or the frame-transfer architecture. The application of a TDI is characterized by very high resolution in combination with high sensitivity. Scanning with a TDI, however, is complicated by the mechanical scanning in one direction combined with fully synchronized clocking of the imager. In the opposite direction the scanning is fulfilled fully electronically.

9.5. Conclusions

Several performance optimizations in the solid-state imaging world can be achieved by adapting or even completely redesigning the architecture of the devices. Several examples of both minor changes in the device concept and completely new inventions are reported in the literature. A few of them have been described in this chapter.

First, there is the matter of the limited horizontal resolution due to the layout of the horizontal output register. Increasing the horizontal pixel density further has resulted in the creation of a multiplexing technique applied to the horizontal output register of two-dimensional image sensors. This technique was applied in earlier devices to overcome the problem of resolution limitation imposed by the horizontal register layout. However, it is used nowadays to restrict the speed of the horizontal registers, resulting in lower on-chip power dissipation, simpler driving electronics and a lower bandwidth requirement for the output amplifier. Besides the above-

mentioned advantages, however, the multiplexing technique has to cope with transport problems. These can be overcome by :
- optimized design of the transfer gates;
- gate tapering;
- extra channel implantations;
- compound channel construction;
- a combined method.
All methods have the same objective : to increase the fringing fields in the lateral direction in the horizontal registers in order to speed up the transport of charge packets when they have to travel across the horizontal registers.

Examples of an adapted clocking scheme or optimized device architecture aimed at enhancing the performance of vertical CCD shift registers have been studied namely :
- accordion clocking to increase the vertical resolution of a two-dimensional image sensor. The device has its complete driving circuitry built on-chip in dynamic CMOS logic;
- the charge-sweep device to optimize the aperture ratio of the photodiodes at the expense of the width of the vertical shift registers. Nevertheless, the imager adapts its clocking scheme to the "shortcomings" of vertical shift registers;
- metal-strapped polysilicon CCD gates to lower the RC constant of the CCD gates and speed up the driving capability of the vertical CCD registers in (frame-) interline-transfer and frame-transfer devices. This is a very powerful technique applied in large-area devices which have to operate at high frame rates (e.g. HDTV imagers).

Electronic Still-Picture applications have to do with capturing a progressively scanned image with the image sensor and displaying it in an interlaced pattern. Special architectures for frame-transfer CCDs are designed to meet the requirements of the ESP option. Although interline-transfer type sensors always have the capability to capture the images progressively, they also have to store the second field in the photodiodes during the read out. Nevertheless, a shutter in the camera is needed for ESP and the electronic still-picture application has to cope with nonuniform dark-current effects due to the storage-time differences between the two kinds of fields.

Time-delay and integrating image sensors have an architecture of a kind in between linear arrays and full-frame or frame-transfer concept. In one direction they have a relatively high number of pixels to enable the capture of images with a very high resolution. In the other, only a few lines are located on the chip. The strength of the TDI concept lies in clocking the device in the latter direction, fully synchronized with the mechanical scanning mechanism used to scan the object in this way. Thus, the application of a TDI is characterized by very-high resolution in combination with high sensitivity. But, as already mentioned, scanning with a TDI is complicated by its mechanical scanning in one direction and fully synchronized clocking of the imager. In the opposite direction scanning is accomplished fully electronically.

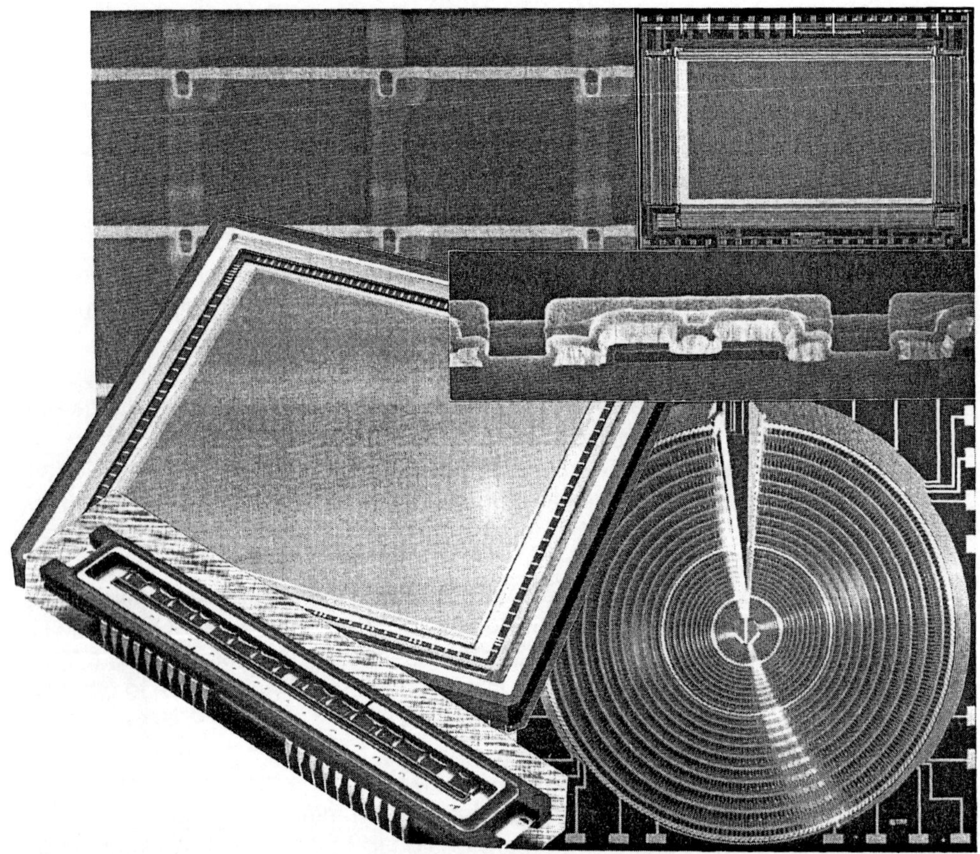

Top left : microphotograph of the pixels from an IR imager based on the CSD concept. The pixels of 26 µm x 20 µm are characterized by a very large aperture ratio and a very small CCD shift register (courtesy of Mitsubishi). Top right : the device shown is basically an MOS-XY addressable imager, but the pixels have an amplification function included. The imager has an aspect ratio of 16:9 (courtesy of Olympus Optical Co.). Middle left : a full-frame CCD having 25M pixels of 12 µm x 12 µm (courtesy of Dalsa). Middle right : cross section of an interline-transfer CCD. At the left and the right side, two photodiodes can be recognized, and between these two structures, a single layer of poly-silicon forms the gate of the vertical shift register. The poly-silicon gate is contacted or strapped by the metal light shield to decrease its resistivity (courtesy of NEC). Bottom left : a linear CCD for color applications having three parallel pixel rows (R, G, B) each containing 8000 elements (courtesy of Kodak). Bottom right : a CCD image sensor imitated the human eye or the "Foveated Retina Sensor". A small interline-transfer CCD in the center is surrounded by several linear image sensors built as concentric arrays (courtesy of IMEC).

Chapter 10

NONCONSUMER IMAGING

In this chapter some features of the solid-state imagers will be discussed which are not directly related to consumer-type applications of these devices. However, as stated earlier, some of the techniques originally invented in relation to non-consumer image sensors are transferred also to the mass-produced imagers and are nowadays commonly used. For some device options, it is rather arbitrary to catalogue them in the consumer or nonconsumer range of the business.

This nonconsumer range can be divided in scientific imagers, smart imagers, imagers for nonvisible light detection, high-speed devices and contact-type image sensors. Scientific imagers are mainly characterized by their extremely large number of pixels, combined in most applications with a very long integration time. To overcome the low-yield problem of these types of devices, buttable structures are proposed which could form a focal plane from several individual image sensors. Developments to enhance the performance of large-area devices are the notch CCD (to improve the transport of small charge packets) and the skipper CCD (to lower the readout noise). Dark currents can be greatly reduced by driving the devices in a pinned-phase mode. This method of operation resembles virtual-phase technology.

Smart imagers combine part of the electronics (driving circuitry or analog video pre-processing) on-chip in the same technology as the imager itself. In this section only a few typical examples will be studied : ASIC vision, a three-dimensional integrated imager, and a foveated-retina sensor.

Solid-state imagers for nonvisible light detection can be classified according to the wavelength of the incident light. In this chapter attention will be paid to infrared imagers, UV detectors, and X-ray sensitive devices. For the latter two imager types back-side illumination or thinning of the silicon substrate is quite often applied.

High-speed devices make use of special architectures or "high-speed material" by substituting the silicon by gallium-arsenide substrates.

Contact-type imagers are very useful in facsimile-type applications where a lens has to be omitted in order to make the overall construction of the imaging system as compact as possible. Because of their setup and architecture, contact-type imagers are constructed using amorphous silicon structures.

10.1. Scientific imagers

Some of the features described in this section have been typically developed for scientific applications, in which high performance of the devices outweighs the price of the silicon. On the other hand, some of the device characteristics described here are still on course to be transferred to the consumer area, e.g. the all-phase pinned CCD. The boundary between consumer and scientific applications is in fact becoming much less clear-cut.

10.1.1. LARGE-AREA DEVICES

Imagers which really live up to their name as scientific devices in all respects are the large-area devices. The demand for these big chips is being primarily driven by applications involving astronomical observation using telescopes incorporating a solid-state imager. At the end of the 70's, the first large-area devices had 500 x 500 pixels; later at the beginning of the 80's, 800 x 800 pixels (Blouke 83) and 1000 x 1000 (McGrath 83, Beal 87); and later, 2k x 2k (Blouke 85) were announced; while, at the end of the 80's, 4k x 4k (Janesick 90) appeared on the scene. In the early 90's, even 5k x 5k image sensors were reported (Chamberlain 93). Note that the early large-area devices of about 1k x 1k were smaller than the sensors used nowadays in HDTV applications with 2 million pixels. As time goes on, the definition of "large area" is being progressively extended.

In addition to their high pixel numbers, large-area scientific CCDs are generally characterized by square pixels (equal vertical and horizontal pitch), full-frame organization and low field rates. Square pixels are attractive in cases where the video information has to be post-processed by a computer. For instance, scaling can be done by the same factor in both dimensions and rotation becomes quite simple with square pixels whereas it is rather difficult with rectangular pixels. If high resolution is the key parameter of the imager, all the silicon available is devoted to the imaging section and no space is left for storage. For this reason, scientific imagers are almost all of the full-frame type. Low field rates are associated with the enormous number of pixels which have to be shifted out of the CCD. If the pixel output rate is high, the field rate will still be low. On the other hand, this low field rate is also application-specific. In the world of astronomy, small object changes over long periods of times have to be observed. Consequently, the integration time for these imagers is long and the number of fields will be low. To cope with long integration times in combination with reasonable readout time, large-area devices have quite often to be cooled.

Not surprisingly, fabrication of these big devices always involves a "battle" with CCD technology. Chips which far exceed the "consumer standard" as far as number of pixels and silicon area are concerned, are extremely difficult and costly to produce.

The large areas of gate oxide and interpoly oxide make these devices extremely sensitive to the quality of the isolating SiO_2. For that reason, the design of these large areas is to a great extent focused on this problem, the aim being to keep the amount of isolating SiO_2 as low as possible within the fixed format of the imager or to make the device less sensitive to oxide leakages. The first aim can be achieved with a virtual-phase imager : only half of the gates are made with polysilicon, the other half being formed by implantations in the substrate (McGrath 83). The second means to make the device less sensitive to defects consists in constructing the CCD imagers as three-phase devices in three different polysilicon layers : one for each phase. Leakages or even shorts between gates made out of the same polysilicon layer have only a minor effect or none because they belong to the same CCD phase. Nowadays, all scientific devices make use of this latter type of technology.

10.1.2. BUTTABLE DEVICES

Buttable devices are intended to tackle the main problem of large-area imaging devices : the low production yield in connection with the area of silicon needed. The principle of buttable imaging devices is fairly simple, namely to try to build a large imager by combining a number of smaller ones. This joining of several smaller devices together to construct a bigger one, also called butting, can be done in either one or two dimensions, with linear (Aoki 85) or area imagers. Figure 10.1 gives an example of how a two-dimensional construction of four buttable area sensors

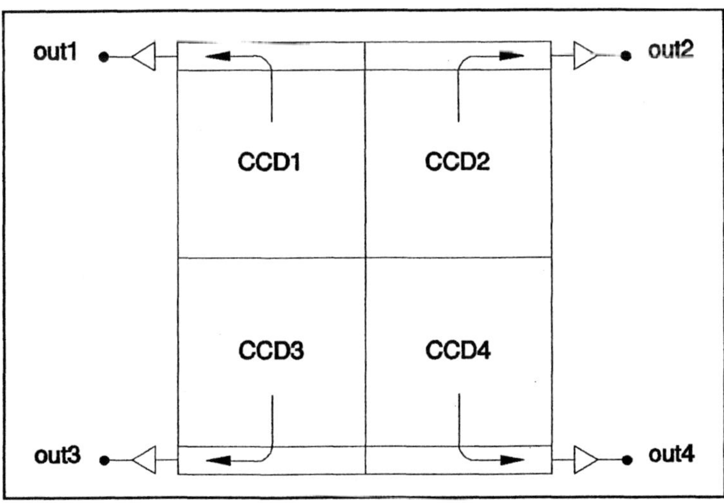

FIGURE 10.1. Organization of a two-dimensional construction of four buttable area sensors in a two-by-two matrix.

in a two-by-two matrix can be organized. Note that the four basic imagers can all be of the same type of device, mirrored around a vertical and a horizontal axis, so as to construct the two-by-two matrix. Each device has its own driving circuitry and its own output stage. Special attention has, however, to be paid to :

    - the layout of the basic imager to make the device suitable for butting : no interconnection can be made along the butting sides of the device. There is no possibility on the chip itself to drive the gates from the butted edge and no bondpads can be located along these chip sides;

    - the technology used in the devices to suppress dark-current generation near the saw edges of the dies. Stress in the material and crystal defects at these locations can give rise to local dark-current nonuniformities which have to be suppressed;

    - die bonding of the matrix because there will always be at least one pixel missing between two imagers of the butted array. To keep the loss of information between two imagers as small as possible, accurate positioning of the devices in their package is needed. In most practical applications of buttable devices the mismatch between pixels which are adjacent but belong to two different chips can be kept very small;

    - the video information processing : the missing pixel(s) on the butting borders has (have) to be replaced by means of external video-processing units. Moreover, if the devices of the two-by-two matrix are all the same and rotated through 90° in their butted configuration, two of them will read out the information along horizontal lines and two along vertical columns. A complete reshuffling of the video information is needed to reconstruct the captured image.

Although it was stated at the beginning of this section that buttable devices can tackle the issue of the amount of silicon for large-area devices, it will be questionable which solution is best for a large-area imaging problem : a single chip fabricated with a low yield, or a matrix of several separate devices with their individual butting problems ?

10.1.3. NOTCH CCD

A natural consequence of the large pixel number in large-area scientific CCDs is the huge amount of charge transports in these devices. In the worst case for a 2k x 2k, the number of transports from one gate to another can amount to as many as several thousands. For instance, in the aforementioned device with 4M pixels, single output and three-phase CCD shift registers, the worst case situation means 2k x 3 transports in the vertical CCDs of the image section and 2k x 3 transports in the horizontal readout register. The net result is 12k transports. With a transport inefficiency of 10^{-5} an amount of charge equal to 12 % is lost. To lower this quantity of transfer loss, the technology has to be further optimized, but there are limits of course.

Another solution may be the notch CCD, which is a device dedicated to handling small charge packets with high transfer efficiency (Bredthauer 91). Bearing in mind that, in cases of incomplete charge transfer, the last charges to move are always small charge packets, the notch CCD also transfers larger charge packets with great efficiency. The basic configuration of a notch CCD is shown in Figure 10.2. The illustration sketches schematically part of the connection between the vertical CCDs and the horizontal output register. In the design of the image sensor, all CCD channels have a narrow (typically 3 μm) strip in which the buried-channel

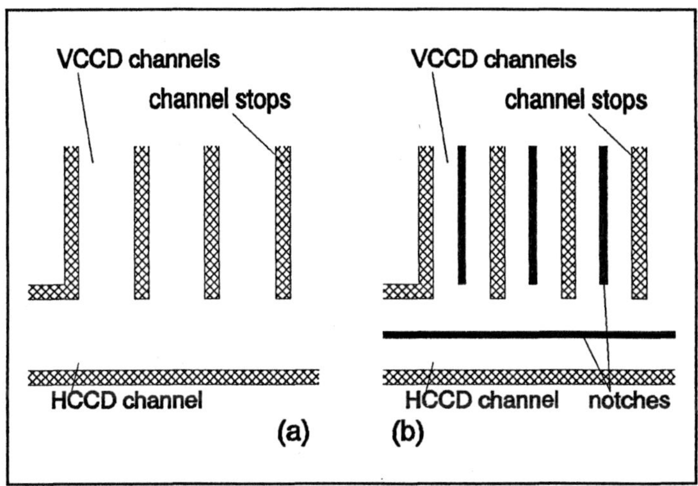

FIGURE 10.2. Basic configuration of the original (a) and the notch CCD (b) : all CCD-channels are provided with a smaller, deeper channel by means of an extra implantation.

implant concentration was increased, resulting in a channel within a channel which has a potential a few volts "deeper" than the normal CCD channel. Small signal quantities would presumably be confined to this trench and would thus encounter fewer traps than if the charges were to spread out across the entire channel width. A factor of 10 increase in transport efficiency is possible with this simple notch construction (for a 40 μm wide channel with a 3 μm notch). Especially for wider channels the incorporation of a notch can be effective; the gain in transport efficiency is less for smaller channels. A first guess for the gain can be the ratio between the width of the original channel and the width of the notch itself.

10.1.4. SKIPPER CCD

With today's technology and new device architectures, charge packets of only a few tens of electrons can successfully transported even in very large devices. The

CCD might, however, be unable to read the charge accurately because of the relatively high noise floor inherent in the sensor's on-chip amplifier.

The skipper CCD was invented to circumvent the 1/f-noise problem and effect a square-root reduction in noise with increasing sample time, thereby allowing even sub-electron noise floors to be achieved. The principal function of the skipper architecture is to permit nondestructive measurement of the charge contained in a pixel and this operation multiplies several times using a floating-gate amplifier. More details are shown in Figure 10.3 where the basic principle is sketched (Janesick 90). The charge packets are transferred underneath the floating gate of the CCD register (movement 1) by means of the horizontal clocks and gates V_{G1} and V_{G2}. The floating gate is connected to a MOSFET source-follower amplifier and to a MOSFET reset switch used to preset the gate to a reference voltage before the signal charge is dumped. During the sensing operation the horizontal clocks are inhibited for the duration of the skipper cycle. During the actual sensing the voltage at the source node of the amplifier is sampled off-chip and results in a first sample of the pixel. Then the charge packet is fed back to where it came from (movement 2) by a reverse transport via gates V_{G1} and V_{G2}. In a following sequence the floating-gate structure is reset and the same packet as before can be restored underneath the floating gate and sensed a second time (repetition of movement 1). The operation whereby samples are transported back and resampled is repeated several times. After a well-defined number of samples, the charge packet is transferred further on into the CCD register by an appropriate operation of gates V_{G3} and V_{G4} (movement 3). In its final stage in the horizontal register, the charge packet can be sampled again by a classical floating-diffusion amplifier and discarded through a second reset transistor.

The various samples collected for a given pixel are averaged off-chip, reducing the random noise of the on-chip amplifier by the square root of the number of samples taken. For example, if a pixel is sampled 25 times, the random noise associated with the on-chip amplifier is diminished by a factor of 5. An example of the skipper action on the random-noise level is illustrated in Figure 10.4 where the photon transfer curve of a real imager is measured (Chandler 90). The photon transfer curve is equal to the noise level expressed in the number of electrons if given as a function of the charge content. Note that for high values of the charge content the noise floor is dominated by the shot noise on the photon-generated signal. In these situations the noise equals the square root of the signal itself. But, as the signal values become lower, the on-chip amplifier noise becomes dominant. If the skipper is not operative (number of samples n = 1) the number of noise electrons is limited to 4.8 electrons rms. With the skipper active and the number of samples per pixel n = 16 chosen, the noise equivalent signal drops as low as 1.2 electrons rms. This accords fully with the theory which predicts a gain of a factor of four in noise performance.

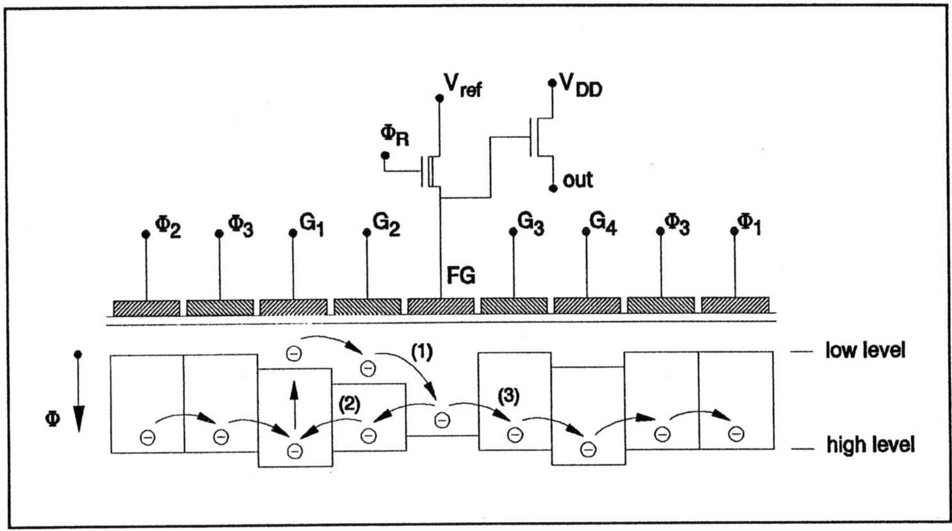

FIGURE 10.3. Schematic diagram of the skipper CCD built around a multiple nondestructive readout by means of a floating-gate amplifier.

Observe that the enhanced noise performance has to paid for with a slower pixel rate and more complicated processing of the video signal. The pixel rate is lowered by a factor equal to the number of skipper readouts, while the more complicated signal processing is due to the averaging action needed to lower the noise.

10.1.5. PINNED-PHASE CCDS

For scientific applications, extremely long integration times are no exception. Instead of milliseconds, the devices are held in their integration mode for seconds, also their readout cycle takes that much time. The parameter causing trouble in these situations will be the dark current. Even if the dark current is acceptably low for consumer applications, enlarging the integration time by a factor of 1000 can cause a full well just by dark-current generation alone. One solution to keep the generation of dark current as low as possible may be to cool the device. As described earlier in section 3.3, decreasing the device temperature by 8°C causes the dark current to decrease by a factor of 2.

An alternative or additional method to decrease the dark current of the sensors is to keep the interface $Si-SiO_2$ inverted. Until normal operating conditions for a CCD, the interface is kept free of carriers and the carrier-generation rate of the interface states has its maximum value. However, the generation rate decreases by orders of magnitude if a high density of free carriers exists at the interface as in accumulation or inversion. Consequently, the technique employed is to bias the interface of the buried-channel imager into inversion, thereby suppressing dark-

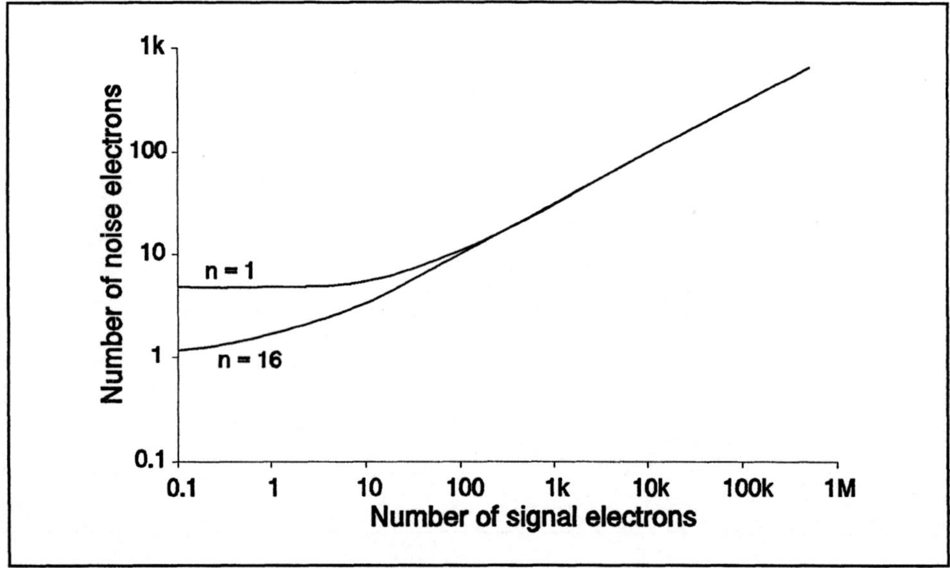

FIGURE 10.4. Photon transfer curve of a CCD provided with a skipper-output configuration, with a single readout (n=1) and a multiple readout (n=16).

current generation by the interface states (Saks 80).

Figure 10.5a illustrates the electrostatic potential in-depth through the center of the n-buried channel structure. The cross section of the structure under study is illustrated in Figure 10.5b. The channel itself is located between two p^+ channel stopper implants. The gate bias is shown as a parameter in Figure 10.5a. For the situation shown, the channel potential has its maximum at a depth of about 0.6 μm in the silicon. Minority carriers generated will be collected at this location. The surface potential at the Si-SiO$_2$ interface is positive for gate voltages higher than - 4 V. If the gate potential drops lower than - 4 V, the interface will be pinned to the p^+ channel-stopper potential, which is equal to 0 V in this example. This pinning results from the inversion layer of holes at the interface. The holes are supplied by the p^+ stoppers. For voltages on the gate, decreasing lower than - 4 V, the density of holes in the inversion layer will increase, as also does the electric field across the oxide, but the width of the space-charge region will remain constant at its minimum value.

Operation of the CCDs with the interface inverted has three advantages :
- shielding of the interface states;
- reduction of the space-charge region, decreasing the generation of dark current inside this smaller volume;

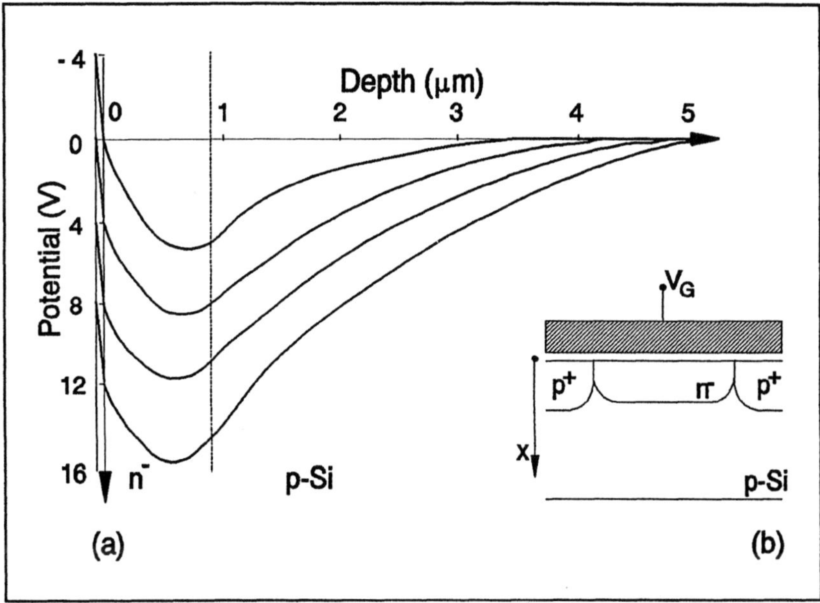

FIGURE 10.5. Illustration of the electrostatic potential in-depth (a) through the center of the pixel, whose cross-section is shown (b), and with a grounded Si substrate.

- reducing the generation of dark current at the boundary between the p^+ channel stop and the active CCD channel because the depleted p^+-n^- transition region no longer exists there.

Pinned-phase CCD technology is based on the aforementioned effect : shielding the interface with an inversion layer to avoid the generation of dark current by the interface states. This technique, originally applied only to scientific CCDs, has now also found its way to the consumer world. In consumer devices it is not always applied in exactly the same way as is described here. In this section three pinned-phase structures will be explained : the open-phase pinned, the multi-phase pinned and the dynamic-pinned CCDs.

<u>Open-phase pinned CCDs</u>
The construction of the open-phase pinned CCD closely resembles that of the virtual-phase CCD : both have one of their phases not covered with gate material but only defined by means of implantations. The main difference between the two structures is their polysilicon part of the gate definitions : only one phase for the virtual-phase device, and two for the open-phase pinned CCD. A cross section of the CCD cell along the direction of transport of the charge packets of the open-phase pinned CCD is shown in Figure 10.6 (Janesick 89). Three gate regions can be discovered : two underneath polysilicon and the third under the open-phase implantation. The shallow p implant serves as the pinning layer to equal the interface

potential to the channel stoppers. The deeper n implant locally increases the channel potential for signal-charge collection. In the open-phase pinned technology, the transport of charge packets is entirely controlled by the external clocking, unlike in the virtual-phase technology where the direction of charge transport is frozen by the local implants under all gates.

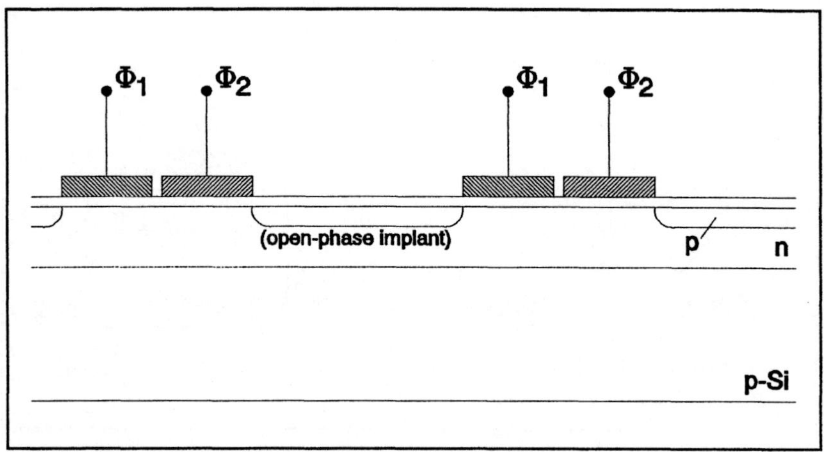

FIGURE 10.6. Cross section through the CCD channel of the open-phase pinned CCD.

In comparison with virtual-phase CCDs, the open-phase pinned devices show high charge-transport efficiency (like all three-phase structures), and have high flexibility of clocking. Additionally, this new sensor architecture is capable of operating totally inverted for low dark-current generation and has a high quantum efficiency, due to the open gate areas. Observe that, in most imaging applications, the gate structure can be optimized for high sensitivity by choosing the ratio of the open-pinned phase area and the total CCD cell area as high as possible.

To illustrate the basic working principle of the open-pinned phase system, the potential plots of an open-pinned phase pixel are reproduced in Figure 10.7 (Janesick 89). This shows the potentials in depth under a clocked phase at a high gate potential (curve 1), a clocked phase at a low gate potential (curve 2) and an open phase (curve 3). Note the pinning of the potential to the 0 V of the substrate bias for the open phase (curve 3) and the clocked phase at a low gate voltage (curve 2).

<u>Multi-phase pinning CCDs</u>
A logical evolution of the open-phase pinned system is the multi-phase pinning technology. The objective is the same as before : to shield the interface states from any interaction with the CCD channel. In multi-phase pinning technology, however, the idea is adapted to a CCD which is fully covered with polysilicon gates,

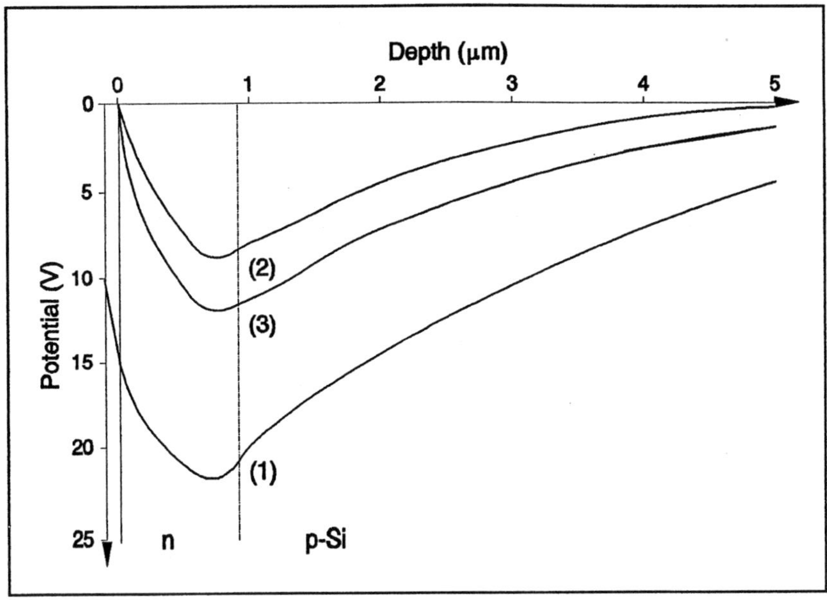

FIGURE 10.7. In-depth potential plots of an open-pinned phase pixel : under the clocked phase at high gate voltage (1) and low gate voltage (2), under the open phase (3).

instead of a combination of polysilicon gates and open phases. The adoption of multi-phase pinning ensures a higher charge-handling capability in CCD shift registers than the open-phase pinned structure does.

A schematic cross section of the structure is shown in Figure 10.8 (Lee 90). In Figure 10.8 a two-phase CCD with two gates per CCD cell is examined. The direction of transport is determined by the extra-shallow p implant. The n-type channel implants are chosen such that, during the integration of charges, the gates can be changed to their negative-voltage state without losing the internal potential wells to collect the charges. The Si-SiO$_2$ interface will be completely pinned to the substrate voltage. This pinning process is, of course, not possible during the transport operation of the CCD, but, if the transport time is kept considerably shorter in comparison to the integration time, the dark current can be kept as low as possible. For example : with conventional clocking, the dark current of the device under consideration is 0.5 nA/cm^2 at room temperature, but can be reduced to 10 pA/cm^2 if a clocking mode according to the multi-phase pinned technology is applied. A measured curve for the dark current as a function of temperature for an imager based on a multi-phase pinned structure is included in Figure 10.9.

Multi-phase pinning is a typical example of a technique which has been introduced into the scientific imaging world to decrease dark current, but which is being adopted to general imaging applications in order to suppress dark current and dark-current

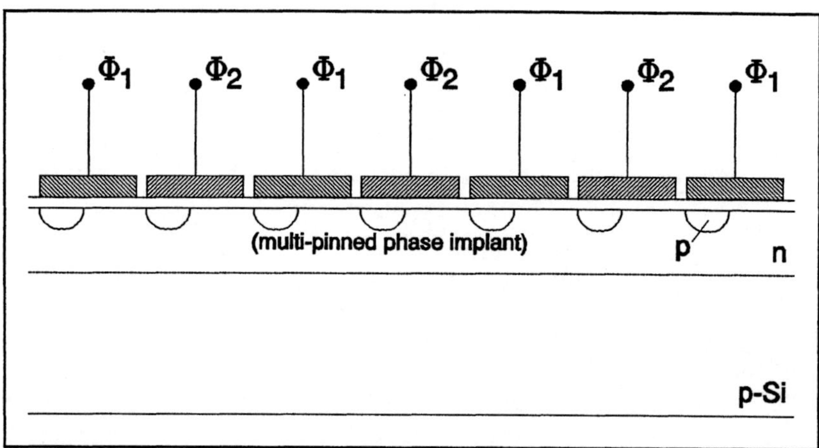

FIGURE 10.8. Cross section through the CCD channel of a multi-phase pinned CCD.

nonuniformities. Although the step from the original application field to the later one might seem simple, in reality it is quite complicated, because a general-purpose imager has to provide antiblooming and charge reset or electronic shuttering facilities. In combination with the doping profiles necessary for multi-phase pinning, this is a hard job to do, but devices have nevertheless been reported (Okada 91). Dark currents around 200 pA/cm² at 60°C for a small (65 mm²) frame-transfer device have been measured.

<u>Dynamic pinning</u>
As already explained in the general description of phase-pinning CCDs, an inverted surface can be applied to decrease the surface-generation current virtually to zero. This technique can be successfully used to suppress the surface-generated dark current in buried-channel CCDs. In n-channel devices the gate potential is lowered to a value that brings the surface potential close to the substrate potential. At this point the holes fill the surface region, inverting the surface and pinning the surface potential to that of the p-type substrate or p-type channel stoppers. Even with the surface pinned to zero a potential well exists in the buried channel to allow electrons to be stored. However, most three- and four-phase CCDs are constructed with the same buried-channel implant and gate insulator thickness under all phases. For these devices the buried-channel potential would be the same under all gates if all were simultaneously biased for surface inversion. Charge packets could, therefore, not be isolated from one another as they are in open-phase pinned and multi-phase pinned technologies.

In dynamic pinning, a method has been developed for complete suppression of the interface-state dark current. It relies on the dynamic properties of interface states and can be applied to buried-channel CCDs without extra implants or other

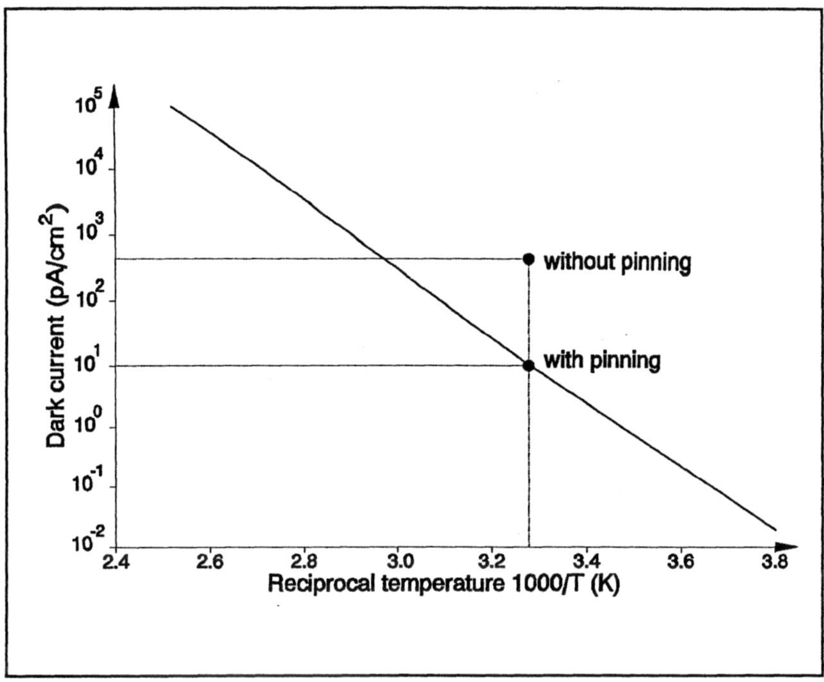

FIGURE 10.9. Dark current versus reciprocal temperature for a multi-phase pinned CCD, as shown in Figure 10.8.

physical modifications of the device (Burke 91). The technique is based on the time-dependency of the generation rate of interface states after a surface has been switched from inversion to depletion. The generation of free carriers at a depleted surface is a two-step process : first, the excitation of electrons from the valence band to levels near the mid-gap (or the emission of holes to the valence band, which is the same); second, the excitation of the electrons to the conduction band. Inversion of an n-channel BCCD forces the mid-gap levels to be emptied of electrons by recombination with holes supplied by the inversion layer underneath the low-biased CCD gates. When the surface is depleted again, these levels must be repopulated before the generation of free electrons resumes. If the surface is inverted again before the mid-gap levels are repopulated, then the generation process is effectively quenched.

This theory of dynamic pinning is applied to a three-phase n-channel BCCD (Burke 91). Clock levels of the imagers were chosen as + 3 V for the high and - 7 V for the low level at which inversion occurs. At the start of the integration period, two gates out of the three (e.g. Φ_1 and Φ_2) are biased low, while the third (e.g. Φ_3) is used to collect the electrons. At regular time points in the integration period, one of the two low-biased gates (e.g. Φ_2) is switched to the high level, while the former

high-biased gate (Φ_3) is switched low.   This sequence has the following consequences :
> - it brings the interface underneath gate Φ_2 into depletion while keeping the interface states filled with holes and the surface-generation current is kept zero "for a while";
> - it brings the interface underneath gate Φ_3 into inversion, keeping the dark current locally to zero as far as surface generation is concerned;
> - the interface underneath gate Φ_1 is not changed and remains inverted.

If this type of clocking is repeated all over again with a well chosen sequence, the charge packet will be stored under one gate and transferred back and forth inside one CCD cell. If the repetition rate is chosen short enough (e.g. a few μsec), all the interface states are kept filled with holes and are inactive during the process of dark-current generation.

The excitation process of holes from the interface states is highly temperature-dependent. At higher temperatures the time constant of hole excitation becomes shorter (only a few msec at room temperature), while at lower temperatures the opposite is true (about 3 hours at - 80°C). The repetition rate of the dynamic pinning method can, therefore, depend on the operating temperature of the imager.
In the mentioned example of the three-phase CCD with - 7 V and + 3 V clock levels, operating at room temperature, the dark current can be decreased from 230 pA/cm^2 without dynamic pinning to 20 pA/cm^2 with an optimized dynamic pinning clocking cycle (Burke 91).

WORTH MEMORIZING

"Scientific imagers" are classified in different groups :
> **- large-area devices with pixels ranging in number from 4M to 25M, and characterized by their quite low frame rate and very high price;**
> **- buttable devices which are sensors especially designed for use in a matrix configuration to construct a larger imaging array;**
> **- notch CCD with its buried charge-transfer channel adapted for optimized transport efficiency not only for large charge packets but also for small ones;**
> **- skipper CCD with its multiple readout of each charge packet so as to lower the read-out noise;**
> **- pinned-phase CCDs to eliminate the effect of interface states as far as dark-current generation is concerned.**

10.2. Smart image sensors

In general terms, solid-state imagers with part of the driving electronics or video-processing electronics on-chip can already be classified as smart image sensors. In semiconductor technology, new developments based on a "marriage" between CCD and CMOS technology were in fact published a number of years ago. So it should not be surprising to learn that CCD imagers including some electronics on-chip have been reported (Theuwissen 84, Theuwissen 88, Hirama 90). These kinds of devices simply combining CCD and CMOS technology are not dealt with in this study of smart image sensors. This section discusses a few topics concerned with image sensors which are application-, technology- or user-specific. The survey is in no way intended to be exhaustive but only to give an idea of the kind of research or development topics figuring in the scientific literature.
Other topics such as integrated circuitry aimed at processing signals, including video information, but without an imager on-board, are also omitted, though several of these devices use signal processing in the charge domain, e.g. wire transfer of charge packets, comparators, differentiators, and integrators (Fossum 91).

The following image sensors will be described in this chapter : an imager for visible light, built in ASIC technology, a three-dimensional integrated image sensor plus processor, and a foveated-retina sensor.

10.2.1. ASIC VISION

What is meant by "ASIC Vision", which is also the title of a paper (Renshaw 90), the combination of a solid-state imager of the MOS-XY type, with its complete set of electronics on a single chip, is shown schematically in Figure 10.10. The basis of the integrated circuit is a CMOS technology, in which digital and analog CMOS circuits can be built but, for this specific application, a matrix of photodiodes is also included. The sensor array comprises 312 x 287 pixels, together with its entire addressing and sensing circuitry, including two logic processors : one to implement the synchronization with a standard timing format and the other to control the exposure setting by varying the integration time adaptively. All the digital electronic elements have been designed using typical ASIC design tools. The device requires only a fixed aperture lens, a 6 MHz clock, and a single 5 V power supply to implement a complete camera function.

This technique applied to a custom-image sensor with all electronics on-board, makes a very quick turnaround of camera design possible when performance is not the key issue, but very fast response to the needs of the customer is of primary importance.

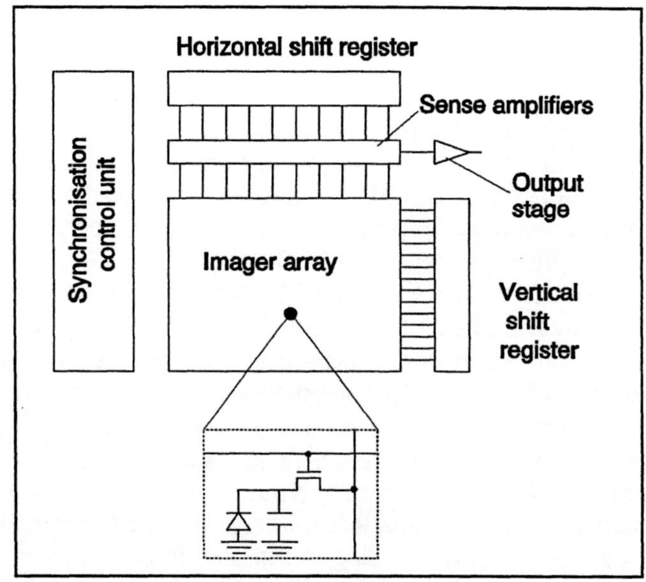

FIGURE 10.10. The basis of ASIC vision : the sensor, its driving electronics, and its video processing, all built in one technology on a single chip.

10.2.2. THREE-DIMENSIONAL INTEGRATED IMAGE SENSOR

In reality the human eye is the most intelligent and complete image sensor ever studied. Its complexity and extremely high quality are made possible by the massively parallel-working elements. This architecture can be duplicated by future image-sensor systems built on silicon. Vertically stacked layers of active solid-state devices are suitable for implementing massively parallel systems on a single die. The logical structures of parallel processing are implemented in the physical structures by using the vertical direction for data flow. High-performance stacked devices will be able to bring about an advanced image sensor.

In the example described in this section, 3D-LSI technology is used to build a single-chip character-recognition system. A possible architecture of such a device is shown in Figure 10.11 (Kioi 88, Kioi 90).

The three-dimensional character of the image sensor can easily be recognized :
    - the top layer contains the photo detectors. In the example shown, a two-dimensional array of photodiodes is fabricated in SOI-technology (Silicon-On-Insulator);

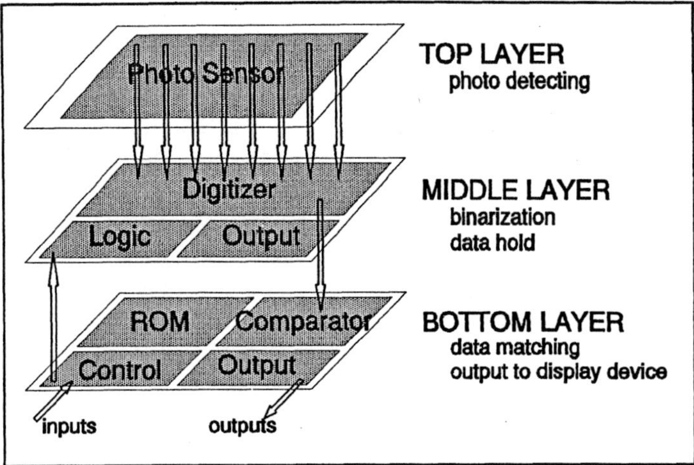

FIGURE 10.11. Architecture of a 3D-image sensor with image processing on-board : top layer with detectors, the middle layer with binarization and the bottom layer with data matching.

- the middle layer contains the binarization unit. Also this circuit is manufactured in the same technology as the top layer : SOI. All photodetectors in the top layer have their own connection to the middle-layer converter. This configuration makes it possible for all video information to be passed from the top layer to the middle layer in only a few microseconds. In the binarization unit the analog information is reshaped into digital pulses;
- the bottom layer consists of the data-matching unit, the control system, and the interface to the outside world. All this circuitry is built into the bulk of the semiconductor and can be made in straightforward CMOS or nMOS technology. The digital data stream generated by the binarization unit of the middle section is passed on in parallel to the comparator, the data are compared simultaneously with match data taken from the ROM memory. If the match data are found, the expectation data are transferred to the output section of the device. Communication between the lower-two layers is bidirectional : control signals are fed from the bottom to the middle section and data are transferred from the middle layer downwards.

A device whose the architecture is shown in Figure 10.11 makes it possible to identify a sensed image as a stored word in about 3 μsec. This time is estimated to be about several thousand times faster than that with a conventional serial data processor. Although the parallel-processing features make these kinds of imager architectures attractive, it will be a few years longer before the return on such devices can be a reasonable one. Strain, thermal stability, and other difficulties in the fabrication process are obstructing the commercial introduction of these three-dimensional devices as image sensors.

10.2.3. FOVEATED-RETINA SENSOR

If rapid detection, localization and tracking of an object are of greater importance than exact reproduction, nonconventional sensing systems can be developed. The architectures of these devices can be totally different from those of classical sensor arrays described so far. The foveated-retina sensor is an example of how a nonconventional sensor system, built in a conventional CCD technology, can be optimized in the direction of scene-analysis performance rather than in that of simply detection (Van der Spiegel 89). In situations where real-time coordination between sensory perception and motor control, for example, is needed, this type of sensor could be of great help.

The foveated-retina sensor is partly based on the biological visual system. Research of the anatomy of the eye has revealed that the photoreceptors are not uniformly distributed over the retina. The cone density shows a peak in the center and decreases towards the periphery. The sampling structure of the retina-like imager is loosely modeled on the early stages of the biological visual system so as to capture the logarithmic mapping as the eye does. Instead of using a uniform square grid, the retina sensor has a highly nonuniform sampling grid. The center fovea has a constant resolution while the peripheral sensors are organized in the manner of a circle whose size increases linearly with eccentricity. A schematic representation of the device is shown in Figure 10.12a. Every randomly chosen point in the retinal plane can be described by its polar coordinates (r,θ) :

$$z = r.e^{j.\theta} \ . \tag{10.1}$$

After mapping of the retinal plane into a cortical representation as shown in Figure 10.12b, the new coordinates become :

$$w = \ln(z) = \ln(r) + j.\theta = u + j.v \ . \tag{10.2}$$

This complex logarithmic transformation has interesting properties for pattern recognition. Its main characteristics are size and rotation invariance. These properties are also characteristic of the human visual system : an object does not change its perceived shape when it is rotated or scaled. Rotation of the image impinging on the sensor will result in a linear translation along the v axis in the cortical plane. Similarly, an enlarged image will be represented in the cortical plane as a translated version of the original image.

The center fovea performs the function of a high-resolution sensor once an object has been tracked by the peripheral parts of the sensor.

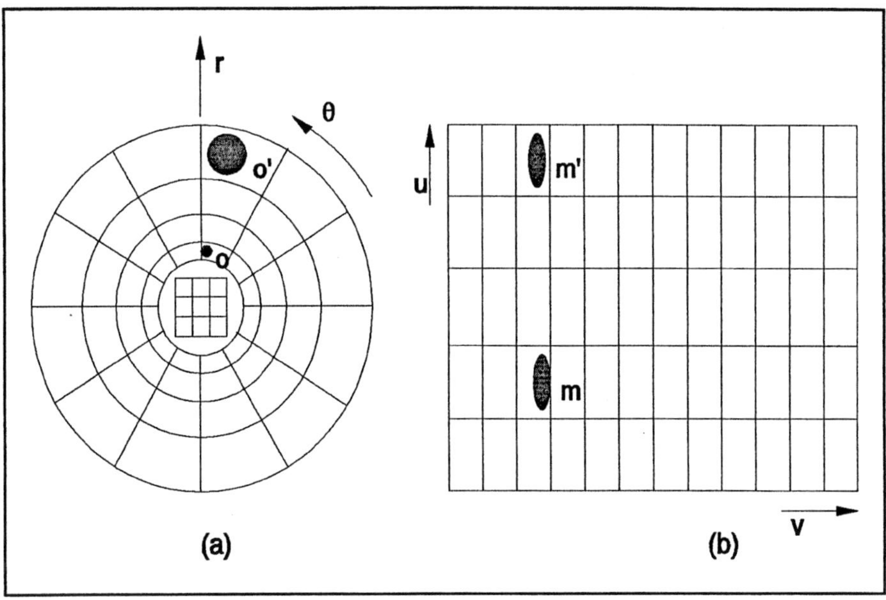

FIGURE 10.12. A schematic representation of the foveated-retina sensor (a) and its mapping of the retinal plane into a cortical representation (b).

This foveated-retina-like sensor is a typical example of how a certain amount of algorithm can be built into the sensor itself. Of course, the device loses its ability to reproduce an image, but it provides the camera with the facility to perform for instance, real-time tracking of moving targets.

WORTH MEMORIZING

Smart image sensors combine an imager with on-chip processing facilities. The latter can be implemented in dedicated electronics, in the incorporation of image-processing techniques, or in a dedicated image sensor design.

10.3. Nonvisible imaging

Nonvisible imaging includes detection of energetic particles which have wavelengths outside the range of 0.4 μm to 0.7 μm. In principle, a classical solid-state imager as described in the previous sections is more or less insensitive outside this wavelength range. As regards solid-state imaging by means of silicon devices, wavelengths can be divided into three ranges :

- wavelengths which are longer than 1.0 μm : the near infrared and the infrared spectrum. For these wavelengths the corresponding absorption coefficient in silicon is too low and almost all photons pass through the structure without being absorbed;

- for wavelengths shorter than 0.4 μm and longer than 0.01 μm the opposite is the case. The absorption is even so high that all photons are absorbed in the top layers above the CCD;

- if the wavelength of the incoming particles is lower than 0.01 μm, the range of X-rays has to be considered. For these low wavelengths, the absorption coefficient is again comparable to that for visible light. In this case, however, the energy of the particles is so high that they can damage the solid-state detector.

The three different wavelength ranges are shown in Figure 10.13 (Bosiers 85), which also includes their penetration depths in silicon (for the definition of penetration depth, see section 5.1). For each of these three groups of wavelengths, typical solutions have been developed which adapt the silicon detectors to make them suitable for the specific application. In this section some of the solutions for nonvisible imaging will be highlighted, namely for infrared imaging, UV-sensitive devices, and X-ray imaging, in that order.

10.3.1. INFRARED IMAGING

Infrared imagers fabricated in the well-established silicon technology are almost exclusively of the Schottky-barrier type. Pure silicon is incapable of absorbing infrared photons. For this reason infrared photons are converted into electrons by means of a Schottky-barrier structure in which the pixels make use of PtSi detectors. An array of these detectors is coupled to a charge-coupled readout system in a linear organization, or a two-dimensional system.

The working principle of the Schottky barrier detector in its simplest form is illustrated in Figure 10.14 (Kosonocky 87). A platinum-silicide layer (PtSi) is deposited on a p-type silicon substrate. The detector is back-side illuminated and is connected to a silicon CCD readout (not shown in Figure 10.14). The infrared radiation with a photon energy lower than the bandgap of silicon (E_g = 1.1 eV) is transmitted through the substrate without any absorption loss. Absorption of the infrared radiation in the silicide layer results in the excitation of a photocurrent across the Schottky-barrier by internal photoemission. The holes are injected into the silicon substrate if they have enough energy to overcome the Schottky-barrier formed between the silicide and the p-type silicon and with a height equal to the metal-semiconductor workfunction Φ_{MS}. Detection of the infrared optical signal is completed by capturing the negative charge from the silicide electrode and transferring into a CCD readout structure.

The spectral energy window of a back-side illuminated Schottky-barrier detector on silicon is :

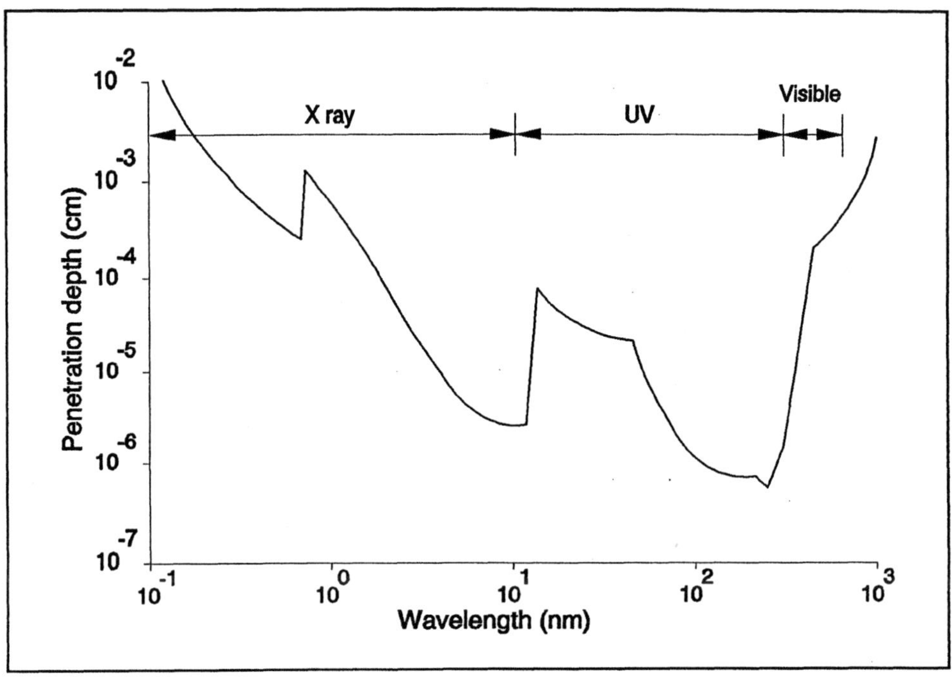

FIGURE 10.13. Penetration depth of silicon for the different ranges of UV, visible and infrared wavelengths.

$$\Phi_{MS} < h.\nu < E_g \, . \qquad\qquad [10.3]$$

The right-hand side of expression [10.3] shows that the photon has to pass through the silicon and its left-hand side that it has to be absorbed in the Schottky-barrier detector.

To ensure maximum responsivity of the detector, the PtSi-Schottky barrier is constructed with an "optical cavity" as illustrated in Figure 10.15. The infrared signal is introduced from the back (bottom side in the illustration in Figure 10.15) of the silicon substrate. A PtSi film about 2 nm thick is formed on the p-type silicon substrate and is separated from an aluminum reflector by an SiO_2 dielectric layer about a quarter-wavelength thick. This construction maximizes the optical absorption by setting up a peak of an optical standing wave at the thin PtSi film. An antireflective coating deposited on the back of the silicon substrate increases the coupling of the infrared radiation into the Schottky-barrier detector by about 30 %.

The present state of the art for high-performance PtSi Schottky-barrier detectors for thermal imaging can be represented by the responsivity shown in Figure 10.16.

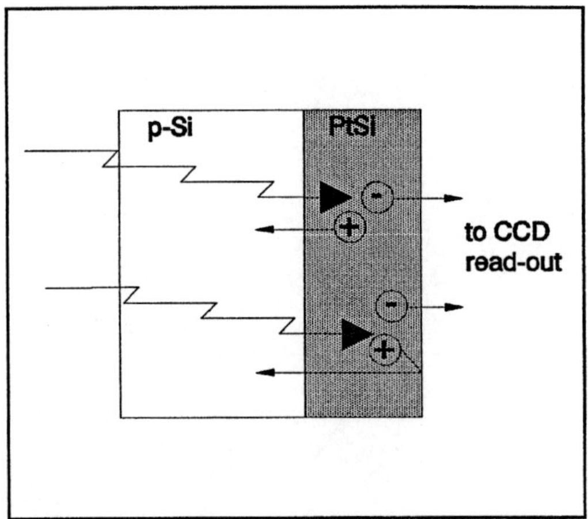

FIGURE 10.14. Cross section of a Schottky-barrier infrared photon detector : a platinum-silicide layer deposited on a p-type silicon substrate.

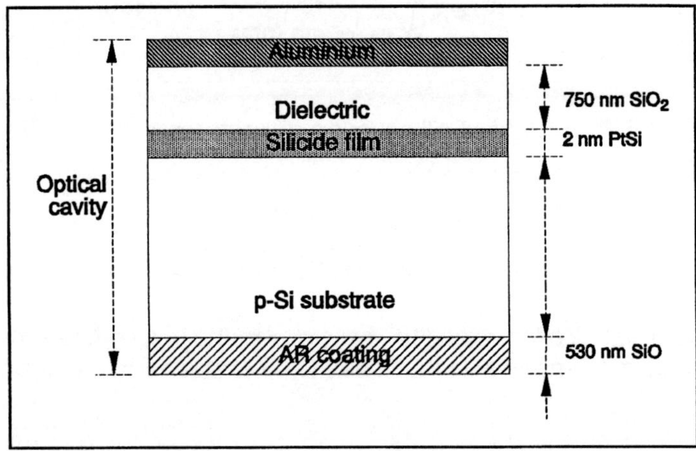

FIGURE 10.15. Cross section of a Schottky-barrier infrared photon detector with an optical cavity.

This responsivity R can be modeled by :

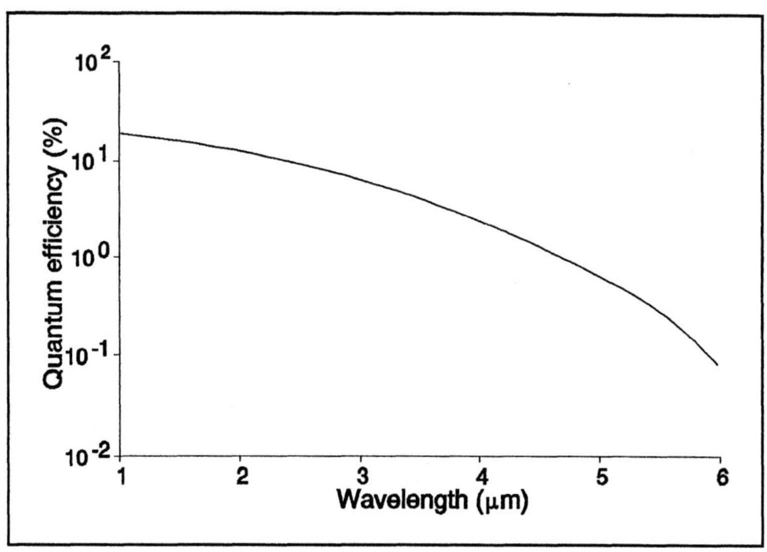

FIGURE 10.16. Responsivity of a high-performance PtSi Schottky-barrier thermal infrared detector.

$$R = C_1 \cdot \left(1 - \frac{\Phi_{MS} \cdot \lambda}{1.24} \right)^2 \qquad [10.4]$$

with C_1 as the quantum efficiency coefficient. For PtSi detectors, C_1 equals 0.267/eV. Using equation [10.4] for the responsivity, the cut-off wavelength λ_c can be expressed as :

$$\lambda_c = \frac{1.24}{\Phi_{MS}} \cdot \qquad [10.5]$$

For a typical PtSi Schottky-barrier detector, with Φ_{MS} = 0.22 eV, the value of the cutoff wavelength is equal to 8.7 μm. This cutoff wavelength can be extended with electric-field-induced barrier lowering obtained by increasing the substrate doping using a shallow implant. However, a more successful approach to extending the long-wavelength response to λ_c has been achieved with IrSi Schottky-barrier detectors. A barrier height of 0.152 eV corresponding to a cutoff wavelength of 8.2 μm has been reported.

As far as detector organizations are concerned, linear sensors for one-dimensional and interline-transfer devices or MOS-XY addressable imagers for two-dimensional scanning purposes have been announced. Designs based on quite conservative design rules have resulted in state-of-the-art structures with 4096 PtSi Schottky-barrier pixels on a linear basis and 512 x 512 IrSi or 1040 x 1040 PtSi detectors in a matrix organization (Yutani 91). The infrared detectors themselves are coupled to CCD readout structures which make use of advanced CCD technology or CCD architectures : Charge-Sweep Device and Transversal-Signal Line techniques are both applied in solid-state thermal image sensors. Infrared imagers compatible with standard TV readout were recently announced, namely 648 (H) x 487 (V) PtSi interline-transfer CCD on a chip area of 16.5 (H) mm x 13 (V) mm with square pixels of 21 μm side. The imager operates at a pixel rate of 12.3 MHz. The CCD-channel doping of this device is optimized to operate with high transport efficiencies even at liquid-nitrogen temperature. Normally at low temperatures, donors in a buried-channel CCD are frozen out and behave as charge traps during the charge transfer. To counteract this phenomenon, the donor concentration should be decreased. Together with the donor concentration in the buried CCD channel, the charge-handling capability is also lowered. However, the charge capability is more demanding in a vertical CCD in the image area of the interline-transfer device than in a horizontal CCD. The transfer efficiency is more demanding for a horizontal CCD than for a vertical CCD because the signal-transfer frequency is higher in the former. The Schottky-barrier infrared sensor should therefore be composed of a heavily-doped vertical CCD and a lightly-doped horizontal CCD (two-dopant concentration CCD) for a low-temperature CCD image sensor (Konuma 91).

10.3.2. UV IMAGING

As briefly mentioned earlier in this section on nonvisible imaging, the absorption of UV photons in silicon can be high. Not only silicon but also silicon dioxide has a high absorption coefficient for energetic particles of this kind. It will be recalled that SiO_2 is used as the gate-isolating material in all MOS structures. It will consequently be very difficult to make a solid-state detector for UV photons because they hardly reach the detector substrate before getting absorbed in the top layers.

To avoid this absorption in the multilayer structure on top of the solid-state imager, several solutions are possible :

- deposition of a UV-sensitive phosphor on top of the active area of the imager. By an appropriate choice of the down-converting phosphor, the UV information can be converted to a wavelength matching the spectral response of the CCD. A typical example of such a phosphor is coronene : when excited by ultraviolet radiation of wavelength less than 0.4 μm, coronene fluoresces in the green portion of the visible spectrum, with its peak near 0.5 μm. The sensitivity of a CCD imager covered with coronene is extended from a wavelength of 0.1 μm up to the near infrared. The photoresponse of a virtual-phase CCD before and after covering with the phosphor is shown in Figure 10.17 (Blouke 79);
- illumination from the back-side (Burstein 80). This technique is not new and, as already mentioned, it is also applicable to visible imaging (see

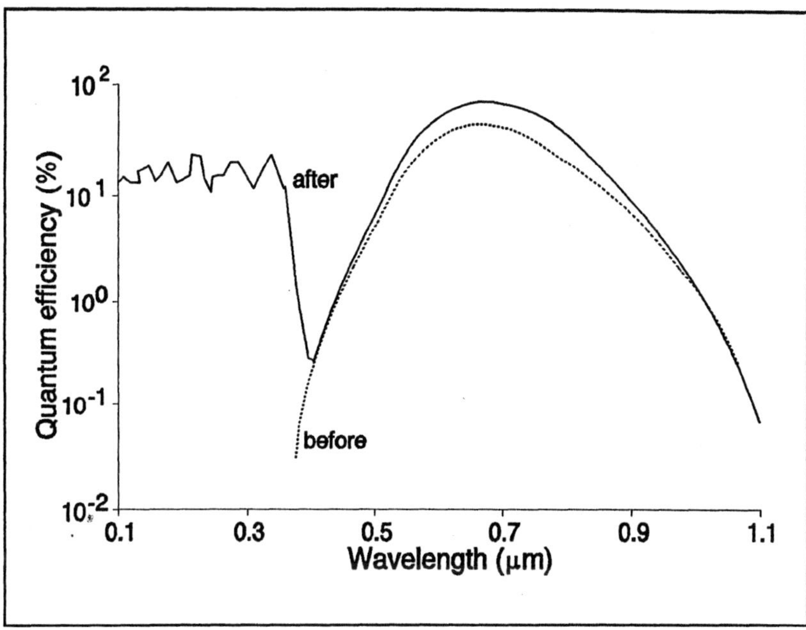

FIGURE 10.17. Photoresponse of a virtual-phase CCD before and after covering with a phosphor which converts UV into green light.

section 7.4). However, to bring the generation sites of the charge carriers as close as possible to their collection sites, the substrate of the detectors has to be thinned, typically to about 10 μm. Due to this extra processing and the subsequent delicate handling, this method is too expensive for consumer imaging in the visible spectrum, but can be suitable for UV detection.

The etched back-side can cause problems with recombination of generated charge carriers. The electron-hole pairs are generated somewhat closely to the back surface where the surface is not passivated and is consequently characterized by a disordered crystal structure. Recombination will be high at this surface and many generated charge carriers will recombine before collection at the front surface of the detector. To obviate this drawback, the back-side of the detector can be implanted with a very shallow p$^+$ layer to generate an extra electric field so as to drive the generated electrons toward the front surface before getting lost by recombination. A potential technological problem is this implantation after the thin etching because further high-temperature processing is not possible with finished wafers. An alternative to implantation is the growth of a delta-doped silicon layer using a molecular-beam epitaxy technique (Hoenk 92). In these delta-doped layers the dopant atoms are nominally located in a single monolayer of the crystal, reducing the width of the

back-side to considerably less than 1 nm. Because of this ultrathin layer, recombination of UV-generated electrons is almost negligible;
- deep-depletion CCDs (Bosiers 85) : in deep-depletion CCDs, a very-lightly doped, high-resistivity substrate is used, such that the depletion region under the CCD gates is extended to the back of the wafer. Minority carriers generated by the UV photons illuminating the rear of the CCD are swept to the front side by the electric field in this depleted region. This approach to deep-depletion CCDs avoids not only the absorption in the polysilicon gates but also the thinning required for back-side illuminated CCDs with the conventional doping density. A second important advantage of the depletion approach is that high-temperature processing of the rear of the wafer is possible and a variety of back-side passivating structures can be obtained.

A cross section of a deep-depletion CCD is shown in Figure 10.18. Note the back-side implanted (and annealed) p^+ layer to improve the quality

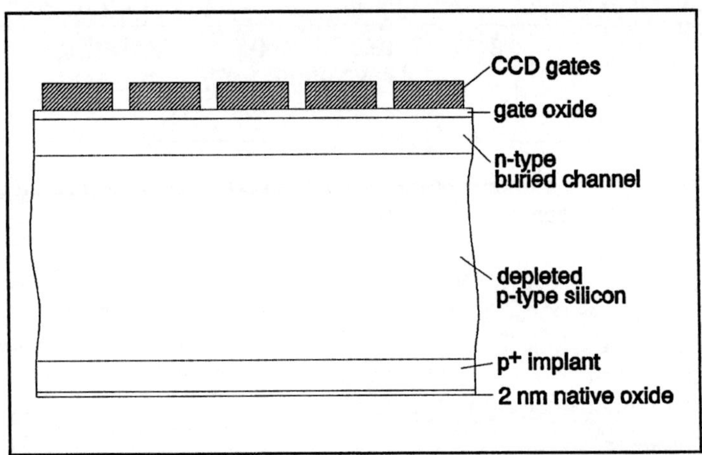

FIGURE 10.18. Cross section of a deep-depletion CCD built on a high-resistivity silicon substrate.

of the back-side of the CCD, by decreasing the dark current and increasing the quantum efficiency of the device. The thickness of this deep-depletion substrate is about 150 μm, while the resistivity of the substrate is in the range from 4 kΩ.cm to 10 kΩ.cm.

A drawback in the approach to deep-depletion CCDs is the dark current, its level increases linearly with the volume of the space-charge region. Dark current is significant at room temperature, but drops dramatically as the temperature is lowered. Cooling can be easily accomplished in most scientific UV applications and therefore dark current should no longer be a problem.

As far as pixel numbers were concerned, solid-state infrared and UV imaging was always a few steps behind visible-imaging technology, but infrared sensor research moved along, and quite recently a very interesting device was announced which combines visible imaging with UV and IR detection in one focal-plane array (Tsaur 90). A cross section of the device is shown in Figure 10.19. The front-illuminated focal-plane array is essentially identical to those previously described for back-

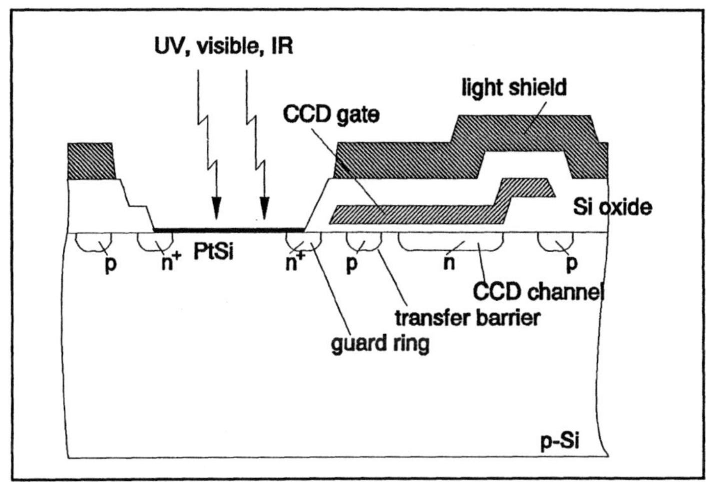

FIGURE 10.19. A cross section of a detector able to be used for visible imaging as well as UV and IR detection in a single focal-plane array .

illumination operation except that for the front-side the optical cavity structure has been eliminated. Electrons generated by UV, visible and IR photons are accumulated on the PtSi electrode and subsequently transferred to the buried CCD channel. Figure 10.20 shows the quantum efficiency of the device, measured between wavelengths of 0.29 μm and 7 μm and when it is cooled down to about 50 K and biased to a reverse voltage of 5 V. For wavelengths beyond about 1 μm, corresponding to photon energies below the bandgap of Si, the detector response is produced by carriers generated by absorption in the PtSi film. The wavelength dependence in this region is similar to that for back-illuminated Schottky-barrier detectors.

When the photon energy is increased above the bandgap of Si, the quantum efficiency increases drastically, as shown in Figure 10.20, because the radiation transmitted through the thin PtSi film is absorbed in the Si substrate and generates carriers that contribute to the detector response. The collection efficiency for these carriers is very high since they are generated very close to the PtSi-Si interface, and this interface is of excellent crystalline quality.

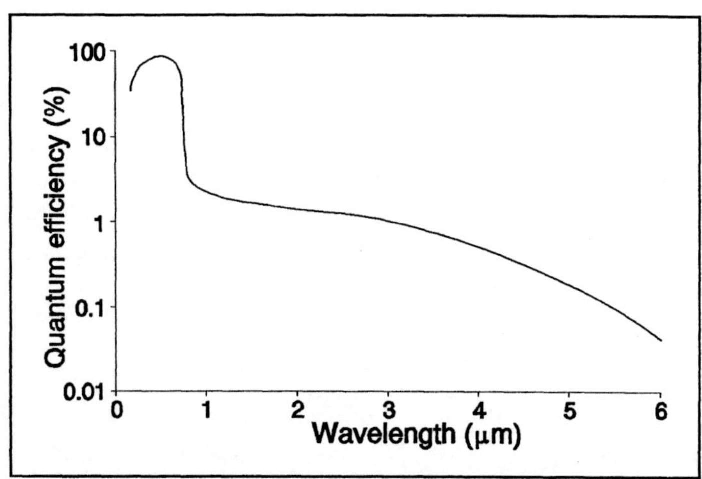

FIGURE 10.20. Quantum efficiency of the device shown in Figure 10.19.

This array can be extremely useful in remote sensing applications and imaging conditions where its multispectral capability is advantageous.

10.3.3. X-RAY IMAGING

The field of X-ray imaging can be divided into two parts : medical imaging and spectroscopic imaging. Fairly high-energy X-rays (> 25 keV) are used in medical imaging, while soft X-rays (< 10 keV) are used in spectroscopic applications. In fact, there is another characteristic which separates the field into two parts : in medical applications, the imaging result is reproduced in terms of spatial resolution and intensity. In other applications, in addition to the intensity of the signal and its spatial frequency contents, the spectral information of the X-ray source is also of great importance. This last item places some extra and new constraints on the imaging concept of CCDs, namely how to resolve spectral information.
The two fields will be described briefly in this section : nonmedical imaging and the solid-state sensors for the medical-application field.

<u>X-ray imaging for spectroscopic purposes</u>
In the recent past, data and results were published for developments of CCDs designed specifically for the detection of X-ray radiation, particularly in the 0.4 keV to 10 keV energy range (Wadsworth 84). CCDs of this kind will have applications in the fields of crystallography, plasma research, and X-ray astronomy, where both spectral and spatial information can be obtained simultaneously.

The basic performance characteristics of a CCD X-ray imager are good simultaneous spatial and spectral resolution in combination with high quantum efficiency. Spatial resolution has to do with to the accuracy with which the location of an X-ray event can be determined. Spectral resolution relates to the accuracy with which the energy of an X-ray event can be determined. The energy of an incident X-ray photon can be determined since the number of electrons generated upon absorption is directly proportional to the photon energy.

The quantum efficiency at low photon energies is limited by the "dead layer" of the device. For front-side illumination, there is a high possibility that low energy X-ray radiation will be absorbed in the polysilicon and oxide layers covering the active imaging area of a pixel. Absorption in the dead layer is a serious problem in these devices. A solution to this problem is to thin the back surface of the CCD to the depletion layer and to illuminate the device from that side. This method is also used in UV detection and even in visible-light sensors.

The two parameters, spatial and spectral resolution, are interrelated since the energy of the incident photon determines the number of electrons in an event. A loss of spatial resolution is equivalent to a loss of spectral resolution. The major cause of spatial and spectral resolution degradation is charge diffusion, as described by the MTF_D in section 5.5.1. The diffusion of generated charges is perceived as image smearing and as a decrease in the apparent energy of the X-ray event.

These diffusion effects can be minimized by incorporating a deep depletion layer into the CCD. In practice this is accomplished by fabricating the device on high resistivity material. Most techniques applied in X-ray imaging technology look very similar to these already applied to UV imaging with CCDs.

A totally different alternative can be used to minimize diffusion effects when devices are fabricated on epitaxial material. This technique utilizes the reflective properties of interfaces. The interface between a p-type epitaxial layer and a p^+ substrate will act as a reflector when charge impinges upon it.

A second cause of spatial and spectral resolution degradation is the charge loss when X-ray photons are absorbed at the boundary of two or more pixels but within the depletion layer. If the event is generated for instance at the boundary of two pixels in the same column, charge splitting between these two pixels can occur. On the other hand, when an X-ray is absorbed in or around the channel-stop region, charge splitting and even charge loss can occur. This charge loss can be the result of recombination processes in the highly-doped stopper region. These sources of degrading of the spatial and spectral resolution cannot be dealt with easily. The most practical means of reducing these boundary effects is to increase the pixel area at the expense of spatial resolution.

<u>X-ray imaging for medical purposes</u>
At first glance, the problems and issues involving in imaging with a solid-state imager for X-ray applications remain the same, irrespective of the application, whether medical or not. High-energy X-rays bother CCDs at least in the same way as soft X-rays do (Peckerar 79). The main issues are again :
>   - loss of quantum efficiency due to recombination of carriers released below the depletion region, and due also to radiation passing through the device without absorption;
>   - degradation of the modulation transfer function due to diffusive spreading of carriers.

Solutions to these problems are, of course, identical to those already described before : high-resistivity substrates and illumination from the back-side.

Where the application fields of X-ray imaging in the two sections (high and low energies) differ, are in the effects the X-ray introduced has in the imager itself. High-energy particles can severely damage the semiconductor structures of the MOS type. CCDs and all other imagers are MOS devices and, consequently, they will be partly damaged when exposed to X-rays in the higher energy range.

The main damage caused by high-energy X-rays is located at the Si-SiO_2 interface and in the SiO_2 dielectric. This radiation damage may be translated into the generation of an extra number of interface states and an increased number of oxide charges. The net overall results for the CCD are an increase in dark current and dark-current nonuniformities, threshold-voltage shifts, and an increase in transport inefficiency. Ultimately, the device can even be destroyed by the high energy particles. Many of these radiation effects have been closely studied and are well understood, so that preventive measures can be taken in the fabrication of these devices. Several processing steps during the fabrication of the sensors have a direct or indirect effect on the radiation hardness or their ability to withstand X-rays. In some respects it is possible to fabricate imagers so that they can tolerate part of the radiation, but, to date, it is still impossible to design or fabricate CCDs totally impervious to damage caused by high-energy impinging particles.

All in all, it is very important to keep high-energy X-rays as far as possible away from the CCD, and direct imaging of these energetic particles does not seem to be very practical. A few indirect imaging techniques have been developed for that reason : the incoming signal to be detected may be converted into an intermediate signal which can be detected by the CCD without any risk of damage.
This intermediate signal can be generated by means of a phosphor sensitive to X-rays, using exactly the same technique as already discussed for UV imaging. The phosphor can be deposited on top of the imager surface, permitting detection of the converted X-rays without any further optical aids. A possible drawback may

be the penetration of nonabsorbed X-rays into the CCD with the known consequences.

To ensure there are no residual X-rays on the detector surface, the sensor has to be located out of possible beams from the X-ray source. In this case the converting medium has to be placed on a carrier (e.g. glass) or may even be the phosphor screen of an X-ray imaging tube. The intermediate optical signal has to be coupled to the CCD by means of a lens or a fiber-optic coupling. This possibility of fiber-optic coupling of a solid-state imager to an imaging tube is shown schematically in Figure 10.21.

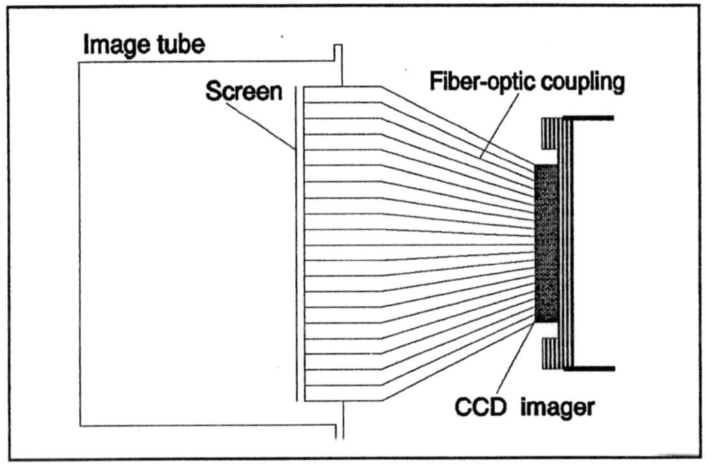

FIGURE 10.21. Fiber-optic coupling of a solid-state imager to an imaging tube.

With reference to applications of this kind, the lens coupling is often bulky compared to the fiber-optic solution. On the other hand, fiber-optic coupling also has its drawbacks : because of the large number of pixels in the solid-state imager and the technology of fiber-optic plates, perfect coincidence or matching of the pixels to the fibers is not possible. To minimize resolution losses, fiber diameters are usually chosen small compared to pixel pitches, and as a rule, several fibers are bonded to each single pixel.

Besides defects in the fiber and geometric distortion due to fiber taper, more basic interferences between pixel and fiber periodicities lead to aliasing or Moiré patterns. If some of these various precautions are taken, it is almost impossible in practice to detect any measurable resolution loss due to fiber optics.

WORTH MEMORIZING

If CCD imagers are used in applications outside the visible spectrum, special precautions have to be taken to make the device sensitive to the wavelength concerned. If the wavelength of the incoming light is too short (less than 0.4 μm), most photons are absorbed in the layers on top of the silicon; if the wavelength of the incoming light is too high (greater than 1.1 μm), the photons pass through the silicon without any absorption at all.
Infrared imaging is based on photon absorption in a Schottky-barrier photodiode : a PtSi layer on a p-type silicon substrate. UV imaging can be done using a conversion phosphor on top of the CCD (which converts the incoming UV light to the visible spectrum) or by backside illumination combined with a CCD of the deep-depletion type. In most applications involving X-ray imaging, an intermediate step to convert the X-rays into a visible picture is also taken. Especially for X-ray detection, the CCD itself has to be kept out of the X-ray beam because photons of this kind can drastically damage the charge-coupled device.

10.4. High-speed imagers

Conditions needed for high-speed devices or high-speed pixel rates can be created by :

> - using frame rates higher than 50 or 60 Hz. Examples can be found in the capturing of fast moving objects;
> - standard frame rates but with an extremely high number of pixels within one frame.

Both categories are characterized by high pixel rates at the output stage. CCD transport speeds are limited, as also is the bandwidth of the output amplifier. To overcome these boundary conditions, the output construction of the CCD does not have to be limited to a single stage, but can be extended with more stages, or the fabrication technology used for the device can be changed to a substrate different from silicon to upgrade the intrinsic speed limit of the CCD.

10.4.1. IMAGERS WITH MULTIPLE OUTPUTS

In fact the devices described in some of the previous sections (9.1) are already making use of two or three parallel output stages in the consumer application area. For very high-speed operation and/or capture of very fast-moving objects, simple demultiplexing of a horizontal output register may not be enough. Devices with

6 x 32 output stages are reported to capture images with a frame rate of 2000/sec (Lee 81).

10.4.2. GALLIUM-ARSENIDE CCD IMAGERS

Schottky-barrier-gate GaAs CCDs possess properties which make them extremely attractive for special imaging applications and optical signal-processing applications where very high speed is important. GaAs CCDs with transparent conducting indium-tin-oxide gates which can operate up to a frequency of 1 GHz are reported (Sahai 83).

WORTH MEMORIZING

High-speed imagers can be made through its architecture (by means of multiple outputs) or by fabricating the devices on special substrate material, e.g. GaAs.

10.5. Contact-type linear imagers

A typical example of solid-state imaging completely different from the objects discussed so far are the contact-type linear image sensors, for instance, used for facsimile applications. Their architecture and working principle have nothing to do with CCDs or MOS-XY sensors, but the application field is so important that it is worthwhile to include a description of them here.

Contact-type linear imagers have been developed in order to produce more compact facsimile equipment. Characteristic of these devices is their length, which is equal to the document width, so that the apparatus does not require an optical lens system. This makes it small and simple. The imager virtually contacts the document. A schematic setup of the facsimile scanner using a contact-type linear imager is shown in Figure 10.22 (Kaneko 82). The document is scanned by the sensor in the direction perpendicular to the illustration; in the direction indicated by the arrow, the document is scanned mechanically. Illumination is done by solid-state LED arrays, and the coupling of the information carrier to the sensor surface is realized with a fiber array. Observe the extremely compact structure of the construction : only 2 cm in height, which is the typical advantage of the document scanner with a contact-type linear imager.

The working principle of the imager is based on the effects of resistivity changes in photoconductor materials. Typical photoconductors used in the contact-type linear imager technology are CdS-CdSe layers (Komiya 81) and amorphous silicon (Kanoh 81, Kaneko 82). The working principle of the device can be easily explained with the aid of an equivalent electric circuit as shown in Figure 10.23 (Komiya 81).

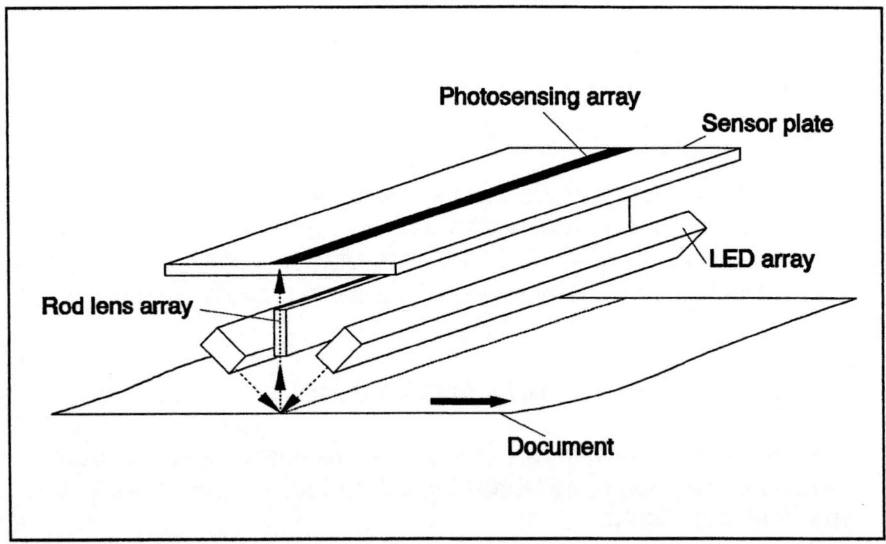

FIGURE 10.22. System concept using a contact-type linear image sensor in a facsimile scanner.

The imager itself has a 2048 element CdS-CdSe photoconductor array deposited on a glass substrate. The photoconductors are connected to the matrix circuit to reduce the number of output terminals required. This matrix can include, for instance, 64 common electrodes and 32 individual electrodes. Each terminal is connected to an external switching device, or it is possible by appropriate circuiting of the matrix to connect each individual photoconductor serially to the output preamplifier. In a succeeding operation, time-sequential video signals are obtained through the output stage. As can be seen from Figure 10.24, the design of a contact-type linear imager is based on two key items : the individual photoconductor array and the scanning circuit.

The aperture size can be typically designed to be around 100 μm by 55 μm, and the scanning circuit is based on combinations of MOS transistors because of their high-speed operation and low power dissipation.

Note the relatively great length of such a facsimile sensor : 2048 pixels with a pitch of 100 μm account for over 20 cm. For this reason two of the main issues with contact-type linear imagers are, not surprisingly, uniformity of the pixel response and possible image lag. A typical value for signal nonuniformity is about 10 %, which can be considered satisfactory for black and white facsimile application. The nonuniformity of the pixel response is due to factors such as (Komiya 81) :
- heat-treatment conditions during fabrication of the device;
- aperture-shape accuracy or geometric concerns;
- photoconductor film thickness uniformity;

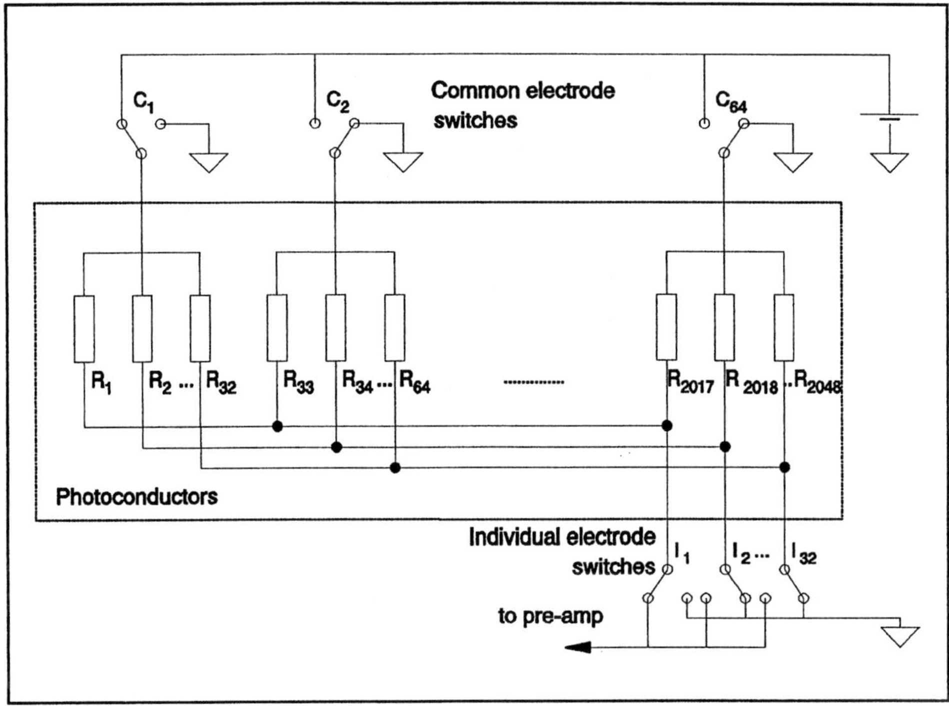

FIGURE 10.23. Equivalent electric circuit of a contact-type linear imager.

- contacts between the photoconductor and matrix electrode.

Image lag can be kept down to 10 %. This value means there are no severe limitations as far as facsimile applications are concerned. A typical voltage-response has a rise time of 2.5 μsec, which is equivalent to a 400 kHz operating frequency.

WORTH MEMORIZING

Contact-type linear imagers are sensors capable of capturing their objects by very close contact instead of using an optical system. A typical application field is the facsimile domain. The basic construction of the contact-type linear sensor is a one-dimensional array of photoconductors, read out one after the other by a digital scanning circuit.

10.6 Conclusions

In this chapter on nongeneral imaging applications special attention has been paid to imagers of this type, which are not usually found in mass-production environments.

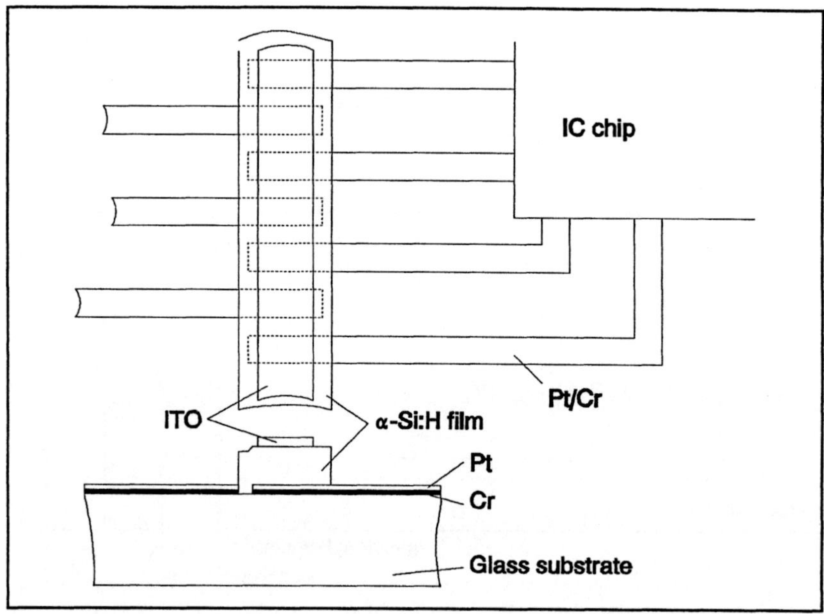

FIGURE 10.24. Design concept of a contact-type linear imager based on the individual photoconductor array and the scanning circuit.

Typical examples are scientific imagers, smart image sensors, detectors for infrared and UV light, CCDs for X-ray imaging, and contact-type imagers.

In the scientific world there is very great interest in extremely large imaging arrays. Pixel numbers into the mega range are common in this kind of application. But production of imaging devices of this type is accompanied by some typical problems : huge chip size, low production yield, and extremely high price. The fabrication technology and the device architecture are adapted to each other to make the yield of the production process as high as possible. Scientific imagers are mainly characterized by square pixels, full-frame organization, low field rates, and a high pixel number.

To circumvent the problems associated with the huge silicon area, a single large-area scientific device can be replaced by a matrix of smaller devices built up from subarrays. In the case of this alternative the image sensors of the subarray can be specially designed for butting to each other. Busbars are located at only one side of the imager and precautions are taken to eliminate any dark-current effect from pixels located very close to the edge of the CCD chip.

Developments to upgrade the performance of scientific devices are described : the notch CCD and the skipper CCD. The former incorporates an extra-deep CCD channel in the existing buried channel to enhance the transport of small charge packets. The skipper CCD is intended to lower the output-amplifier readout noise by a multiple readout cycle and external averaging of the measured multiple samples of the output signal.

Phase-pinned CCDs turn their Si-SiO$_2$ interface into the inversion state to eliminate dark-current generation by the interface states. Phase pinning can be done in several ways : by open-phase pinning technology (closely resembling the virtual-phase technology), multi-phase pinning technology (in which all the gate regions are pinned during integration), and a dynamic pinning clocking system (in which the regions are pinned only for a shorter period of time).

Smart image sensors in this chapter were limited to devices which have part of their processing circuitry on-board, completely integrated in CMOS technology, or completely integrated in a three-dimensional IC, or by their concept or architecture.

In the scientific, technical, and medical world, imaging of nonvisible optical information is a very important application area. In this chapter attention was paid to infrared imaging, UV imaging, and X-ray imaging.
The infrared field is covered by CCDs with Schottky-barrier photodiodes because pure silicon is unable to absorb the infrared photons. Platinum silicide is commonly used as Schottky-barrier material but, depending on the wavelength concerned, other materials can also be chosen. In today's infrared sensor world, devices with full standard-television resolution are reported. Because most of the devices have to be operated at lower temperatures, specific issues have to be tackled in connection with the transport phenomena at liquid-nitrogen temperature.
Direct UV imaging is very difficult to do with a silicon CCD because most of the photons are absorbed in the layers on top of the active silicon. To overcome this problem, the UV detectors are illuminated from the back-side and the silicon devices are fabricated on high-resistivity material. The depletion regions underneath the CCD gates (located on top of the substrate) are thus extended to the rear of the detector. If this technique is too cumbersome, a conversion phosphor can be deposited in front (or even on top) of the front-side illuminated CCD.

Techniques similar to those applied to silicon CCDs to make them UV-sensitive are also used to make the devices sensitive to X-rays. But in the case of X-ray imagers, a difficult problem emerges : the Si-SiO$_2$ interface is extremely sensitive to damage introduced by the X rays. In other words, the kind of radiation to be detected destroys the device itself, especially if these are high-energy X-rays. For this reason, in the case of X-ray detection with a solid-state image sensor, a conversion medium is used in almost all circumstances to change the X-ray information into visible light which can be easily detected by the CCD without any risk of damage.

High-speed devices are made using a clever design or an adapted architecture, e.g. the use of several output stages in parallel. Another possibility is to use "high-speed material" : GaAs instead of Si as substrate material.

The last section of this chapter was dedicated to contact-type linear image sensors. These devices are used in almost direct contact with the object which has to be detected. This makes them somewhat large, but the overall construction of the imager and the object is very compact. The imager itself is built from photoconductors made of amorphous silicon. All individual photoconductors are addressed and read out via a digital scan circuit.

Left : a small frame-transfer CCD (360,000 pixels on 38 mm^2) based on the accordion transport mechanism. At the left-hand and right-hand side of the chip the drivers built in CMOS technology can be recognized (courtesy of Philips). Right top : a color sensor intended for electronic still picture applications is shown. Mounting in a plastic package lowers the overall cost of the imager (courtesy of Texas Instruments). Right bottom : a four inch wafer containing frame-transfer CCDs. The wafer is backside thinned and the devices are backside illuminated to increase their sensitivity (courtesy of Thomson CSF).

Appendix 1

HOW CCD IMAGERS ARE MADE

This appendix gives an overview of a possible route which can be followed to make a CCD imager. The fabrication technology is purely an example and the vehicle used to describe the various steps is also a hypothetical device.

CCD processes proceed from complicated, MOS-based technologies and processes. The complexity is a result of the relatively large number of technological steps and the consecutive relatively large throughput time in the production facilities. Fabrication involves several implantation steps and the deposition of several layers of, for instance, silicon oxide, silicon nitride, polycrystalline silicon and at least one layer of metallization. If the imager has to be used in color applications, color filters layers have also to be added. Microlenses are another possible option. The minimum number of masks needed in the production trajectory is about 10 to 12, but, for very highly sophisticated devices (such as HDTV imagers with color filters), as many as 25 masking steps are no longer an exception. Throughput times for CCD imagers processes range from a fortnight to several weeks. In other words, there are probably as many different fabrication processes as there are CCD designs because the technology is quite often optimized with the main application and the device architecture in mind.

The architecture on which this overview is based is that of a full-frame device highly suited to describing a minimal route. Figure A1.1 shows a top view of the device under study. The full-frame imager has a light-sensitive part, a horizontal output register and an output amplifier connected to the floating diffusion. All these regions are designated in the illustration. The cross section of interest is indicated by the dotted line A'-A. Starting from point A', the cross section passes first through the light-sensitive area, then through the horizontal output register and floating diffusion, and finally through the reset transistor. In the remaining part of this appendix, the individual technological steps of the fabrication process will be described, successively and their consequences for the CCD itself will be illustrated along the cross section A'-A.

Some characteristics of the full-frame device under study :
- the technology is based on a triple-polysilicon gate process;
- a single level of metal is used, both for the interconnects and for the light shield;
- vertical antiblooming has to be catered for by a single p-well implant;
- CCD transport has to take place in a buried CCD channel;

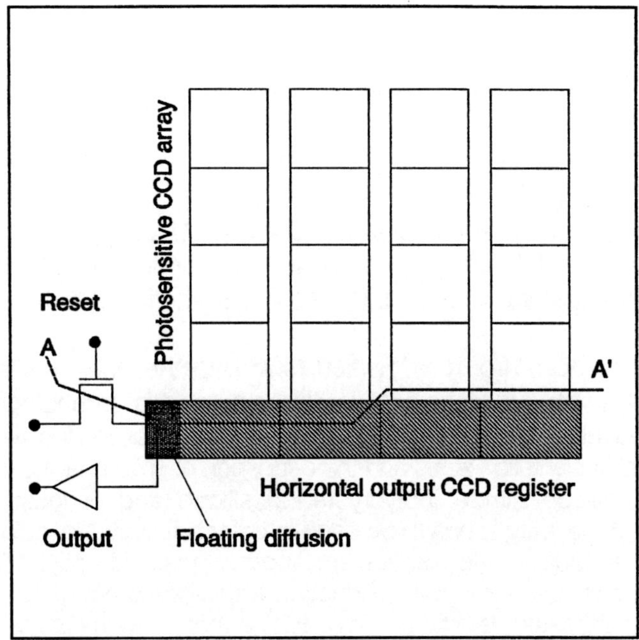

*FIGURE A1.1. Illustration of the full-frame imager which is used
as the technology vehicle to describe the fabrication process of
a charge-coupled device.*

- the CCD has to be capable of making color images by color separation on-chip
  with complementary colors;
- the light-sensitive array has be provided with microlenses (this specification
  item is only for the sake of completeness because, as will be explained
  later, it has only theoretical value).

Figure A1.2 shows the codings of the various regions used to build up the CCD
imager. These codings will be retained in the remaining part of this section.

The following are some general semiconductor fabrication techniques which are
also widely used in the CCD fabrication process are :
- lithography : photoresist is applied to the silicon wafers using spin-on
  deposition methods. After a short curing cycle the photoresist is exposed
  to UV light through a mask and subsequently developed. Positive
  photoresist will remain in the nonexposed areas of the wafer after
  developing and will disappear from the exposed areas during the
  developing process;
- implantation : a well-defined number of ions are implanted using a well-defined
  amount of implantation energy;

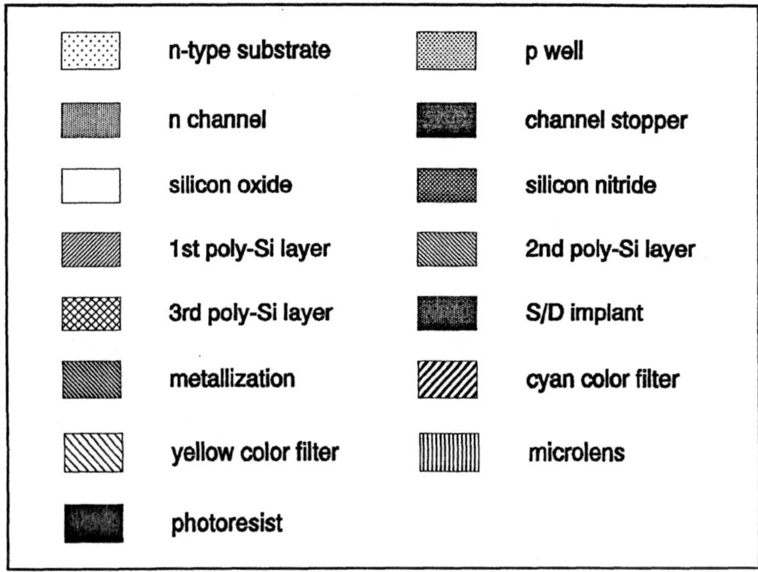

FIGURE A1.2. Coding of the various regions of the CCD.

- etching techniques : material can be removed from the wafer by wet chemical etching (isotropic etch) or by dry etching, which can be a combination of chemical and physical removal of the material (isotropic or anisotropic);
- deposition techniques : several methods exist to provide the wafers with a uniform coating of thin films : spin-on deposition, sputtering, low-pressure chemical vapor deposition, plasma-assisted chemical vapor deposition, and evaporation;
- annealing steps : heating the wafers to a well-defined temperature in a conditioned atmosphere (inert, oxidizing, reducing) for a well-defined time.

Any hypothetical full-frame imager can be fabricated using a more or less complex combination of the above techniques and a complete set of masks.

A1.1. Substrate preparation

To achieve "perfect" CCD imagers, the most important requirement is to start with "perfect" silicon wafers. Any impurity or crystal defect in the wafer which can interact with the active region in which the CCD is made, can cause pixel defects or dark-current non-uniformities. Another very tight parameter of the wafer specification is the concentration and uniformity of the doping. To meet the first condition for defect-free active regions, the wafers undergo a series of very closely controlled annealing procedures before the actual processing starts. The annealing processes can in some cases be carried out on the vendor's premisses. The aim of these steps is twofold : first, to make the top layer of the wafer free from oxygen, thus creating the so-called denuded zone and, second, to create oxygen precipitates at the center of the wafer. This renders the active layer of the wafer free from impurities because the oxygen precipitates at the center of the wafer attract and getter any contaminant. The starting wafer material may be of the n^+ type with a 10 μm thick epitaxial n^- layer grown on it. The latter process is characterized by excellent control of the doping concentration and uniformity. When the wafers enter the clean room, they are coded, cleaned, and thermally oxidized. The result at this point is shown in Figure A1.3 : the n-type substrate with a thin thermal silicon oxide layer on both sides of the wafer.

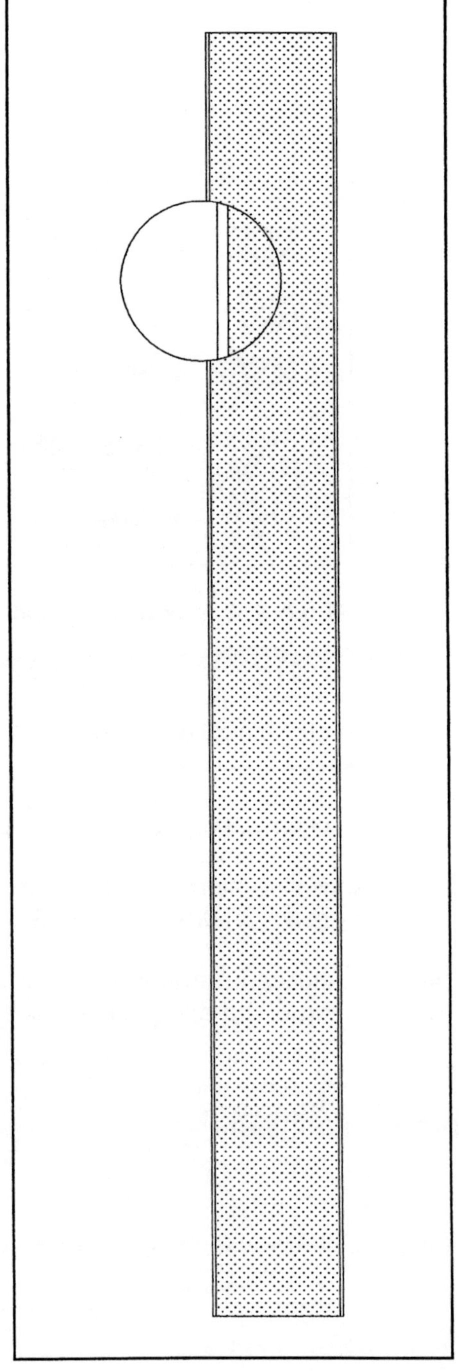

FIGURE A1.3. Preparation of the substrate before starting the first masking step.

A1.2. Implantation of the p well

In the case of vertical antiblooming, definition of the p well is quite crucial, especially in the light-sensitive part of the device. As explained in section 6.3.3, the doping profile of the p well should be more or less sinusoidal shaped to create a saddle point in the center of the pixel. The exact location in depth of this saddle point defines the antiblooming effect.

To achieve the desired doping profile, the p well is not uniformly implanted in the image section but into a striped pattern in which the stripes are parallel to the CCD channels and the direction of charge transfer. These stripes are defined in the photoresist which is spin-coated and processed prior to the implantation. The photoresist is used locally as a masking step against boron ions.

Outside the light-sensitive areas, where anti-blooming is not concerned, the p well is uniformly implanted.

The situation at this point is shown in Figure A1.4 : the pattern in the photoresist is chosen such that the implantation through the holes (or stripes) in the imaging section can result in the sinusoidal doping profile. The energy and the dose of the implanted boron ions are both low : less than 100 keV and around $10^{12}/cm^2$, respectively.

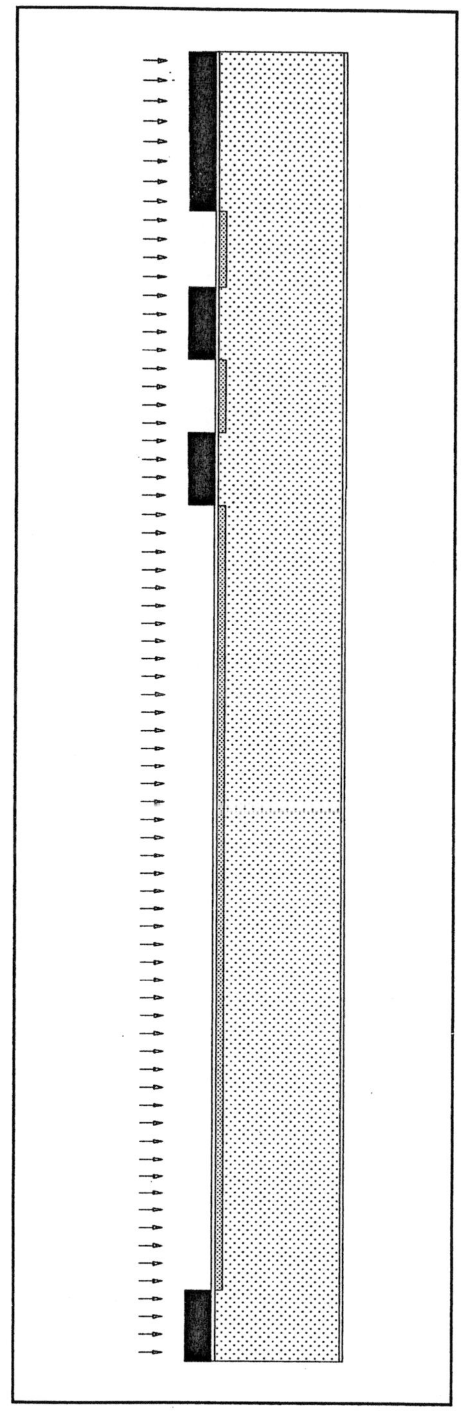

FIGURE A1.4. Implantation of the boron ions to define the p well.

A1.3. Drive-in of the p well

After the implantation of the p well the wafers are annealed to diffuse the implanted boor ions deeper into the silicon. Outdiffusion of the ions takes place not only in depth but also laterally. In general, the length of the lateral diffusion (also called underdiffusion) is about 75 % of the depth of the in-depth outdiffusion.

Note that this temperature step is for sure not the last one in the overall fabrication process of charge-coupled devices. The p well will outdiffuse even further during the temperature steps which still have to follow this annealing step. During determination of the flow chart (= the sequence of the various fabrication steps which follow in the clean room), the technologist has to take into account the temperature budget along the entire fabrication route to arrive the ultimate definition of the p well. The annealing step at this point in the flow chart is only part of the temperature budget as a whole, although this phase of the process takes place at very high temperatures (e.g. > 1100°C) and over a very long time (e.g. 10 hours).

The result at this point is shown in Figure A1.5 : the photoresist has been removed, and the wafers cleaned and stored at higher temperatures to drive in the implanted boron ions. Note that the implanted stripes in the image section have already diffused so far sideways that a more or less continuous p well has also been generated in the image section.

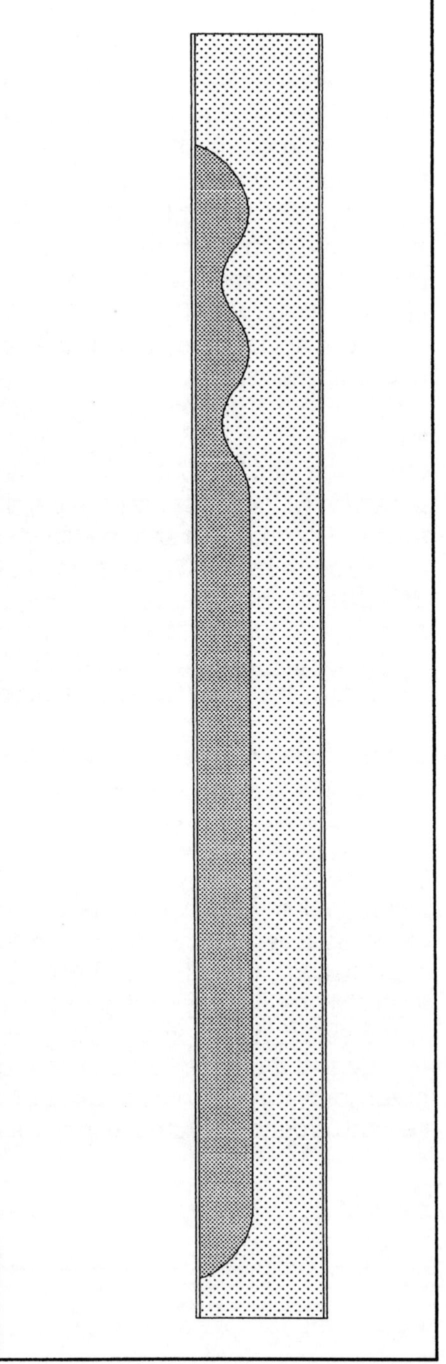

FIGURE A1.5. Situation after the drive-in anneal of the p well.

A1.4. Implantation of the CCD channel

The second masking step is concerned with implantation of the CCD channels. As will be shown in the next masking stage, the actual definition of the channels is a combined effect of this implant and the next one.

Implantation is done through a single opening in the photoresist : the CCD channels in the imaging area, the horizontal output register, and even the reset transistor, all form part of a single implanted channel area.

This implantation is dedicated entirely to the buried CCD channels and its depth is defined by the charge-handling capability and the CCD transport characteristics. In this case the exact doping profile again depends on the implantation energy and the implanted dose, together with the continuing temperature budget.

The result at this point is shown in Figure A1.6 : the pattern in the photoresist is chosen such that it in fact covers the entire active CCD area. The energy and dose of the implanted phosphorus ions are both of the same order of magnitude as the parameters of the p-well implant : less than 100 keV and around $10^{12}/cm^2$, respectively.

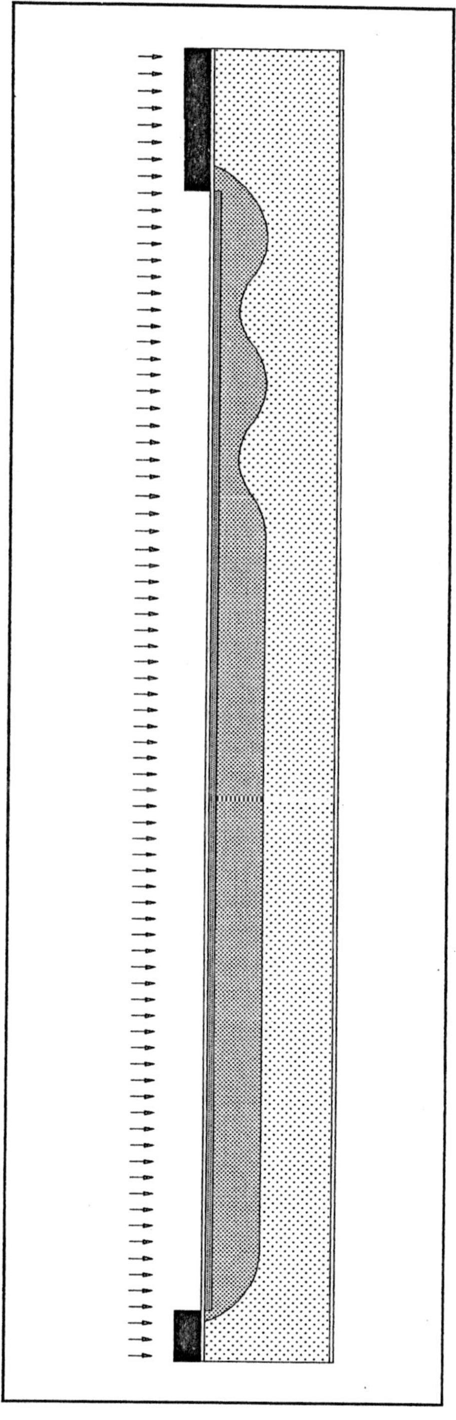

FIGURE A1.6. Implantation of the phosphorus ions to define the n-type buried CCD channels.

A1.5. Drive-in of the channel implant

After the removal of the photoresist, followed by a cleaning process, the wafers are again annealed. In this step the channel implant is diffused deeper into the silicon and, at the same time, so is the profile of the p well. A remark similar to the one made during discussion of the drive-in of the p well, also applies at this stage : the ultimate doping profile of the CCD channels is defined by the overall temperature budget of the flow chart and not just by this drive-in anneal, although by the end of this stage the profiles of the p well and the channels come pretty close to their final state. The depth of the p well in the saddle points and at the deepest points are about 2.5 μm and 3.5 μm, respectively, while the depth of the buried-channel profile is about 0.8 μm.

Compared to the anneal of the p well, the annealing temperature and time are lower, e.g. > 1000°C and only a couple of hours.

The result at this point is shown in Figure A1.7 : after the high-temperature treatment, the implanted phosphorus ions have been diffused deeper into the silicon substrate. In the diffusion of the channel implant, an underdiffusion of 75 % of the in-depth diffusion has again to be taken into account. Note that the all implanted phosphorus ions are surrounded by a p well and fully isolated from the n-type substrate.

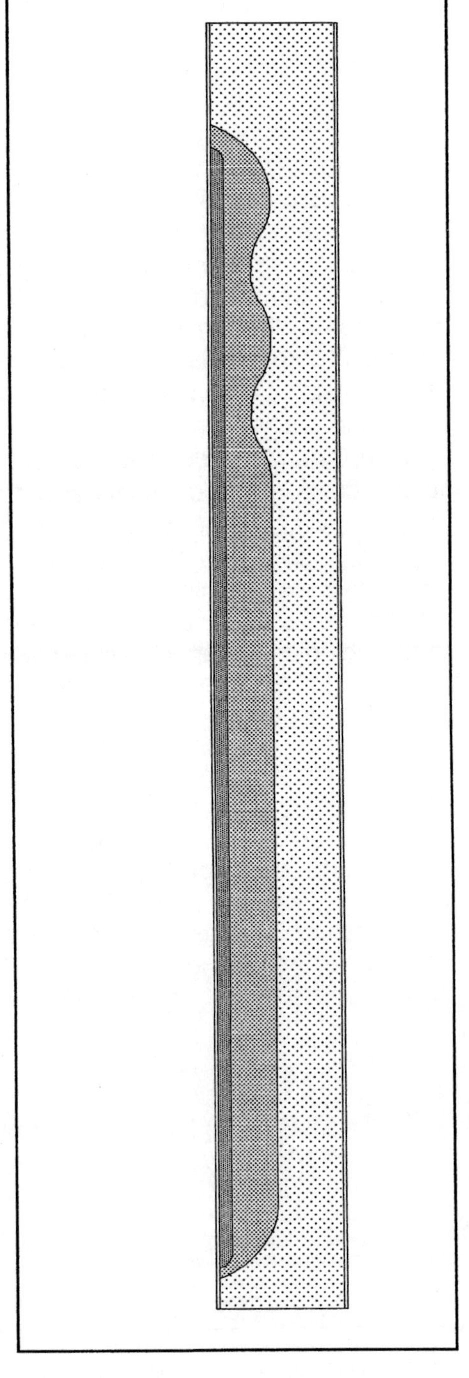

FIGURE A1.7. *Situation after the drive-in of the buried-channel implantation.*

A1.6. Implantation of the channel stoppers

After the preceding annealing step the wafers are again covered with photoresist and the pattern of the channel stop regions is transferred from the mask into the resist. Next comes implantation of the channel separation itself by means of boron ions. To fulfill its function as a channel separator, the energy and dose of the implant should be chosen such that it overdopes the profile of the CCD channels. But the channel stopper has another function, namely to bias the p well electrically. This means that the channel stopper implant has also to be deep enough to penetrate into the p well. (Depth adjustment takes place in the next annealing step.)
The result at this point is shown in Figure A1.8 : the wafers have been implanted with boron ions in very narrow regions defining the separation between different CCD areas. The implantation dose and energy are such that a relatively shallow but more highly doped region can be created with, for instance, values of 50 keV and $2*10^{13}/cm^2$, respectively. The exact value of the dose is not critical so long as it overdopes the channel implant.
In Figure A1.8 the two CCD channels belonging to the image section can already be recognized from the saddle points in the p well but also between the stopper implants.

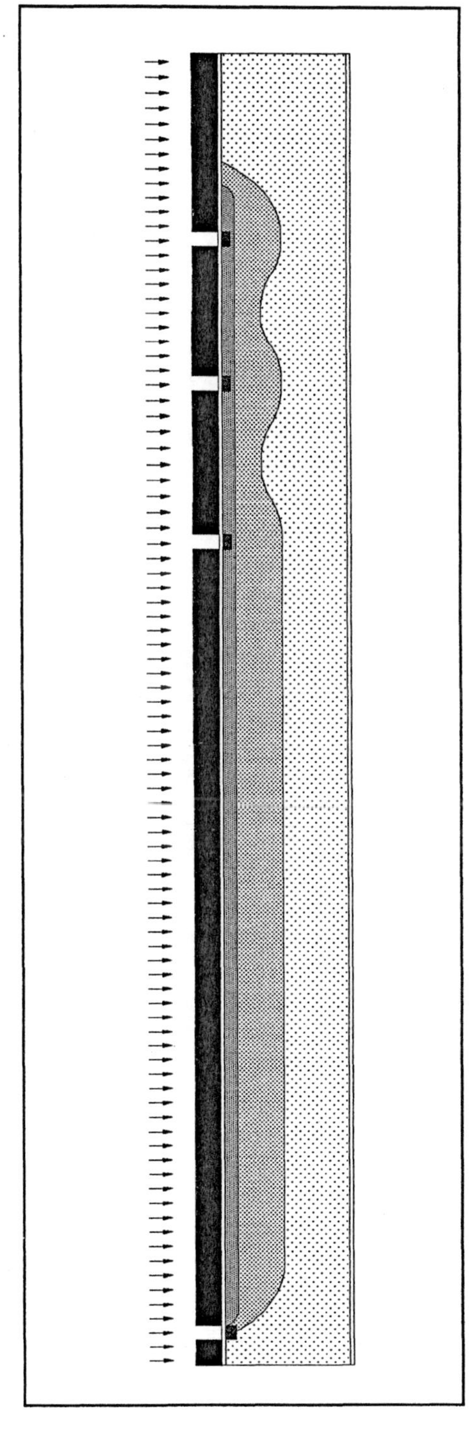

FIGURE A1.8. Implantation of the channel-stop regions by means of boron ions.

A1.7. Gate oxidation

The three implantation steps are now followed by preparations for the formation of the gate material. When the wafer leaves the implanter, the photoresist is removed, but also the original thermally-grown silicon oxide is etched using a classical wet chemical method. The wafers are cleaned and heated in a furnace at 1000°C in order to grow the gate oxide in an oxygen atmosphere. The thickness of the gate oxide layer is 80 nm.

The result at this point is shown in Figure A1.9 : the wafers have been covered on both sides with a thin layer of thermally-grown silicon dioxide, serving as the base layer of the gate dielectric. The channel stopper has been diffused in the meantime to a depth of about 1 μm, overdoping the CCD channels.

An important trade-off is illustrated in Figure A1.9 : the stopper-implantation profile has to be deep enough to contact the p well. But, as well as a "deep" channel stopper, a "broad" channel stopper region has also been defined as a result of the outdiffusion. In the lateral direction, any space taken up by the stopper implantation is at the expense of the CCD-channel width and consequently at that of the charge-handling capability.

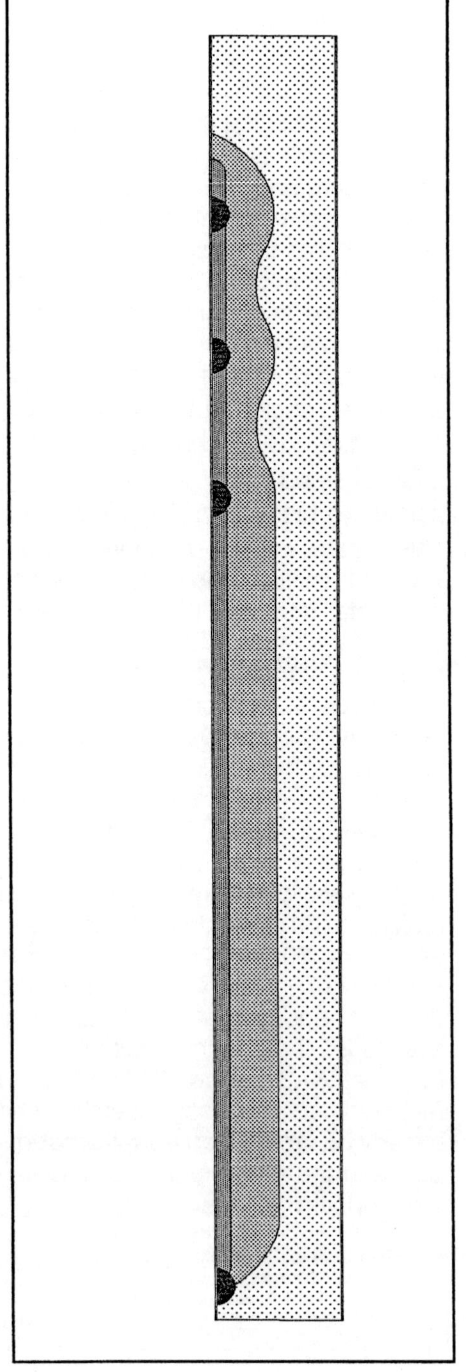

FIGURE A1.9. Situation at the end of the gate-oxidation process.

A1.8. Deposition of the first poly-Si layer

Before the first level of gate material is deposited, the gate dielectric has first to be completed. An additional silicon nitride layer of 40 nm is deposited using an LPCVD technique. The sandwich of silicon oxide (80 nm) and silicon nitride (40 nm) constitutes the entire gate dielectric. The ratio of the dielectric constant of SiO_2 and that of Si_3N_4 is about 2. This results in an overall equivalent gate-dielectric thickness of 100 nm SiO_2.

After the nitride, the first polycrystalline silicon, used as the first gate material, is deposited. The technique applied for the poly-Si deposition, is the same as for the nitride film, namely LPCVD. The thickness of the film is 500 nm after deposition and only 250 nm at the end of the processing, as a whole. The difference is mainly due to the oxidation processes which have to follow.

Immediately after deposition, the poly-Si film is doped. This step can be done either in a heating chamber with ambient $POCl_3$ to dope the film with phosphorus or by implantation of, for instance, arsenic.

The result at this point is shown in Figure A1.10 : the wafers have been covered with silicon oxide, silicon nitride, and a uniform film of polycrystalline silicon. Although not visible in the figure, the poly-Si layer is n-type doped to make the film low in resistivity : about 30 $\Omega/\square$ at the end of processing.

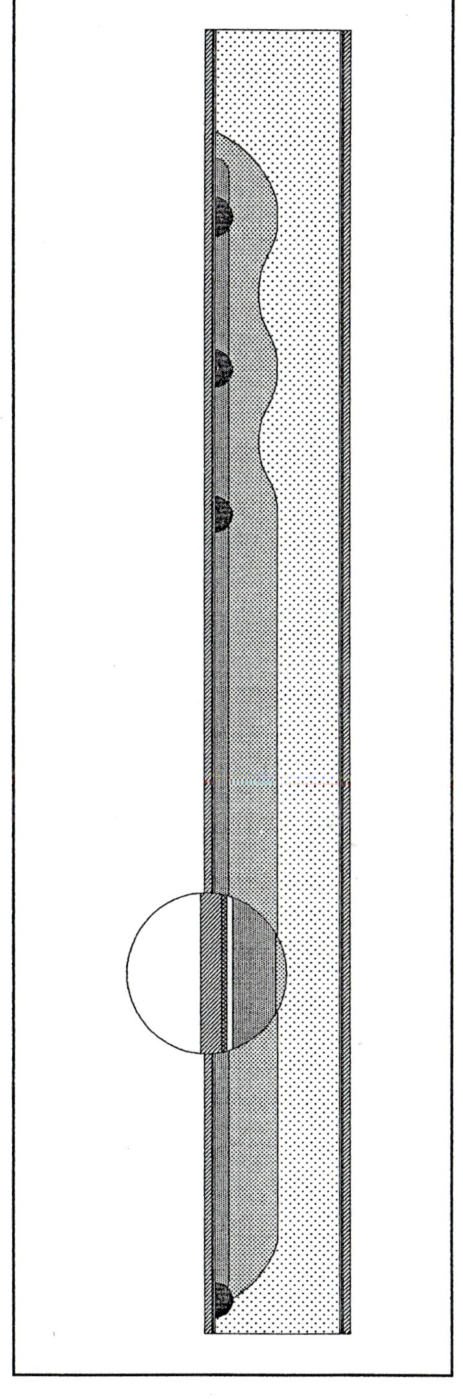

FIGURE A1.10. The substrate covered with the gate dielectric and the first poly-Si layer.

A1.9. Definition of the first polycrystalline-silicon layer

After doping of the polycrystalline silicon layer, the photoresist is deposited and patterned. Next the poly-Si film is etched, preferably using a dry, anisotropic etching technique. The smallest dimensions of gate length and gate spacing can be obtained in this way. The etching process is optimized in such a way that the poly-Si is etched very selectively to the silicon nitride. This means that, once the poly-Si film has been etched, the etchant does not attack the underlying nitride film.

The situation now is shown in Figure A1.11 : the cross section through the image area is drawn perpendicular to the charge transport and as a consequence both pixels are covered by the same gate, made from the first poly-Si layer. Further along the path in the horizontal register the situation is different. The cross section is parallel to the charge transport and only a few pieces of the poly-Si material are left here to form different gates from the horizontal register.

At the very end of the horizontal register a single gate is defined which will serve as the reset transistor's gate.

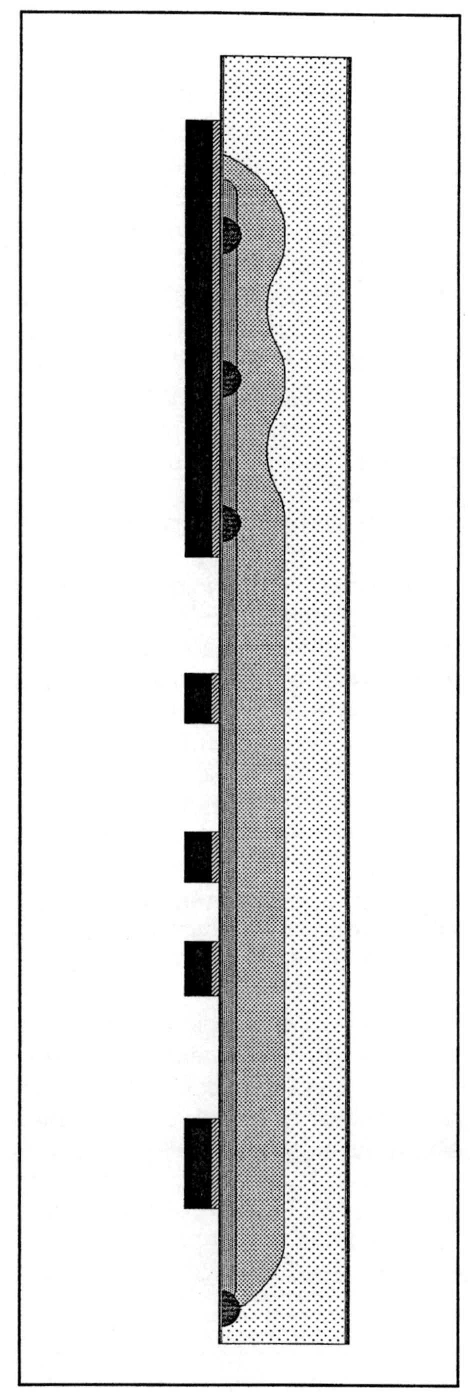

FIGURE A1.11. Situation after the etch process used for the first poly-Si layer.

A1.10. Interpoly isolation

Before a start is made to define the second poly-Si layer, electrical isolation has to be created between the first and the second. Because of the high quality of the oxide-nitride sandwich used for the gate dielectric, the same system is applied between the first two poly-Si layers.

After removal of the photoresist and a cleaning process, the wafers are placed in an oxidizing environment to convert part of the poly-Si gates into silicon oxide. Note that at this point the whole surface of the mono-silicon substrate is still covered with silicon nitride and protected from oxidation. The result is that only the poly-Si gates are oxidized. The oxidation takes place at a temperature around 1000°C and generates about 200 nm of silicon oxide on the poly-Si gates.

The result now is shown in Figure A1.12 : the poly-Si gates have been encapsulated in the thermally-grown oxide. Observe also that the silicon nitride at the rear of the wafer has prevented any further oxidation on that side of the wafer.

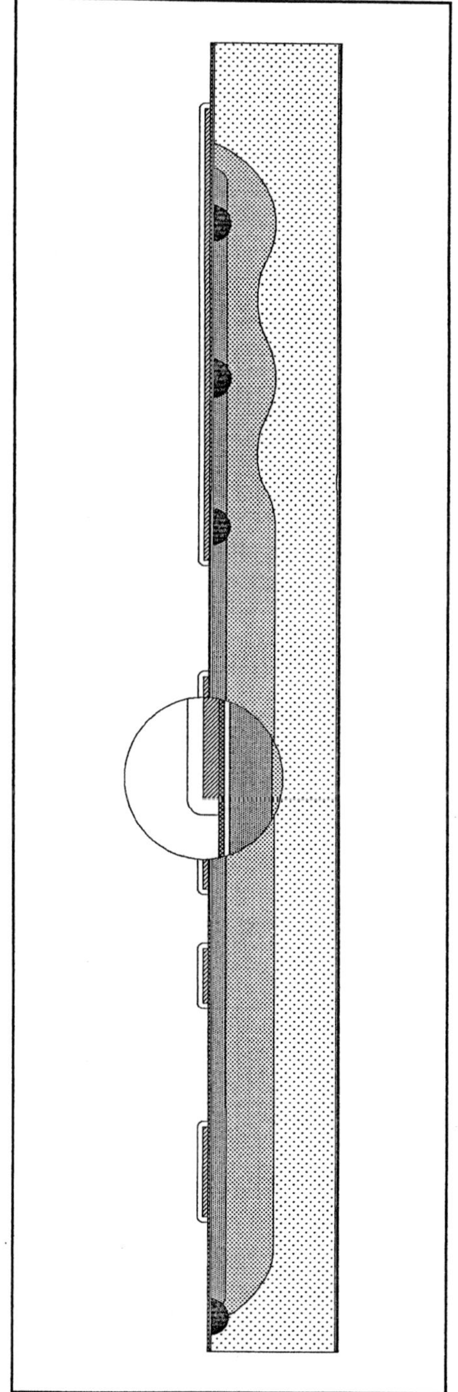

FIGURE A1.12. Situation at the end of the oxidation process used on the first poly-Si layer.

A1.11. Deposition of the second poly-crystalline-silicon layer

The next step is the deposition of the second polycrystalline silicon layer, acting as the second gate material. The thickness of the film is the same as that of the first poly-Si layer, namely 500 nm after deposition. However, only about 350 nm is left at the end of the processing as a whole.

Immediately after deposition, the poly-Si film is doped in the same way as the first poly-Si film was doped.

The resultant situation is shown in Figure A1.13 : the wafer is entirely covered with a uniform film of polycrystalline silicon. The latter is n-type doped to give the film low resistivity : about 35 $\Omega/\square$ at the end of the processing.

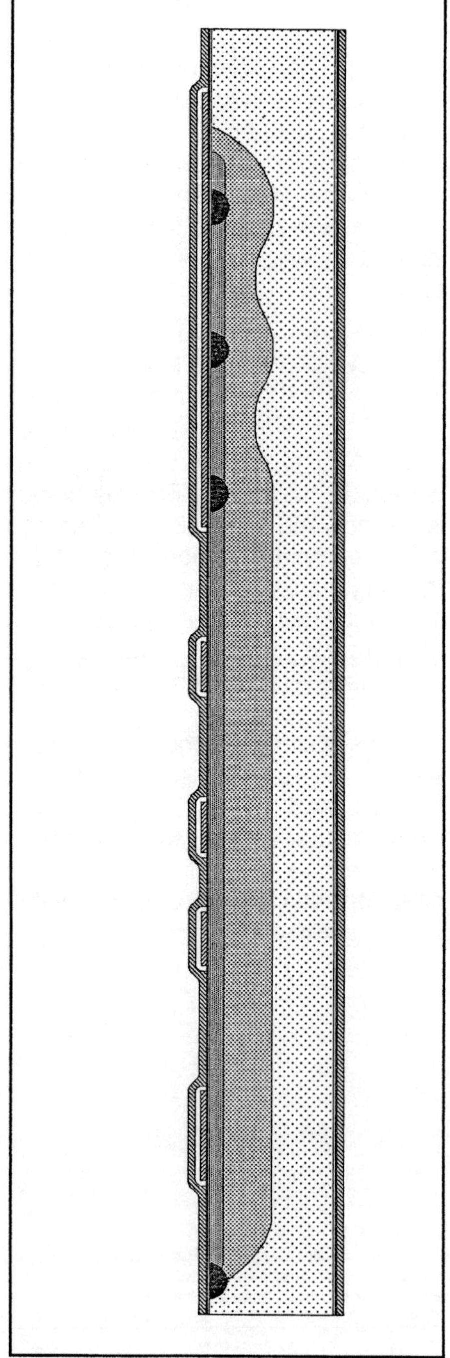

FIGURE A1.13. The wafer covered, at both sides, with the second polycrystalline silicon layer.

A1.12. Definition of the second poly-Si layer

After doping of the polycrystalline-silicon layer, the photoresist is deposited and patterned.   Next, the poly-Si film is etched, preferably using a wet, isotropic etching technique. As can be seen from Figure A1.13, the exact thickness of the second poly-silicon layer is not everywhere the same. This is due to the existing topography at the moment the second poly-Si was deposited. This irregularity of thickness makes it necessary to employ an isotropic etching method. This isotropic etching process, too, is optimized so that the poly-Si is etched very selectively to the silicon oxide and silicon nitride. The situation now is shown in Figure A1.14 : nothing in the image section (or along the cross section considered) has changed : almost the entire second poly-Si layer is removed.   The situation in the horizontal register is different. Here a few pieces of the second poly-Si material have been left to form some extra gates for the horizontal register. Note the overlap between the first and the second poly-Si gates in the horizontal register. This means that the continuous charge transport channel in the silicon can be controlled by the external gates. On the other hand, the overlap capacitances limit the maximum speed of the horizontal register. Overlaps are kept to a minimum for this reason, but are more or less necessary to ensure a smooth CCD transport.

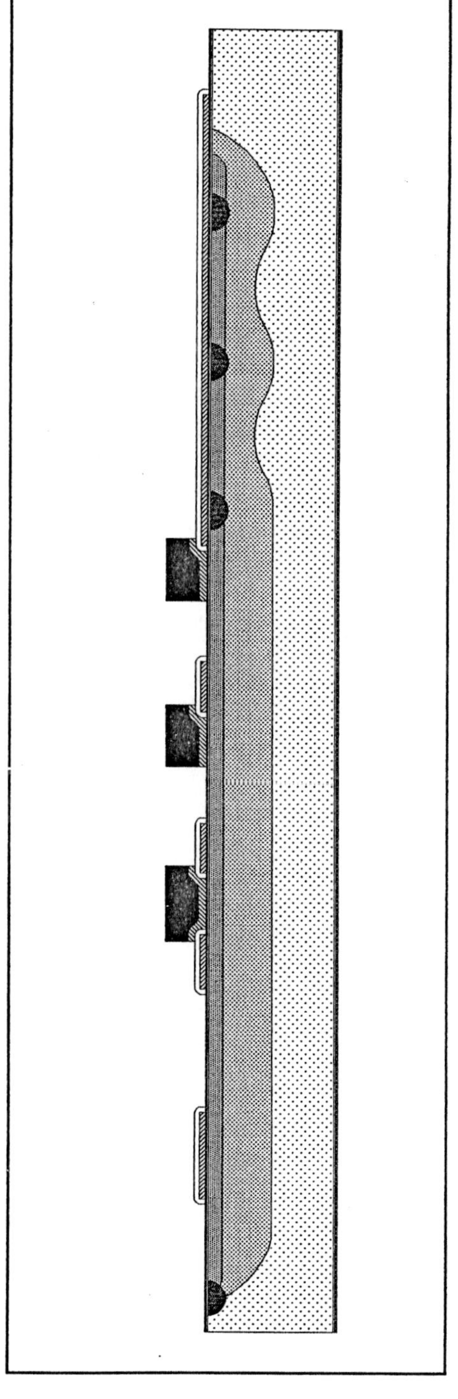

FIGURE A1.14. Situation after the etch process of the second poly-Si layer.

A1.13. Interpoly isolation

Before starting on implementation of the third poly-Si layer, electrical isolation between the second and the third one has to be created. After removal of the photoresist and a cleaning process, the wafers are placed in an oxidizing atmosphere to convert part of the poly-Si gates into silicon oxide. Note that at this point the entire surface of the monosilicon substrate is still covered with silicon nitride and protected from oxidation. The result is that only the poly-Si gates made from the second poly-Si layer are oxidized. Oxidation takes place at a temperature around 1000°C and generates about 200 nm of oxide on the poly-Si gates.

After oxidation, the nitride (from the gate dielectric sandwich) is etched without any mask. The oxidized poly-Si gates act as masks, and all the "free-lying" nitride of the gate dielectric is removed. The result at this moment is shown in Figure A1.15 : the poly-Si gates are encapsulated in the thermally-grown oxide and all silicon nitride has been etched away. Note also that the silicon nitride and the poly-Si layers at the back of the wafer have been removed.

Although not shown in Figure A1.15, growing the silicon dioxide takes up quite a lot of the polysilicon-gate material itself. The creation of 200 nm of oxide accounts for about 100 nm of poly-Si gate material.

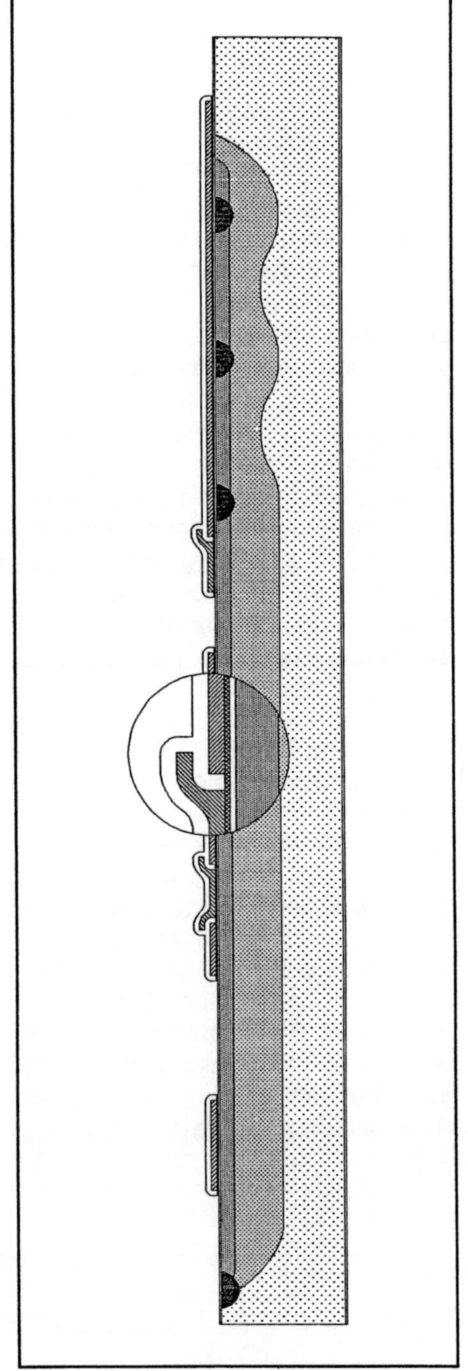

FIGURE A1.15. Situation at the end of the oxidation process for the second poly-Si.

A1.14. Deposition of the third poly-crystalline-silicon layer

Before the deposition of the third level of gate material, the gate dielectric has to be reconstructed.   Instead of acquiring an extra 40 nm layer of silicon-nitride, however, the remaining gate oxide of 80 nm is further oxidized thermally to 100 nm.

At this stage of the process, all gates will be given the same "electrical" gate dielectric, an equivalent of 100 nm silicon dioxide.

After the extra oxidation of the gate oxide, deposition of the third poly-crystalline Si, acting as the third gate material, can take place. The thickness of the film is the same as that of the other poly-Si layers, namely 500 nm, but about 450 nm is left at the end of the entire processing. Immediately after the deposition, the poly-Si film is doped in the same way as the other poly-Si films were doped.

The resulting situation is shown in Figure A1.16 : the wafer is totally covered with a uniform film of polycrystalline silicon. The latter is n-type doped using $POCl_3$ to lower the resistivity of the film to about 25 $\Omega/\square$ at the end of processing.

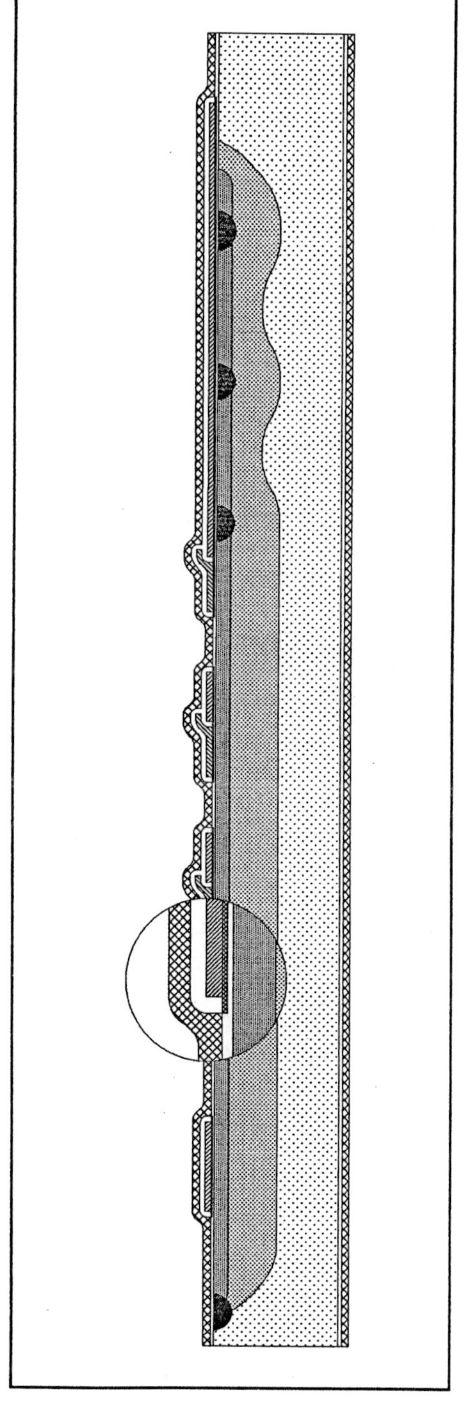

FIGURE A1.16. The wafer covered with the third poly-Si layer.

A1.15. Definition of the third poly-crystalline-silicon layer

After the doping of the polycrystalline-silicon layer, the photoresist is deposited and patterned. Next, the poly-Si film is etched, preferably using a wet, isotropic technique. The reason for the isotropic etching is the same as in the case of the second poly-Si film and can be seen in Figure A1.16 : the existing topography on the wafer makes the thickness of the poly-Si even less perfect than when the second poly-Si was etched. The etching process is optimized in such a way that the poly-Si is etched very selectively to the silicon oxide.

The situation now is as shown in Figure A1.17 : nothing has changed in the image section (or along the cross section considered) and the third poly-Si layer has been entirely removed. In the horizontal register the situation is different. Here only a few pieces of the third poly-Si material are left to form some extra gates of the horizontal register. Note that in the horizontal register the overlap between the first and the third poly-Si gates on one side and the second and the third poly-Si gates at the other side are necessary to ensure a smooth CCD transport. Now that the third poly-Si layer is in position, all CCD gates are available in both the image section and the horizontal output register. The order of the gates in the horizontal CCD register can be chosen randomly (so far as the layout rules or other architectural requirements permit) as shown by the example in Figure A1.17.

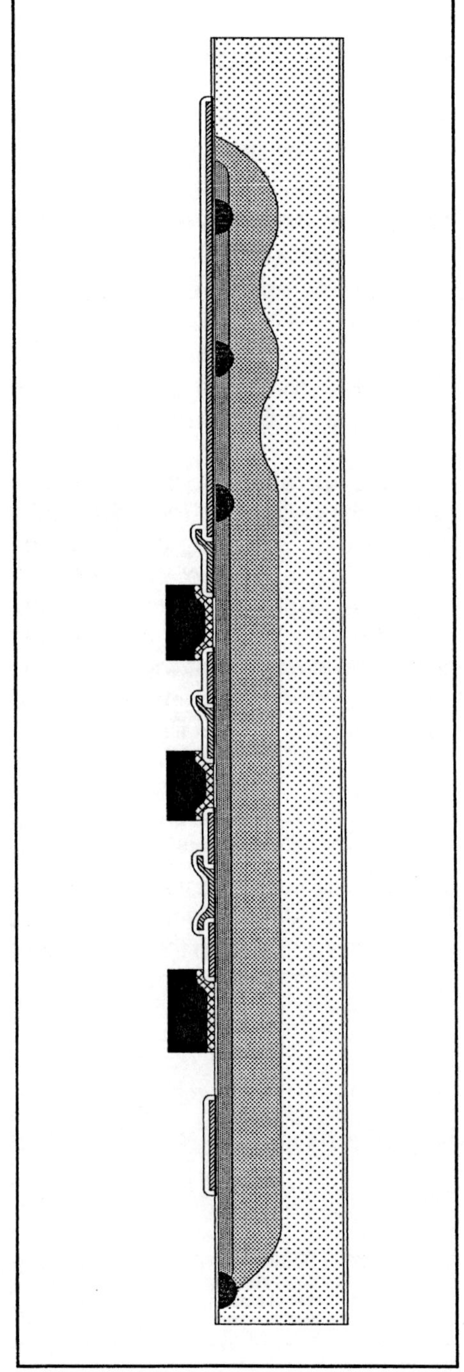

FIGURE A1.17. Situation after the etch process for the third poly-Si layer.

A1.16. Source and drain implantation

Before the metallization processing is initiated the last implantation to define the n^+ regions or the sources and drains of the n-channel MOS transistors has to be done. The resist of the previous step is removed, and new photoresist is applied to the wafers and patterned. Openings are made at those locations where n^+ implants in the monosilicon are needed. All mono-Si not covered with polysilicon is shielded by 100 nm of SiO_2. This is too thick to implant the ions through. Before the implantation, the silicon wafers are etched to decrease the thickness of the SiO_2 film to about 50 nm. Next, the implantation is done with the same resist pattern still on the wafers. A relatively high dose of arsenic ions (of the order of $10^{15}/cm^2$) is implanted. The choice of As instead of P for the n^+ regions has to do with the diffusivity of the As ions in the silicon. To keep the outdiffusion and underdiffusion of the n^+ regions to a minimum (and thus limit the stray capacitances of the sources and drains of the MOS transistors), an ion with a low diffusion constant is needed, namely As. The result at this point is shown in Figure A1.18 : n^+ islands are only needed to create the floating diffusion area, which coincides with the source region of the reset transistor, and the drain of the reset transistor. The floating diffusion is made self-aligned with the aid of the poly-silicon gates (the opening in the resist is greater than the ultimate implanted area in the silicon). Self-alignment at the drain side of the reset transistor is only possible at one edge.

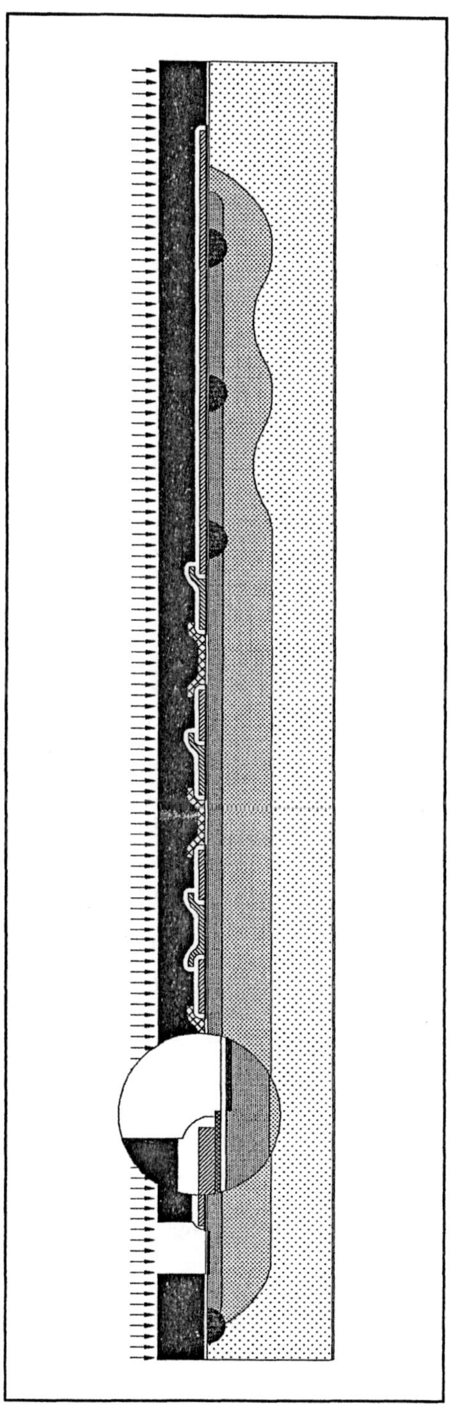

FIGURE A1.18. Situation after the source and drain implantation.

A1.17. Back-end isolation

By definition, the "back-end" of CCD fabrication includes the metallization and scratch-protection processes. Before the metal can be deposited, deposition of the intermetal isolation layer has first to be done. This step consists of two phases : first, the last oxidation of all "free-lying" silicon and, second, the CVD of the isolation layer itself. At this stage of the processing, only the third poly-silicon layer is still free of any oxide. After the last implantation the photoresist is removed and the wafers are cleaned and oxidized at a temperature of about 1000°C. An LPCVD SiO_2 film is deposited onto the wafers next to the thermally-grown-oxide layer. The thickness is chosen around 1 μm and this layer has to serve as the isolation dielectric between the metal layer and the underlying substrate. Because the quality of vapor-deposited silicon-oxide layers is not as good as that of the thermally-grown layers, the LPCVD film is much thicker. This has extra advantages : it makes the topography of the wafers flatter and the parasitic capacitances of the metal wires to the rest of the chip smaller. But, as will be seen during the steps which follow, a thicker intermetal isolation places some exacting requirements on the technique used for the deposition of the metal film. The situation at time point is shown in Figure A1.19 : the third poly-Si layer has been thermally oxidized and the complete structure covered with a 1 μm thick isolating LPCVD silicon-oxide film.

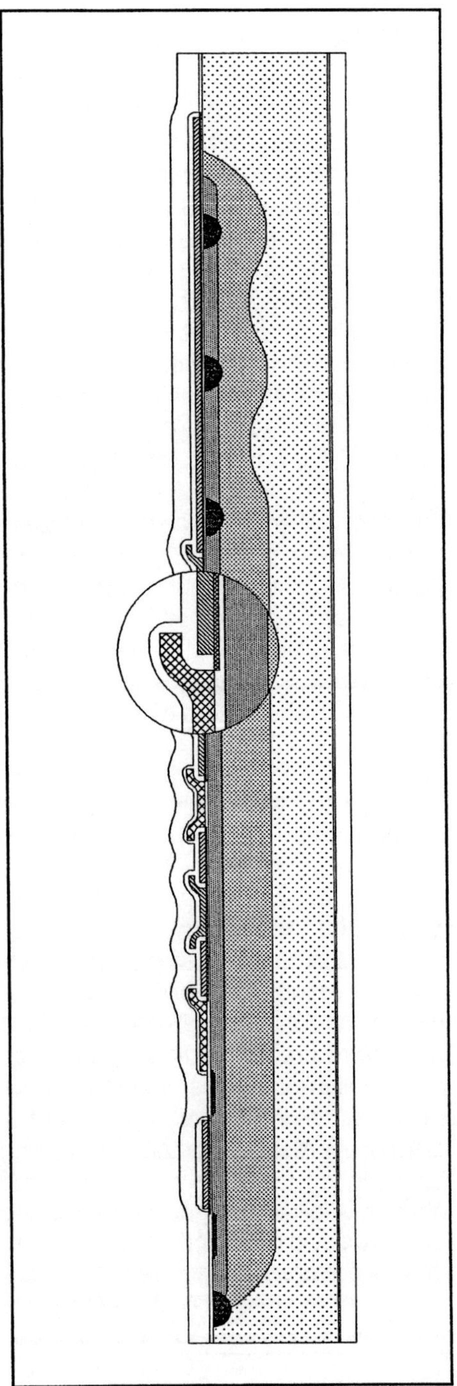

FIGURE A1.19. The wafer is covered with an LPCVD intermetal dielectric, preparatory to entering the back-end processing phase.

A1.18. Contact-hole definition

The first step in the back-end part of the process is definition of the contact holes, i.e. at what locations the metal has to make contact with the underlying structures. The photoresist is applied to the wafers and patterned. The silicon oxide stack (of thermally grown and LPCVD deposited SiO_2) is isotropically etched. This etching process is quite a complicated step because the etchant has to go through different types of layers and the etching rates in these layers can be different from each other. The etching technique itself is optimized so as to give rise to a certain amount of "underetch". The etchant attacks not only the silicon oxide vertically (deeper into the film) but also sideways or laterally. This gives rise to a funnel-shaped contact opening. The benefit of this will become clear in the next processing step.

The situation now reached is shown in Figure A1.20 : the whole structure is covered with photoresist, except for those areas where the metal has to contact the underlying structures (mono-Si or poly-Si). In Figure A1.20 two openings have been made to contact the reset transistor (floating diffusion or its source and drain) and an extra opening has been made to contact the clock phase of the image section made from the first poly-Si layer.

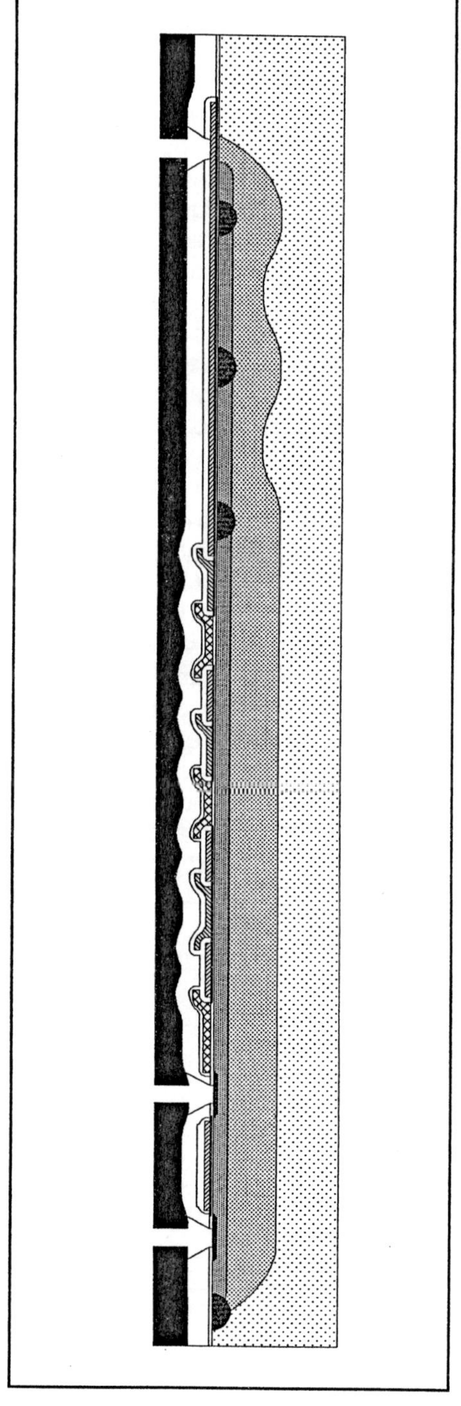

FIGURE A1.20. Location of the contact holes through which the metal has to contact the underlying structures.

A1.19. Deposition of the metal layer

The next step in back-end processing is the deposition of the metal layer, which is done after the photoresist has been removed. The deposition method used to cover the wafers with the metal is a sputtering technique. Its effective step coverage is a particularly attractive feature to deposit the metal in the contact holes, which are relatively deep in comparison to their width.

In most cases metallization of the chips is done by means of an aluminum film doped with 1 % silicon. Aluminum is very attractive for its low resistivity, a typical value being 30 mΩ/$\square$. Doping of the metal film with 1 % silicon is necessary to avoid alloying effects. The latter can be appreciable when pure aluminum is deposited on silicon and alloying takes place due to the aluminum layer "consuming" silicon. Where the aluminum contacts very shallow junctions, the alloying effects can cause the latter to short circuit. The alloying effects can be avoided by doping the aluminum with silicon atoms.

The situation reached at this moment is shown in Figure A1.21 : the complete wafer is covered with a metallization film with a thickness of 1 μm.

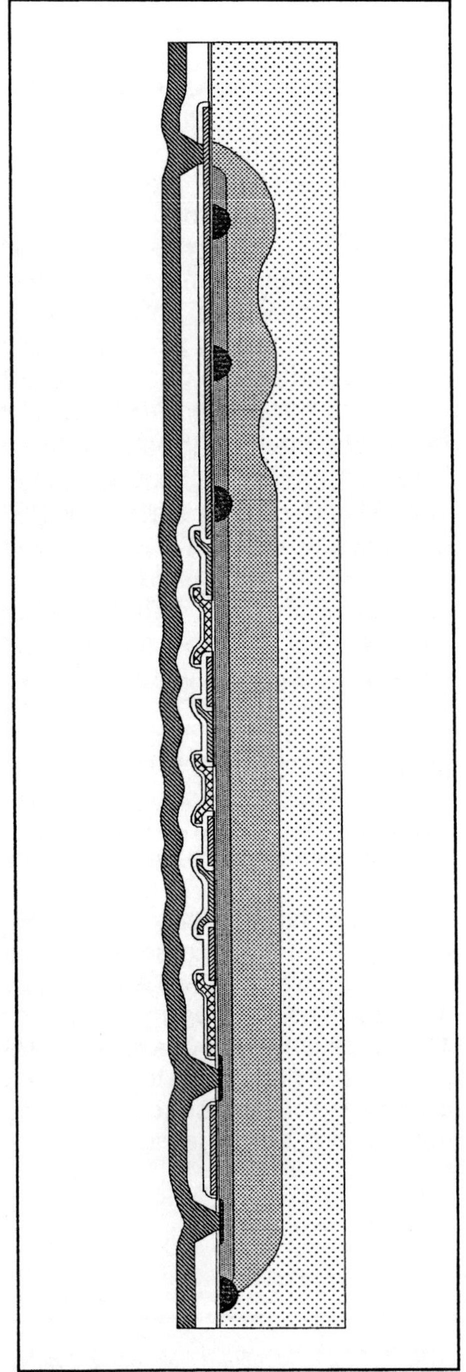

FIGURE A1.21. Deposition of the aluminum layer needed for the metallization of the circuit.

A1.20. Definition of the metallization pattern

Once the metal film has been deposited, the photoresist needed to define the metallization pattern is deposited. The lithographic step is followed by etching of the metal film. Either wet or dry etching techniques can be used, but the etching method should preferably be based on an isotropic process. The main reason for this preference is the underlying topography, which has been mainly generated by the three layers of poly-Si.

The situation thus created is shown in Figure A1.22 : the complete wafer is covered with the photoresist and the film has been isotropically etched. Note the underetching and also the slope of the edges of the metal patterns. In the example shown in Figure A1.22, metal dots are left at the source and drain of the reset transistor and at the contact to the first poly-Si gate layer. In principle these metal dots are also connected (by the metal) to bonding pads (for the reset drain and the poly-Si gate) or to other on-chip parts (for the floating diffusion or source of the reset transistor).

Observe also the aluminum light shield which is left on top of the CCD register. The light shield prevents photons from penetrating into the horizontal readout register and disturbing the video information by generating additional electron-hole pairs.

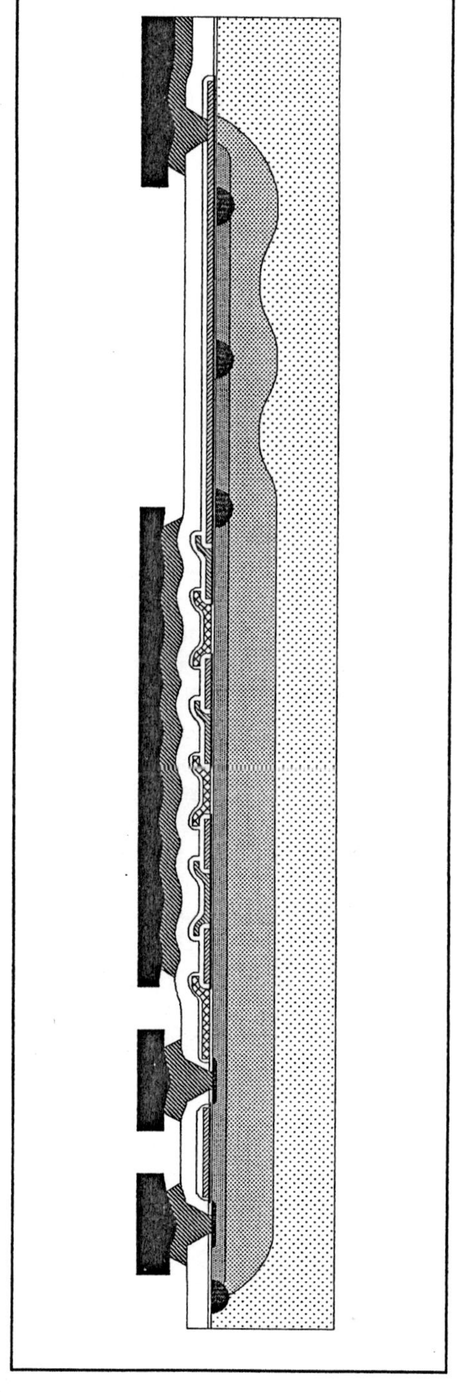

FIGURE A1.22. Definition of the metal pattern : the interconnects and the light shield are included in the same metal layer.

A1.21. Scratch-protection deposition

The scratch protection will complete the processing of a typical black-and-white image sensor. The photoresist of the previous lithographic step is removed and the last film is deposited on the wafer; it is Si_3N_4 and provides scratch protection. A plasma-enhanced CVD method is used to deposit a film with a thickness of 1 μm. After the film has been deposited, the bonding pads have to be cleared of the nitride film. This is done in an additional lithographic step and etching process. Finally, the wafers are annealed at a low temperature, namely about 400°C. This anneal is doubly cured : by sintering the metal-silicon contacts and annealing the interface states available in the active structure. The latter process is made more effective by the very active hydrogen released during the plasma deposition of the nitride film. In the course of heat treatment, the hydrogen will diffuse to the Si-SiO_2 interface and chemically interact with the dangling silicon bonds. Hydrogen can only reach the interface at those locations where the monosilicon is not covered by the thinner nitride layers, because the nitride film prevents the diffusion of any atom. This is why, earlier in the processing, the previous nitride films are defined such that the third poly-Si layer is free of any underlying nitride layer. Hydrogen atoms can then diffuse through these holes (etched in the nitride films). Figure A1.23 shows the situation at this point : the fully finished wafer ready for use as a CCD imager for black-and-white applications.

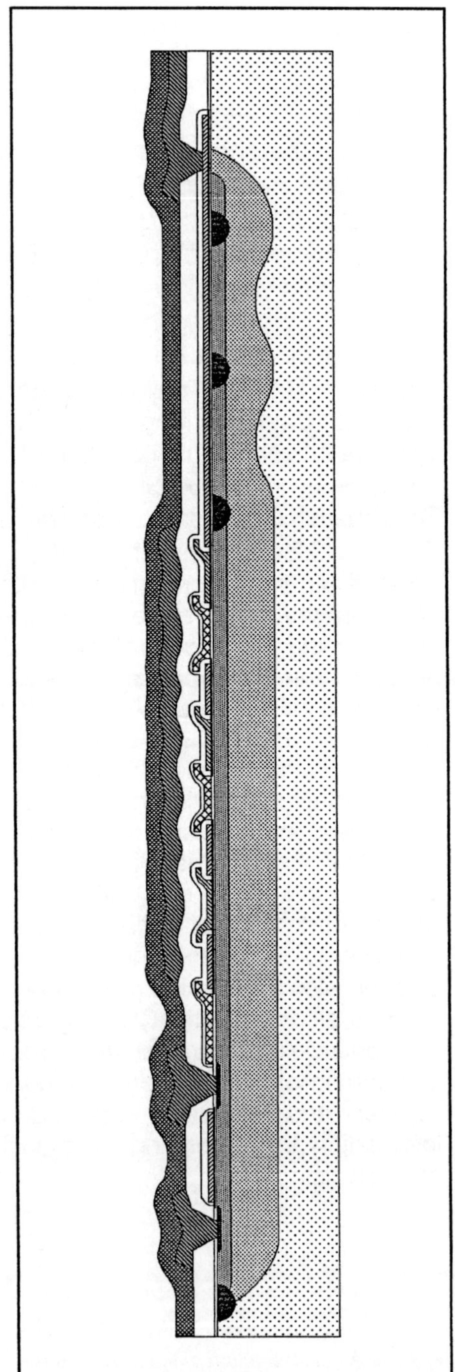

FIGURE A1.23. The wafer at the end of standard CCD processing.

A1.22. Deposition of the first color filter layer

When processing of the black-and-white imager is finished, color filters will be deposited if the device has to be used for color imaging. There are several possible techniques for depositing the filter materials but evaporation of the color pigments in combination with a lift-off lithographic step is described here. Before deposition of the color filter itself, the resist is applied to the wafers and processed. Unlike in classical resist processing, however, the photoresist in the lift-off method is removed at locations where the color filter has to remain on the wafer. Vice versa, at locations where no color filter is needed, the photoresist is left on the wafer. Next comes the evaporation of the first color filter, in this example the cyan pigment. The situation now reached is shown in Figure A1.24 : the photoresist has been applied to the wafer, and "oppositely" patterned, and the color filter has been evaporized on the chips.
Note the typically negative slope of the photoresist; this is needed to avoid the covering of the edges of the photoresist. In Figure A1.24 only one pixel is covered by the cyan filter.

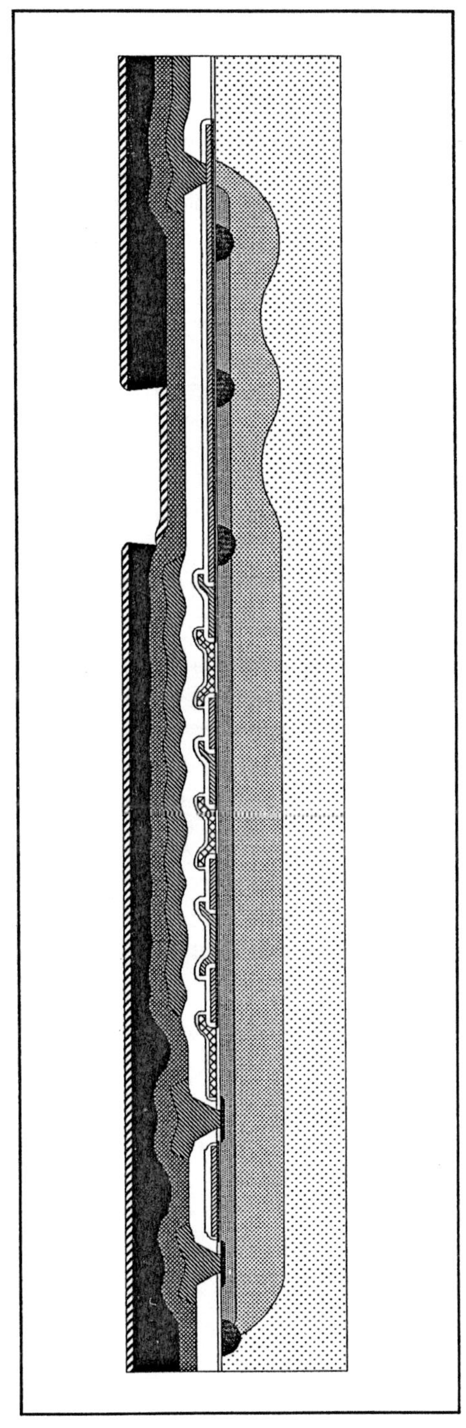

FIGURE A1.24. Deposition of the first color-filter layer in combination with a lift-off technique.

A1.23. Lift-off of the first color filter

To complete the processing of the first color-filter layer, simple removal of the photoresist is carried out. In this operation the color-filter material lying on top of the photoresist is also removed. To effectively remove the photoresist, however, the chemicals which do not attack the cyan filter have to attack the underlying photoresist. To bring the chemicals into contact with the resist, the edges of the resist have to be free from color-filter material. For this reason, the photoresist pattern has, as already explained in the preceding step, to have negative slopes.

The resultant situation at this processing step is shown in Figure A1.25 : after the stripping of the resist, the cyan filter is left only on a single pixel.

The pattern in the photoresist was designed so that the color filter covers the pixel from one center of the stopper implantation to the next. Precise adherence of the specification of these regions is not very critical because the light sensitivity of the stopper regions is quite low. Consequently, the pixel is defined not by the color filter but by the gap between two stopper implantations.

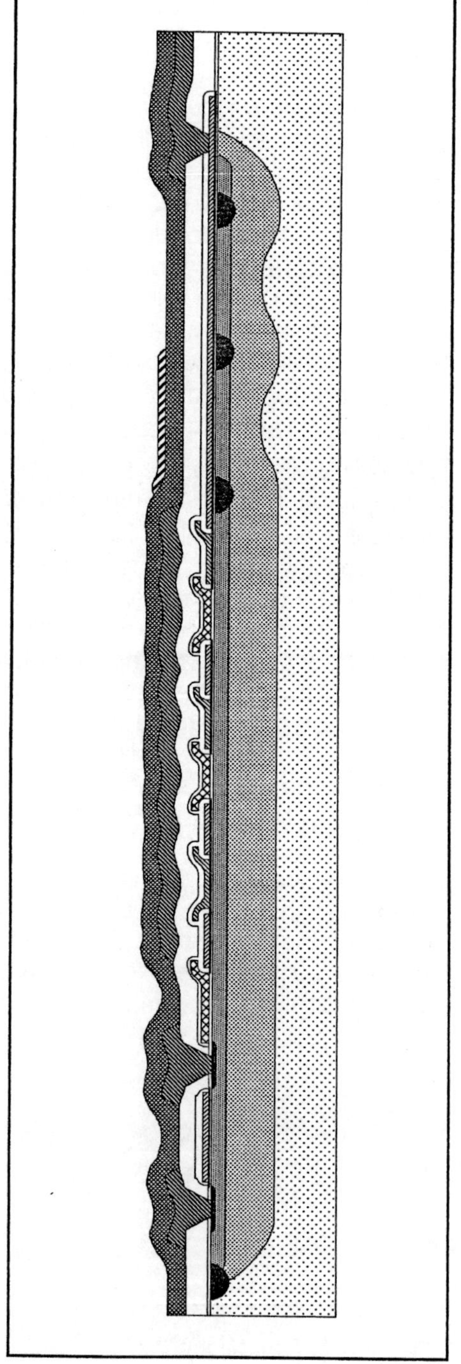

FIGURE A1.25. The situation at the end of the cyan-filter processing.

A1.24. Deposition of the second filter layer

To deposit the second color filter, the same method as explained before is applied : evaporation of the color pigment in combination with lift-off. Depending on the material used, it may be necessary to deposit an intermediate layer between the different color filters. To keep the processing as "simple" as possible, the color pigments are chosen such that they do not chemically interact with each other. Before deposition of the color filter itself, the resist is applied to the wafers and processed. Next comes the evaporization of the second color filter, in this example the yellow pigment.

The result at this point is shown in Figure A1.26 : the photoresist has been applied to the wafer and the color filter is evaporated onto the chips, two pixels covered with the yellow filter. One of the latter was already covered with the cyan pigment from the previous processing step. In combination with the additional yellow layer, the overall color filter for this pixel has become a green one.

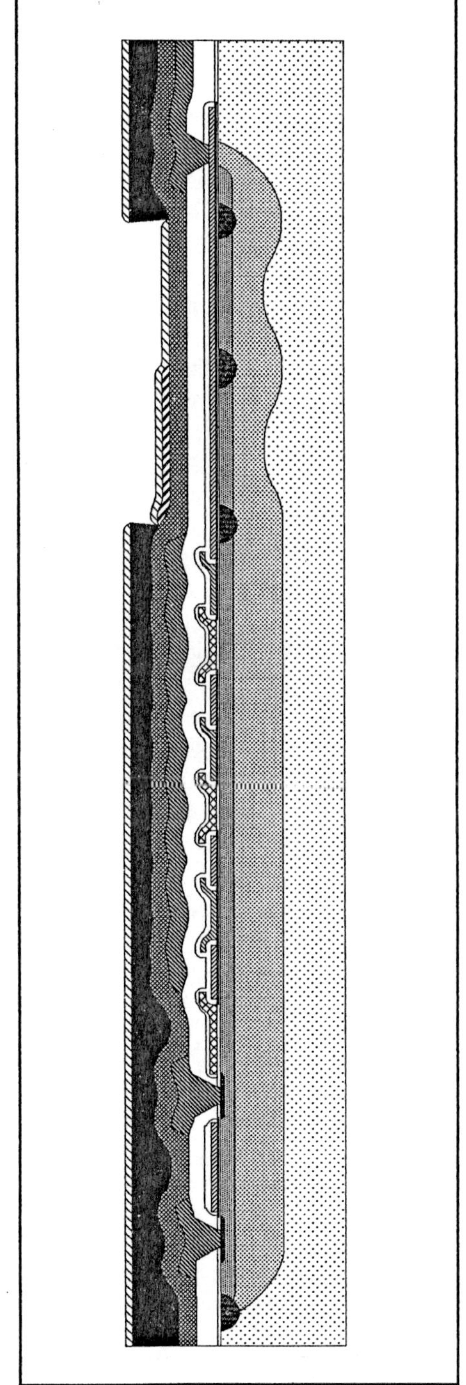

FIGURE A1.26. Deposition of the second color-filter layer in combination with a lift-off technique.

A1.25. Lift-off of the second color filter

To complete the processing of the second color filter layer, simple removal of the photoresist is necessary. The process is the same as that as applied to the first color-filter layer.

The result after this step is shown in Figure A1.27 : after the resist is stripped, the yellow dots are left on the pixels.

As already mentioned, the combination of the previous cyan filter with the new yellow one creates the green filter. In the example, the cyan pixel and green pixel are shown, but single cyan dots are also available at other locations in the active imaging area.

For a typical mosaic-filter pattern, three of the four pigments are available. The magenta color filter is still absent. If the imager has to be provided with the Cy-G-Ye-Mg arrangement, an additional third layer has to be deposited on the wafer. If the filter architecture is based on Cy-G-Ye(-W), the processing of the color filters is finished at this stage. The imager is ready for use in all kinds of color applications.

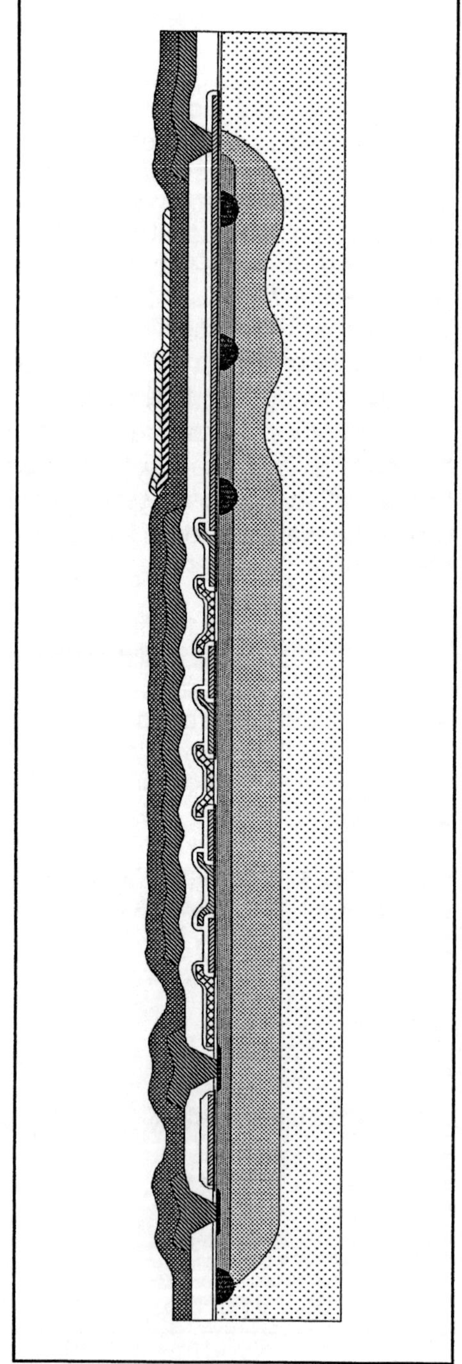

FIGURE A1.27. The situation at the end of the yellow filter processing, and if a Cy-G-Ye filter arrangement chosen, the processing of a full color imager has been completed.

A1.26. Deposition of planarization layers

By this stage of the processing a complete color-imaging device is ready for testing. To optimize the light sensitivity, it may be attractive to coat the imager with microlenses. In the example described in this section, microlenses are not of practical use, but these processing steps are also included in the full-frame device to give some insight into the microlens technology. Processing of the microlenses is initiated with a planarization procedure. The surface of the imager has to be completely flat before microlenses can be incorporated. Planarization is done in two steps : deposition of the planarization layer(s) and etch-back of the structure.

The surface topography can be smoothed by spin-on coating of a planarization layer followed by an optional curing step. Although the planarization film lowers the topography, some surface "roughness" still remains. After deposition of the planarization layer, a relatively thick photoresist layer is added to the wafers. This operation is optimized to ensure a completely flat surface.

The result is shown in Figure A1.28 : the planarization layer added after the color-filter processing. The topography has been reduced but not completely eliminated. A photoresist layer has also been added to the wafers.

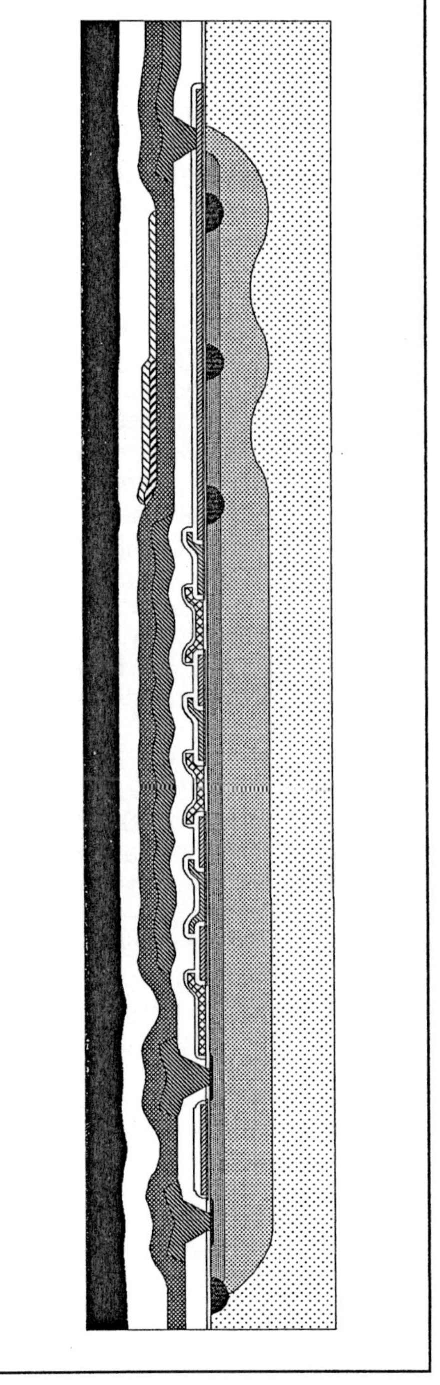

FIGURE A1.28. Deposition of the planarization layers : the spin-on coated planarization film and the photoresist.

A1.27. Etch-back of the planarization layers

After the photoresist has cured, the wafer is etched back. The unpatterned but very smooth surface of the photoresist is etched in a dry etching environment which attacks the photoresist layer very uniformly. Another important characteristic of the etching method is the etching rates of the photoresist and the planarization layer are the same. In this etching process, the photoresist is "peeled off" layer by layer. At a certain moment a situation will be reached in which the photoresist has completely disappeared at locations where the resist thickness was the smallest and where the highest points were located in the topography of the planarization layer. Due to the etch-back mechanism, the etching environment will make no difference between the etching rate of the resist and that of the planarization film.

The flat surface of the resist at the beginning of the etch-back process is completely transferred into the top layer of the underlying planarization film. When the point when the photoresist layer has been totally etched, is reached, the result will be a uniformly flat surface of the planarization layer.

The resultant situation is shown in Figure A1.29 : the photoresist has been totally etched away and its flat surface at the beginning of the etch-back procedure has been transferred into the underlying planarization layer.

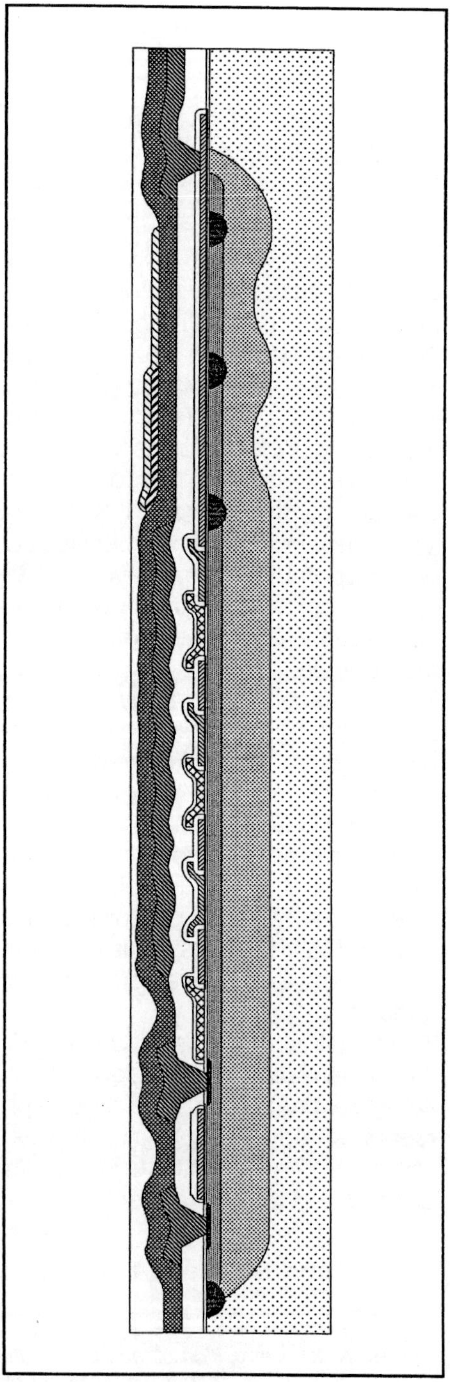

FIGURE A1.29. A uniformly flat top surface of the wafers has been generated by etch-back of the planarization layers.

A1.28. Deposition of the lens material

The lens material chosen is a deep-UV photoresist and it is deposited in complete accordance with the classical method of coating wafers with photoresist. The coating is added to the wafers using a spin-on technique. Due to the fact that the lens material is a photoresist, the patterning which follows is also done in the classical manner. The lens material is exposed to the lithographic process and afterwards developed. This means that both the resist and lens material have now been patterned.

The entire optical characteristics of microlenses are a function of the refractive index of the lens material, of the curvature of the microlenses, of the lens thickness and of the microlens-substrate distance. The latter is determined by the thicknesses of the various underlying layers. In cases where the planarization layer does not add sufficiently to the separation between the microlenses and the substrate, an intermediate, so-called base-resin layer, may be added. This process step is not illustrated in the figures.

For their part, the refractive index of the lens and the thickness of the lens are determined by the choice of lens material and the spin-on method used to coat the wafers with the lens material. The resulting situation is shown in Figure A1.30 : the deep-UV photoresist has been added to the wafers and patterned to create microlenses from it.

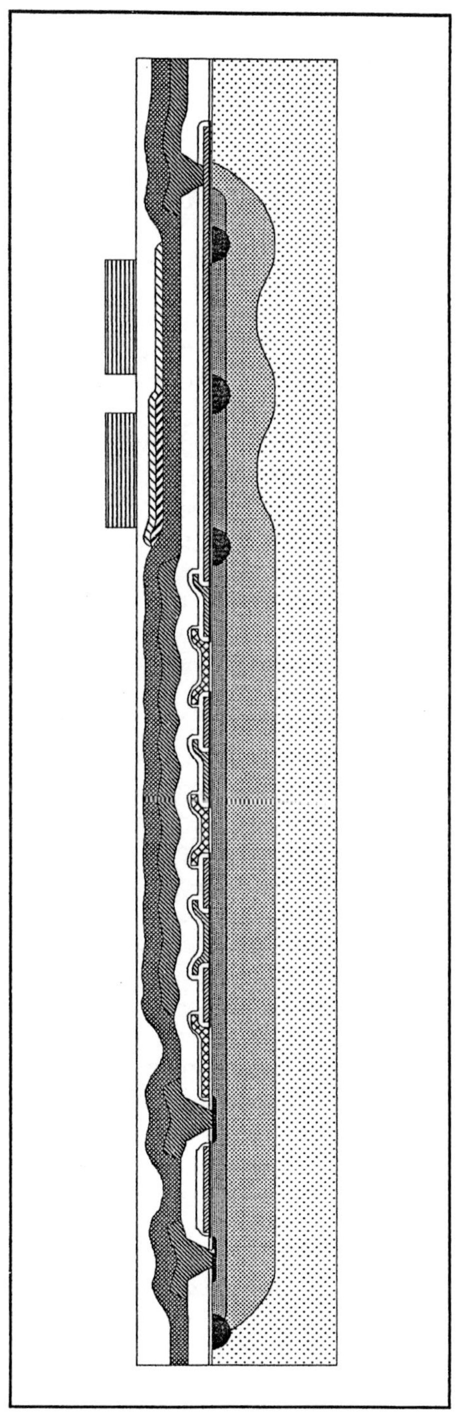

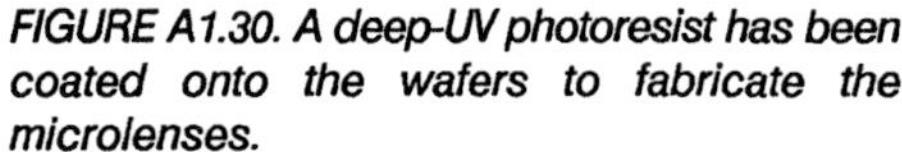

FIGURE A1.30. A deep-UV photoresist has been coated onto the wafers to fabricate the microlenses.

A1.29. Reflow of the microlenses

The patterning of the photoresist in a classical lithographic process is followed by the curing of the resist layer. The same applies to the processing of the microlenses. The curing cycle employed is not the standard one but an optimized process, which does not leave the rectangular resist strips or dots intact. During curing the rectangular elements are deformed into a curved structure which ultimately forms the microlens itself.

The result of the entire processing is shown in Figure A1.31 : the deep-UV photoresist has been cured and microlenses formed from the rectangular dots left on the wafers during the preceding processes. The exact duration and temperature of curing determine the curvature of the microlenses.

The microlenses of the structure shown in Figure A1.31 do not enhance the sensitivity of the imager because, whether they are present or not, the same photons impinge the CCD. The microlenses have no added value in this example and are featured in this discussion only for the sake of completeness. Microlenses only serve a purpose in cases where the pixels have dead areas. Only if they concentrate on the pixel the light, which in the absence of microlenses would have impinged on the dead area, then they can be effective.

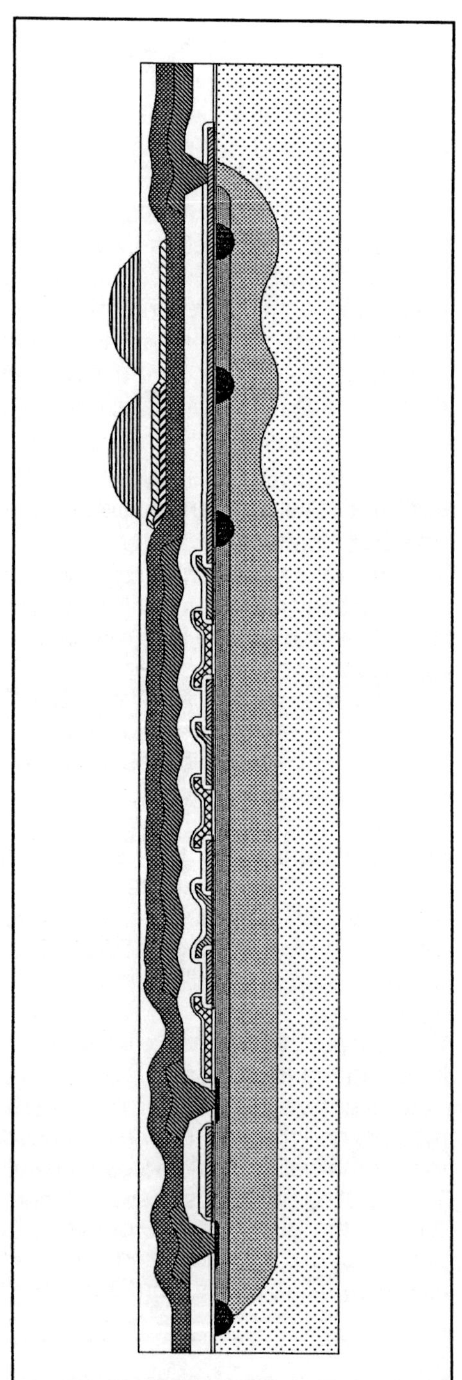

FIGURE A1.31. The situation at the very end of the entire processing of charge-coupled devices for imaging applications, including color filters and a microlens array.

HOW TO INTERPRET CCD ARTIFACTS

A few typical artifacts which can show up on a display when an image is captured by a CCD camera are briefly discussed in this appendix. Characteristics discussed are smear, blooming, charge reset, integration-time shortening, defects in the cover glass, defects due to electrostatic discharge, gate oxide leakages, column defects, white spots, fixed-pattern noise and output amplifier noise. Of course, this list is not exhaustive, but nevertheless it gives a short overview of the things which can go wrong when CCDs are applied.

A2.1. Reference picture

The reference picture is shown in Photograph A2.1 : a metronome, a wooden toy, a lamp, and, at the back, a monitor. The images are captured with a "standard" CCD camera provided with a "standard" frame-transfer CCD. The CCD has 500,000 pixels and is operated in an interlaced scanned mode. The video data delivered by the camera is captured by a memory in single fields, and later displayed in continuous mode on a monitor.

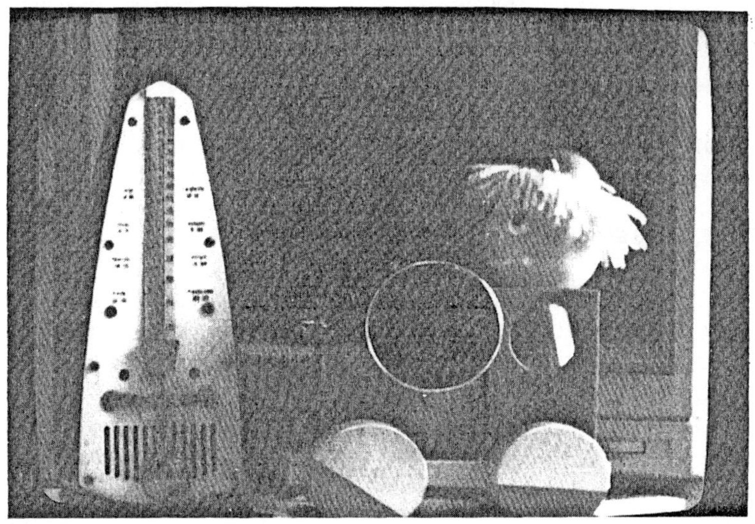

PHOTOGRAPH A2.1. Reference picture.

A2.2. Smear

Switching on the lamp in the scenery will cause local overexposure of the CCD. With proper antiblooming measures, the overexposure will cause no artifacts, but the smear will. Right in the center of the highlight, a vertical smear signal will be generated and the spurious signal can be seen extending from the top to the bottom on the monitor. For the explanation of smear, see section 8.5. The "top part" of the smear is generated before the integration of the video signal; the "bottom part" of the smear is generated after the integration. The net result is shown in Photograph A2.2.

PHOTOGRAPH A2.2. Smear induced by a highlight.

A2.3. Charge reset

Charge reset or electronic shuttering will drain all charges available in the image section of the device and consequently the imager will be completely free of electrons. If this technique is applied after the frame shift of a FT CCD and before the integration starts, all smear signals generated before the integration will also be drained. Photograph A2.3 shows that only the smear after the integration is left on the monitor. More details of charge reset or electronic shuttering can be found in section 6.4.

PHOTOGRAPH A2.3. Smear generated before the integration is dumped by means of charge reset before the integration starts.

A2.4. Blooming

Increasing the intensity of the highlight will raise the level of smear, but at a certain level of light intensity, the antiblooming capabilities of the CCD are able to drain all excess charges. Consequently, they will spill over to neighboring CCD cells and the physical size of the highlight on the monitor will increase (preferentially along the vertical columns). The effect of blooming can be seen in Photograph A2.4 and if it is compared with Photograph A2.2, the increase in spot size and the increase in smear level can be seen.

Both smear and blooming are the result of overexposure. The main differences between them are :

- the intensity on the monitor : the smear has uniform intensity along each column, but its intensity may differ from column to column. The blooming effect has an intensity equal to the top white signal;
- the location : smear is manifested from the top to the bottom on the monitor; blooming disturbs the video information locally.

A2.5. Integration-time shortening

Photographs A2.5 and A2.6 are included to illustrate the influence of the duration of the integration time :

PHOTOGRAPH A2.4. *Blooming effects introduced by local overexposures to such an extent that the antiblooming capabilities can no longer handle the amount of charges generated.*

PHOTOGRAPH A2.5. *"Fast" movements cannot be sharply displayed on the monitor if the integration time of the CCD is chosen too large.*

- in Photograph A2.5, the metronome is "triggered". When the camera is operated with a relatively large integration time, the moving objects in the scenery will be "smoothed" in the displayed image and cannot be visualized sharply;

- in Photograph A2.6, the integration time is shortened to such an extent that the moving object is sampled in a very short time. This method causes it to look very sharp on the monitor.

To obtain the same overall intensity and contrast level on the monitor as in Photograph A2.5, the overall illumination of the scene has to be increased by the same factor by which the integration time is decreased.

PHOTOGRAPH A2.6. Decreasing the integration time of the CCD decreases the sampling period and allows moving objects to be displayed fairly sharply.

A2.6. Column defects

Column defects can be generated by several mechanisms. That shown in Photograph A2.7 has its origin in the transfer from the storage area of the frame-transfer CCD toward the horizontal output register. This effect is due to the nature of the column defects : they all run from top to bottom on the monitor, and each has more or less the same intensity along the column as a whole. Appropriate biasing of the clocks of the horizontal register can help to overcome this kind of artifact.

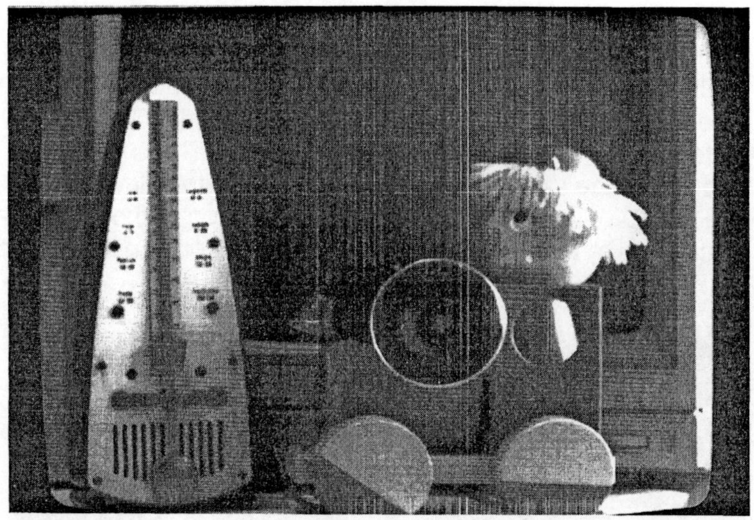

PHOTOGRAPH A2.7. Column defects caused by incorrect charge transport.

A2.7. Cover glass damage

Damage to the cover glass due to particles in the glass is always displayed on the monitor as blurred black objects, as illustrated in Photograph A2.8. This result can be easily explained : the defects are out of focus and limit the amount of light falling on the sensor. If the effect is caused by dust, the cover glass can be cleaned, but if the cover glass is scratched or some particles are enclosed in the material itself, the damage cannot be repaired.

A2.8. Damage by electrostatic discharge

Photograph A2.9. shows an example of damage (at the top of the metronome) caused by electrostatic discharge (ESD).
Charge-coupled devices are built in MOS technology. This fabrication technology is very sensitive to electrostatic discharge phenomena, so also are CCDs. ESD can be introduced merely by handling. Several CCD manufacturers protect their devices by incorporating gate-protection circuits on-chip. Nevertheless, CCDs are still very vulnerable devices. Moreover, ESD damage is destructive and cannot be repaired.
Although the CCD shown in Photograph A2.9 can still operate, in worst-case situations the output amplifier may be destroyed by electrostatic discharge and consequently the device will be unable to deliver any video at all.

PHOTOGRAPH A2.8. Damage to the cover glass.

PHOTOGRAPH A2.9. Damage caused by electrostatic discharge induced by incorrect handling of the devices.

A2.9. Gate dielectric damage

Weak spots in the gate dielectric do not always result in devices which are totally defective, but they can show up on the monitor as local artifacts. An example is shown in Photograph A2.10 : at the top part of the image two white blemishes can be seen. The weak spots in the gate dielectric which generate these kinds of "leakers" can be introduced during the processing of the devices. A small dust particle or some contamination during the oxidation of the wafer can introduce a local breakdown region into the gate dielectric. If such a leaker is of very considerable amplitude, it can eventually generate a white column defects instead of only a white spot.

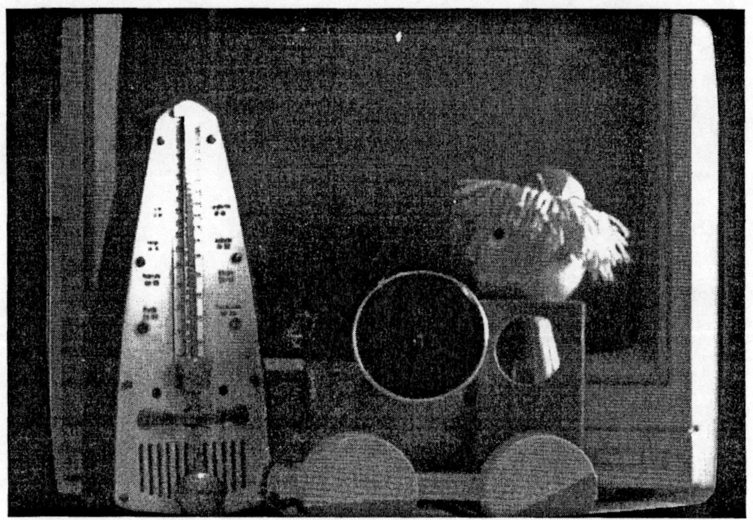

PHOTOGRAPH A2.10. "Leakers" caused by local defects in the gate dielectric.

A2.10. White point defects

Photograph A2.11 shows the effect of excessive dark current nonuniformities caused by the fact that the wafer has been contaminated by a main source of impurities. Examples of such an impurity can be iron, copper, nickel or other metal atoms. These foreign atoms can be introduced in the wafers during the preparation of the starting silicon substrates or during the processing itself (e.g. by a contaminated atmosphere, gas or etchant).
This example shows the importance of ultra-pure starting materials, not only for the silicon substrate but for all gases and chemicals used during the processing of the wafers. Any contaminant present, even in extremely small quantities, can nullify the yield of the production process.

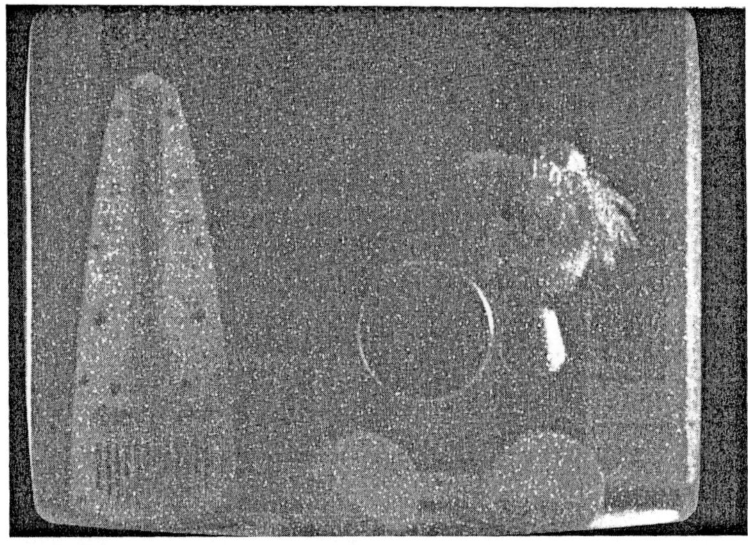

PHOTOGRAPH A2.11. White point defects generated by a single dominant contaminant.

A2.11. Fixed-pattern noise

Even in the absence of a contaminant as described in the previous section, dark-current distribution along the various pixels of the CCD will never be uniform. The reason for this is the fact that the dark current is a random process and the dark-current content of pixels will vary from pixel to pixel. How it looks on a monitor is illustrated in Photograph A2.12 : a fixed-pattern noise which acts like a dirty window through which the camera observes the scenery.
This type of fixed-pattern noise is only visible if the CCD exhibits a fairly high level of dark current, or the device is operated at elevated temperatures. However, the effect of this type of fixed-pattern noise can be lessened by cooling the device.

A2.12. Other noise sources

If the illumination level of the scenery is too low, the gain of the off-chip amplifiers can be increased. Not only the video signal delivered by the CCD will be amplified, but also the thermal noise of the CCD output amplifier. On the other hand, if the illumination level is fairly low, photon shot noise also becomes important. The image shown in Photograph A2.13 contains this amplifier noise, photon noise and probably other noise sources too. Because all these noise types are temporal in nature, they cannot be separated from each other. (In Photograph A2.13 they are all frozen due to the fact that the image is stored in a memory.)

PHOTOGRAPH A2.12. Randomly distributed dark-current generation sites result in fixed-pattern noise.

PHOTOGRAPH A2.13. Thermal amplifier noise, photon shot noise and other temporal noise sources degrade the quality of the picture at low light levels.

Appendix 3

HOW TO COMPROMISE ON CCD SPECIFICATIONS

Specifying a charge-coupled device is a trade-off between the different parameters and characteristics of the imager. Placing greater stress on the optical performance can be at the expense of the electrical characteristics, for instance, or vice versa. Geometric parameters can also play an important role in the trade-off game. This appendix gives a short overview of the interaction between the various parameters, characteristics and geometric values. In each section of this appendix, one of these items is highlighted and its influence on the others is discussed. The following parameters are reviewed : sensitivity or quantum efficiency, resolution, modulation transfer function, antiblooming characteristics, smear, and cost price.

A3.1. Sensitivity and quantum efficiency

Irrespective of the device's architecture, whether it is a frame-transfer or a (frame-) interline-transfer CCD, all pixels have some area which is less sensitive or not sensitive at all (stopper implantation in the case of the FT, vertical shift registers in that of the IL) beside the light-sensitive area. These points are discussed in section 7.2, dealing with the aperture ratio of the IL pixels, and in section 7.3.2, dealing with the amount of polysilicon covering the active pixel in FT imagers. In all situations in which the photodiodes or the light windows in the gates are made larger to increase the sensitivity of devices, the total area available to store the generated charge becomes smaller. Consequently, the dynamic range of the devices is lowered if the pixel area is kept constant.

Conclusion : Increasing the light sensitivity or quantum efficiency by increasing the aperture ratio and keeping the pixel size constant is at the expense of the dynamic range.

A3.2. Resolution

Increasing the number of pixels on the light-sensitive area of an imager and keeping its overall size constant can only be done by decreasing the area of the pixels. As a logical consequence of the theory explained in the foregoing section, however, the sensitive area of the pixel becomes smaller, or the charge-handling capability of the pixel is reduced.

Conclusion : Increasing the resolution of the imager while keeping its total light sensitive area constant is at the expense of the sensitivity and/or dynamic range.

A3.3. Horizontal modulation transfer function

From section 5.5.3, it is known that noncontiguous pixels have the highest MTF value, and the broader the spacing between the active areas of the pixels, the higher the MTF curve rises. The spacing between the active pixels can be influenced by the design of the imager. For an interline-transfer type imager the width of the vertical CCD register determines the dead area between two photodiodes; for a frame-transfer CCD the width of the stopper implantation determines the dead area between columns of the imager. In all cases the designer has a means to optimize the horizontal modulation transfer function, namely increasing the dead area at the expense of the active pixel area and consequently decreasing the sensitivity and/or dynamic range of the imager.

Conclusion : Increasing the horizontal modulation transfer function by broadening of the dead area between pixels and keeping the total width of the pixels constant is at the expense of the sensitivity and/or dynamic range.

A3.4. Vertical modulation transfer function

In the vertical MTF, as is the case of the horizontal modulation transfer function, the absolute value of the parameter also increases if the pixels are noncontiguous. To describe the effects on the vertical modulation transfer function, the device architecture plays an important role too :

A3.4.1. FRAME-TRANSFER CCD

In interlaced scanning the pixels of the frame-transfer imager have an overlapping structure; in progressive scanning the pixels are arranged contiguously. In both cases, the vertical MTF can be increased by appropriate DC setting of the gates during integration of the image section.

Under normal conditions the integrating gates are biased high (e.g. 8 V) and the blocking gates are biased low (e.g. 0 V). Electrons generated underneath the blocking gates will be collected by the neighboring integrating gates through the interaction of the fringing fields. Dividing the electrons equally between the two neighboring pixels causes the pixels themselves to form a contiguous array in each field. This situation is illustrated in Figure A3.1a, where part of the integrating imaging section is shown. The positively biased gates are indicated by (+), the negatively biased by (-). The size of the active part of a pixel is indicated by the solid rectangle in Figure A3.1a.

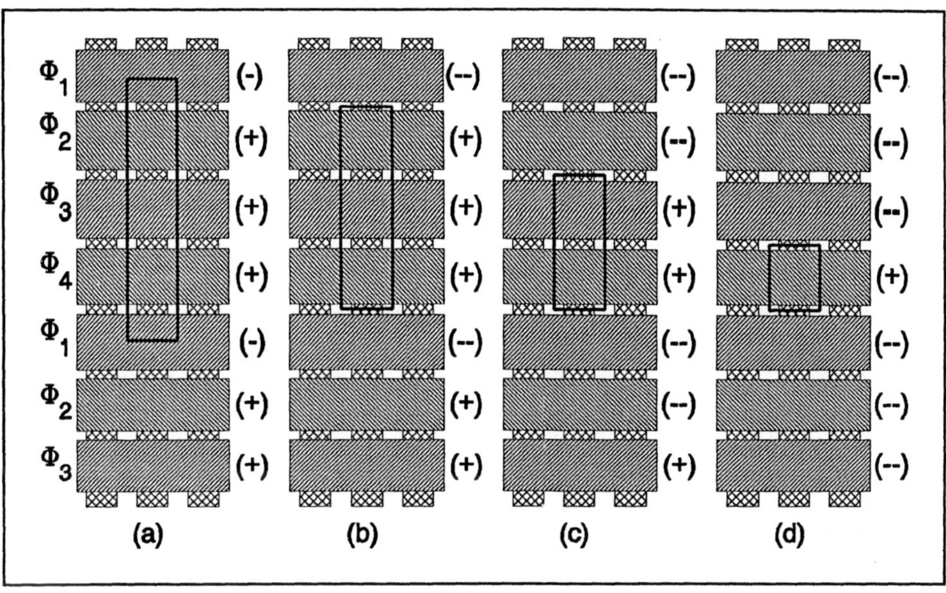

FIGURE A3.1. *Effect of the gate bias on the active area of the frame-transfer pixel for normal conditions (a) and one gate (b), two gates (c), and three gates (d) in a charge-reset mode.*

However, the situation changes when the blocking gates are biased (deeply) negative (e.g. -4 V). In this case the blocking gates are changed to the charge-reset mode and all electrons generated under these gates are drained immediately to the CCD substrate (see section 6.4.1). In other words, the blocking gates are no longer light sensitive and form an extra dead area between two vertical pixels. The new configuration is sketched in Figure A3.1b : the electrons generated under the (--) gate are drained to the substrate and not collected in the charge packet. The active area of the pixel is indicated by the solid rectangle. As can be seen, the organization of the pixels has changed into a noncontiguous array (in the vertical direction).

For a four-phase imaging section of a frame-transfer CCD, a single gate per pixel is normally used for the separation of the pixels. If all gates are of the same length, contiguous pixels (in one field) can be reshaped into noncontiguous pixels with a dead area in between equal to 25 % of the pixel height, as is the case in Figure A3.1b. Making two gates out of the four deeply negative, changes the width of the dead area into 50 %, as shown in Figure A3.1c. In this way the overlapping pixels in the two interlaced fields can be made contiguous. The pixel from a single field is indicated by the rectangle. Ultimately, as many as three of the four gates can be changed to a charge-reset mode during the integration and, even in the interlaced scanning mode, the pixels become noncontiguous. This last condition is depicted in Figure A3.1d.

In no situation do the gates biased into the charge-reset mode collect electrons and decrease the sensitivity of the device. And in the extreme situation of only one gate being positively biased, the dynamic range of the sensor is limited, too.

Conclusion : Increasing the vertical modulation transfer function of a frame-transfer imager by biasing the blocking gates into the charge-reset mode is at the expense of light sensitivity and in extreme situations at the expense of the dynamic range also.

A3.4.2. FRAME-INTERLINE-TRANSFER AND INTERLINE-TRANSFER CCD

In the case of (frame-) interline-transfer pixels, the choice with regard to influencing the vertical modulation-transfer function is not as wide as with the frame-transfer imager. In the latter there are four gates to be biased; in the former each pixel has only two photodiodes in the interlaced mode and only a single one in the progressive scanning. That means that in the case of progressive scanning the vertical modulation transfer function is fixed. Only in the case of the interlaced scanning can the vertical modulation transfer function be influenced.

The devices are operating in the frame-integration mode, the integration time is equal to the frame time : 40 msec or 33.3 msec. Readout of half of the photodiodes (either the even rows or the odd rows) is done each field time : 20 msec or 16.7 msec. Up to this point the readout cycle is completely identical to that described in section 6.1.1 and Figure 6.3. The result is contiguous pixels, as opposed to the overlapping pixels in the field-integration mode. The theory is illustrated in Figure A3.2a for the field-integration mode and in Figure A3.2b for the frame-integration mode. In both cases the area of a single pixel is accentuated by the solid rectangle.

Although the pixels are contiguous in the spatial sense in the frame-integration mode, they overlap in the time domain. The time-resolution characteristics of the imager diminish. To overcome this problem, the imager is used in the frame-integration mode but, after every readout cycle (of, for instance, all odd photodiodes), those remaining (in this example the even photodiodes) are changed to the charge-reset mode. The frame-integrating pixels have a reduced integration time of only half the original time, but the integration times of the odd and the even fields have an offset equal to the field time.

Conclusion : Increasing the vertical modulation transfer function of a (frame-) interline-transfer imager by driving the device in a frame-integration mode in combination with charge reset to maintain the time resolving characteristics, is at the expense of the sensitivity.

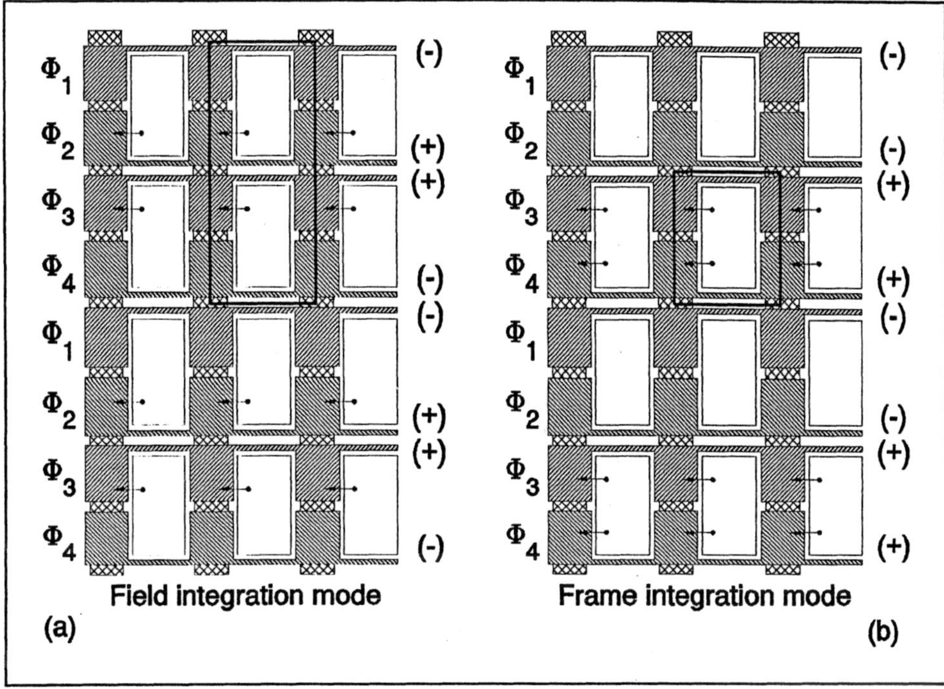

FIGURE A3.2. Dependence of the pixel configuration as function of the integration mode : field-integration (a) and frame-integration (b) mode.

A3.5. Antiblooming

In applications where a higher overload capacity of the imagers is needed, the antiblooming characteristics can be adapted by an appropriate setting of the voltages on the antiblooming drains. Higher voltages on the substrate for vertical antiblooming (section 6.3.3) or higher voltages on the antiblooming implantation for the lateral structures (section 6.3.1) have the same effect : the potential wells which store the charges become shallower, and the overflow of excess electrons is initiated earlier. The spilling of electrons across the antiblooming barrier is enhanced and a higher overexposure level can be handled. This effect is shown in Figure A3.3a for the interline-transfer CCD and in Figure A3.3b for the frame-transfer CCD. In both situations the solid lines represent the potential profiles in the silicon for low or normal substrate bias and the dashed lines correspond to the situation of higher substrate biases. (The terms used in the drawings are the same as in Figures 6.13 and 6.14).

However, due to the decrease of this barrier, the charge-handling capability of the devices is lowered. If the biasing of the substrates changes so much, the location of the barrier may also change and move upwards a little toward the Si-SiO$_2$ interface. As a consequence, the light sensitivity determined by the location of the potential minimum in the p well of the vertical antiblooming structure will decrease. This effect on sensitivity is only a second-order effect.

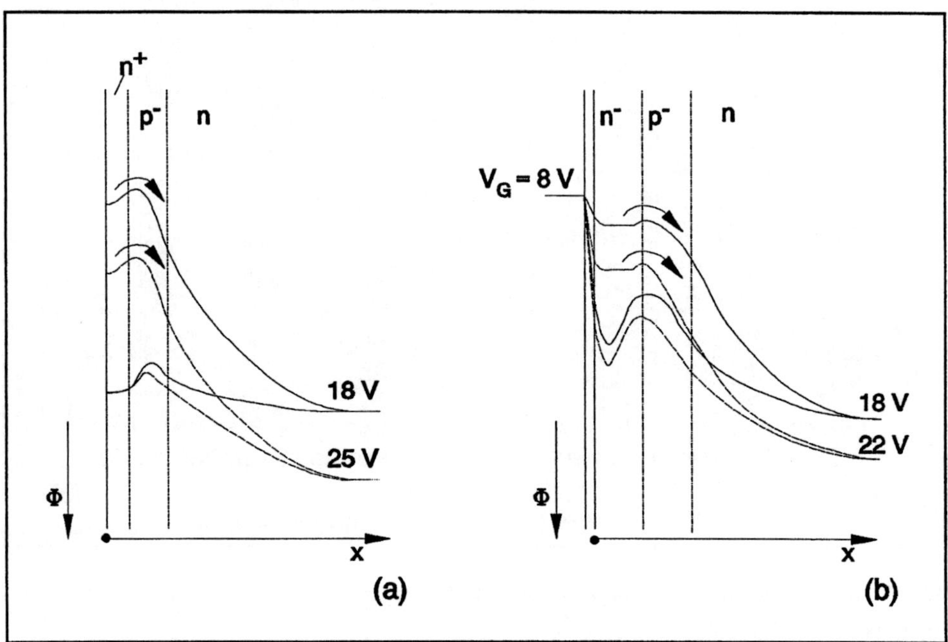

FIGURE A3.3. Effect of the substrate bias on the potential profiles in the silicon : (a) IL CCD, and (b) FT CCD, solid lines for a low bias and, dashed lines for a high bias.

Conclusion : Increasing the overload capability by a higher bias on the antiblooming drains is at the expense of the dynamic range.

A3.6. Smear

In frame-transfer and frame-interline-transfer devices, smear is generated during the time available for the frame shift. If the frame shift is speeded up, the time available for smear generation is reduced. An increased frame-shift frequency might, however, call for more driving power from the (external) drivers. If the drivers are not adapted, the waveform of the individual clocks might collapse at higher frequencies and the effective clock swing available to transport the charges from the image section to the storage section would decrease. With the lower clock swing, the amount of charges which will be transported will also decrease and the charge-handling capability, too, will be reduced.

Conclusion : Decreasing the smear level of frame-transfer or frame-interline transfer CCDs through an increase in the frame-shift frequency may result in a decrease of the dynamic range.

A3.7. Cost price

The cost price of charge-coupled devices is determined mainly by the price of a processed wafer and not of a single processed chip. The number of chips available on a wafer is, of course, directly related to the chip size. On the other hand, the yield of the devices is primarily determined by the probability of having a defect-free CCD. Defects are randomly distributed over the wafers. The yield of the chips is therefore a complex function of the chip size. The relation between both is illustrated in Figure A3.4, with the defect density as the parameter. As can be seen from Figure A3.4, the yield can decrease easily by a factor of 5 for an increase in chip size by a factor of 2, if large areas and large defect densities are concerned.

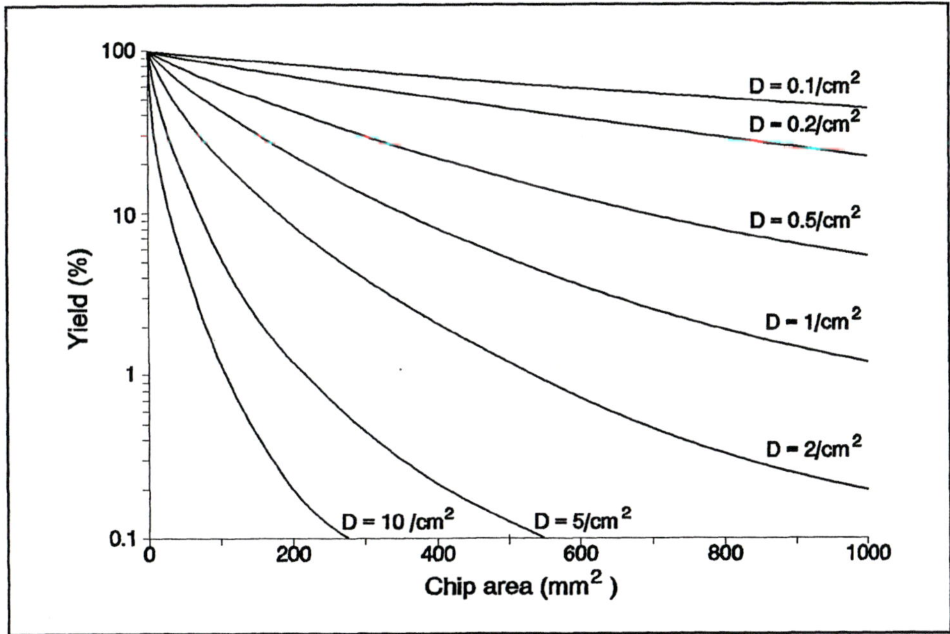

FIGURE A3.4. The relation between the yield of the CCD process and the chip size is shown, the parameter used is the defect density.

Conclusion : Lowering the cost price of the chip by decreasing its area and maintaining the number of pixels equal is at the expense of the dynamic range and the sensitivity.

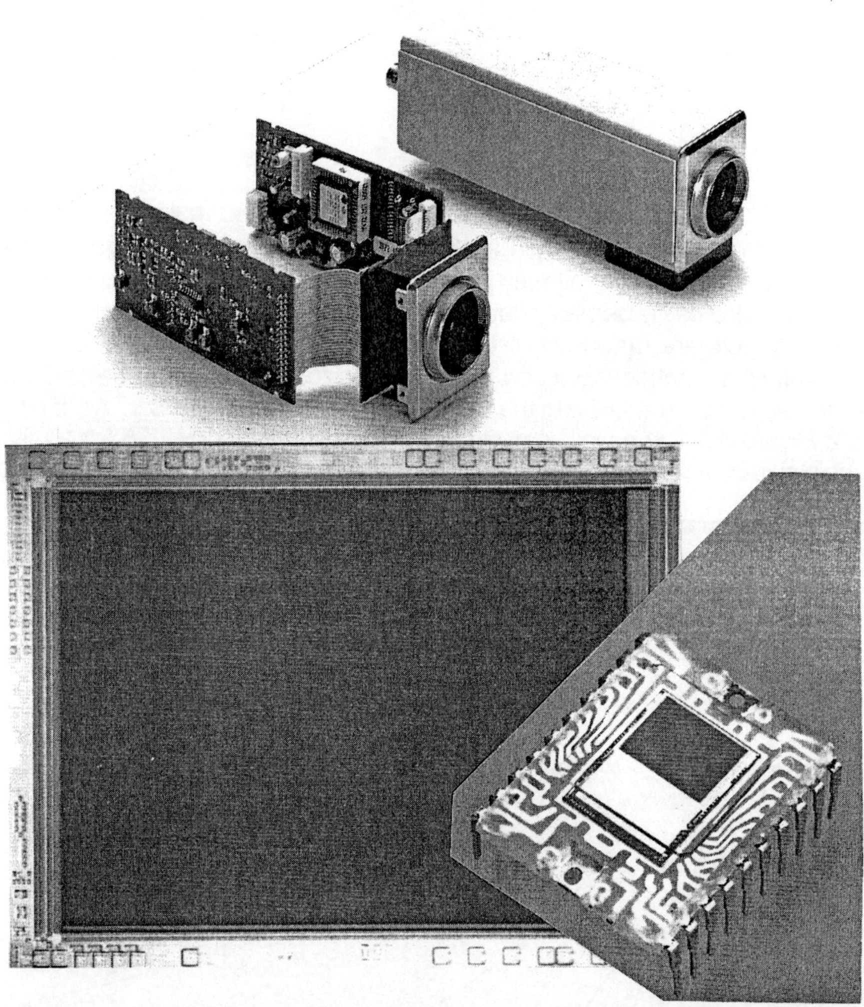

Top : the application of solid-state image sensors in compact camera modules intended for multi-purpose use (courtesy of Philips). Bottom left : an MOS-XY addressable array compatible with a 2/3 inch optical system (courtesy of Hitachi). Bottom right : a 380,000 pixel frame-transfer imager with a 1/2 inch optical format. The low-cost plastic package is covered with a high-quality optical cover glass (courtesy of Texas Instruments).

REFERENCES

Abe 90 : T.Abe, S.Yamaguchi, Y.Takemura, Y.Takeuchi, T.Yamada : "400k pixel full frame reading out FIT-CCD colour pick up system.", IEEE Transaction on Consumer Electronics, Vol. CE-36, No. 4, 1990, pp. 859-865.

Aizawa 85 : I.Aizawa, Y.Todaka, K.Hirose, S.Okada, M.Yamada, M.Masusa : "High resolution electronic still camera with two MOS imagers.", Digest of Technical Papers ICCE, pp. 288-289, Chicago, June 5-7, 1985.

Akimoto 91 : H.Akimoto, H.Ando, H.Nakagawa, Y.Nakahara, M.Hikiba, H.Ohta : "A 1/3-in 410000-pixel CCD image sensor with feedback field-plate amplifier.", IEEE Journal Solid-State Circuits, Vol. SC-26, No. 12, 1991, pp. 1907-1914.

Anagnostopoulos 80 : C.Anagnostopoulos, E.Garcia, G.Lubberts, F.Moser, D.Losee : "Thin polysilicon gate electrodes for frontside illuminated CCD imagers.", Proc. CICC80, pp. 78-81, Rochester, May 19-21, 1980.

Anagnostopoulos 93 : C.N.Anagnostopoulos, H.J.Erhardt, R.H.Philbrick, J.M.Andrus, B.J.Kecskemety : "A single register 8k CCD Trilinear color sensor with unrestricted exposure settings.", Digest Technical Papers ISSCC93, pp. 188-189, San Francisco, Febr 24-26, 1993.

Andoh 90 : F.Andoh, K.Taketoshi, J.Yamazaki, M.Sugawara, Y.Fujita, K.Mitani, Y.Matuzawa, K.Miyata, S.Araki : "A 250,000 pixel image sensor with FET amplification at each pixel for high-speed television cameras.", Digest Technical Papers ISSCC90, pp. 212-213, San Francisco, Febr 14-16, 1990.

Aoki 82 : M.Aoki, H.Ando, S.Ohba, I.Takemoto, S.Nagahara, T.Nakano, M.Kubo, T.Fujita : "2/3 inch format MOS single-chip color imager.", IEEE Transaction on Electron Devices, Vol. ED-29, No. 4, 1982, pp. 745-750.

Aoki 85 : T.Aoki, H.Nakatani, M.Kojima, S.Ohashi, I.Teramoto, H.Mizuno : "A collinear 3-chip image sensor", Digest Technical Papers ISSCC85, pp. 102-103, New York, Febr 13-15, 1985.

Azuma 91 : M.Azuma, T.Nobusada, Y.Toyoda, H.Asada, Y.Saito, S.Hayashi, T.Sugaya, T.Otsuki, Y.Fujita, F.Okano, K.Mitani : "An FPN-free 2/3" 1.3 M-pixel CCD image sensor for HDTV camera system.", Digest Technical Papers ISSCC91, pp. 212-213, San Francisco, Febr 13-15, 1991.

Bakker 91 : J.G.C.Bakker : "Simple analytical expressions for the fringing field and fringing-field-induced transfer time in charge-coupled devices.", IEEE Transaction on Electron Devices, Vol. ED-38, No. 5, 1991, pp. 1152-1161.

Barbe 75 : D.F.Barbe : "Imaging devices using the charge-coupled concept.", IEEE Proceedings, Vol. 63, 1975, pp. 38-67.

Beal 87 : G.Beal, G.Boucharlat, J.Chabbal, J.P.Dupin, B.Fort, Y.Mellier : "Thomson-CSF frame-transfer charge-coupled device imagers : design and evaluation at very low flux level.", Optical Engineering, Vol. 26, No. 9, pp. 902-910, 1987.

Beck 82 : G.A.Beck, M.G.Collet, J.A.A.van Gils, A.J.Klinkhamer, H.L.Peek, W.N.J.Ruis, J.G.van Santen, T.F.Smit, G.T.J.Vandormael : "High density frame transfer Image Sensor.", Proceedings 14th Conference on Solid State Devices, pp. 109-112, Tokyo, 1982.

Beynon 80 : J.D.E.Beynon, D.R.Lamb : "CCD operation, fabrication and limitations." in "Charge-coupled devices and their applications", Ed. J.D.E.Beynon, D.R.Lamb, McGraw-Hill Book Co., 1980, pp. 1-68.

Bisschop 88 : J.Bisschop : "Computer simulations of parallel-to-series conversion in solid state frame transfer image sensors.", Proceedings Simulations of Semiconductor Devices and Processes, Vol. 3, Ed. G.Baccarani, M.Rudan, Sept 26-28, 1988, Bologna, pp. 207-212.

Blouke 78 : M.M.Blouke, J.F.Breitzmann, J.E.Hall : "Three-phase, backside illuminated 500x500 imager.", Digest Technical Papers ISSCC78, pp. 36-37, San Francisco, Febr 15-17, 1978.

Blouke 79 : M.M.Blouke, M.W.Cowens, J.E.Hall : "A UV sensitive CCD detector.", Technical Digest IEDM79, pp. 141-143, Washington DC, Dec 3-5, 1979.

Blouke 83 : M.M.Blouke, J.R.Janesick, J.E.Hall, M.W.Cowens, P.J.May : "800x800 charge-coupled device image sensor", Optical Engineering, Vol. 22, No. 5, Sept-Oct 1983, pp. 607-614.

Blouke 85 : M.M.Blouke, D.L.Heidtmann, B.Corrie, M.L.Lust : "Large area CCD imager sensors for scientific applications.", Electronic Imaging 85, pp. 160-164, Boston, Oct 7-10, 1985.

Bosiers 85 : J.T.Bosiers, N.S.Saks, D.J.Michels, D.McCarthy, M.C.Peckerar : "Deep-depletion CCDs with improved UV sensitivity.", Technical Digest IEDM85, pp. 448-451, Washington DC, Dec 1-4, 1985.

Bosiers 88 : J.Bosiers, B.Dillen, C.Jaspers, A.Kleimann, A.Kokshoorn, H.Peek, M.van de Steeg : "A 2/3" 1188(H) x 484(V) frame-transfer CCD for ESP and movie mode.", Technical Digest IEDM88, pp. 70-73, San Francisco, Dec 11-14, 1988.

Bosiers 91 : J.Bosiers, A.Kleimann, B.Dillen, H.Peek, A.Kokshoorn, N.Daemen, A.van der Sijde, L.van Gaal : "A 2/3-in 1187(H) x 581(V) S-VHS-compatible frame-transfer CCD for ESP and movie mode.", IEEE Transaction on Electron Devices, Vol. ED-38, No. 5, 1991, pp. 1059-1068.

Boyle 70 : W.S.Boyle, G.E.Smith : "Charge-coupled semiconductor devices.", Bell Systems Technical Journal, Vol. 49, 1970, pp. 587.

Bredthauer 91 : R.A.Bredthauer, J.H.Pinter, J.R.Janesick, L.B.Robinson : "Notch and large area CCD imagers.", Proc. SPIE, Vol. 1447, pp. 310-315, San Jose, Feb 25-27, 1991.

Brewer 80 : R.J.Brewer : "The low-light level potential of a CCD imaging array.", IEEE Transaction on Electron Devices, Vol. ED-27, No. 2, 1980, pp. 401-405.

Brown 76 : D.M.Brown, M.Ghezzo, M.Garfinkel : "Transparent metal oxide electrode CID imager.", IEEE Journoul Solid-State Circuits, Vol. SC-11, No. 1, 1976, pp. 128-132.

Burke 76 : H.K.Burke, G.J.Michon : "Charge-injection imaging : operating techniques and performances characteristics.", IEEE Journal Solid-State Circuits, Vol. SC-11, No. 1, 1976, pp. 121-128.

Burke 91 : B.E.Burke, S.A.Gajar : "Dynamic suppression of interface-state dark current in buried-channel CCD's.", IEEE Transaction on Electron Devices, Vol. ED-38, No. 2, 1991, pp. 285-290.

Burstein 80 : P.Burstein, D.J.Michels : "Direct detection of XUV radiation with a CCD.", Applied Optics, Vol. 19, No. 10, pp. 1563-1565, 1980.

Carnes 72 : J.E.Carnes, W.F.Kosonocky, E.G.Ramberg : "Free charge transfer in charge-coupled devices.", IEEE Transaction on Electron Devices, Vol. ED-19, No. 6, 1972, pp. 798-808.

Carnes 72a : J.E.Carnes, W.F.Kosonocky : "Noise sources in Charge-Coupled Devices.", RCA Review, Vol. 33, 1972, pp. 327-343.

Carr 72 : W.N.Carr, J.P.Mize : "MOS device Physics", in "MOS/LSI Design and Application", Ed. R.E.Sawyer, J.R.Miller, McGraw-Hill Book Co., 1972, pp. 1-60.

Centen 91 : P.Centen : "CCD on-chip amplifiers : noise performance versus MOS transistor dimensions.", IEEE Transaction on Electron Devices, Vol.ED-38, No. 5, 1991, pp. 1206-1216.

Centen94 : P.Centen, H.Stoldt, A.Theuwissen, H.Peek, W.Huinink, L.Sankaranarayanan, A.Kleimann, G.Pine, M.Stekelenburg, A.Mierop, F.Vledder, D.Verbugt, J.Oppers, P.Opmeer : "Aspect ration switching with equal horizontal pixel count.", Technical Papers International Broadcast Convention, pp. 1-7, Amsterdam, Sept 1994.

Chamberlain 93 : S.G.Chamberlain, S.R.Kamasz, F.Ma, W.D.Washkurak, M.Farrier, P.T.Jenkins : "A 26.2 million pixel CCD image sensor.", Proceedings SPIE, Vol. 1900, pp. 181-191, San Jose, Feb 2-3, 1993.

Chandler 90 : C.E.Chandler, R.A.Bredthauer, J.R.Janesick, J.A.Westphal, J.E.Gunn : "Sub-electron noise charge coupled devices.", Proceedings SPIE, Vol. 1242, pp. 238-251, San Jose, Feb 12-14, 1990.

Chikamura 84 : T.Chikamura, T.Komeda, D.Ishiko, M.Yoshino, M.Nakayama, K.Yano, Y.Aoki, A.Ueno, T.Yamada, T.Ishihara : "A half inch size CCD image sensor overlaid with a hydrogenated amorphous silicon.", Technical Digest IEDM84, pp. 552-555, San Francisco, Dec 9-12, 1984.

Collet 74 : M.G.Collet, A.C.Vliegenthart : "Calculations on potential and charge distributions in the peristaltic charge-coupled device.", Philips Research Reports, Vol. 29, No. 2, pp. 25, 1974.

Collet 85 : M.G.Collet, J.G.C.Bakker, L.J.M.Esser, H.L.Peek, M.J.H.van de Steeg, A.J.P.Theuwissen, C.H.L.Weijtens : "High density frame transfer image sensors with vertical anti-blooming.", Proceedings SPIE, Vol. 570, pp. 27-34, San Diego, Aug 22-23, 1985.

Collet 86 : M.G.Collet : "Solid-state imagers.", in "Solid-state devices 1985", P.Balk, O.G.Folberth, eds., Elsevier, Amsterdam, 1986, pp. 183-200.

Declerck 83 : G.Declerck, J.Bosiers, J.Sevenhans, L.Van den Hove : "A 3456 element quadrilinear CCD with depletion-isolated sensor structure.", Technical Digest IEDM83, pp. 505-508, San Francisco, Dec 9-12, 1983.

Deguchi 92 : M.Deguchi, T.Maruyama, F.Yamasaki, T.Hamamoto, A.Izumi : "Microlens design using simulation program for CCD image sensor.", IEEE Transaction on Consumer Electronics, Vol. CE-38, No. 3, 1992, pp. 583-589.

De Lang 62 : H.de Lang, G.Bouwhuis : "Colour separation in colour-television cameras.", Philips Technical Review, Vol. 24, No. 9, 1962/63, pp. 263-298.

De Meijer 81 : K.M.De Meijer, G.J.Declerck : "A new method for the two-dimensional calculation of the potential distribution in a buried-channel charge-coupled device : Theory and experimental verification.", IEEE Transaction on Electron Devices, Vol. ED-28, No. 3, pp. 313-322, 1981.

Dillon 79 : P.L.P.Dillon, D.M.Lewis, F.G.Kaspar : "Color imaging system using a single CCD area array.", IEEE Transaction on Electron Devices, Vol. ED-25, No. 2, 1978, pp. 102-107.

Dyck 82 : R.H.Dyck : "Design, fabrication and performance of CCD imagers", in VLSI Electronics Microstructure Science, Vol. 3, Ed. N.G.Einspruch, 1982, Academic Press, pp. 65-107.

Esser 72 : L.J.M.Esser : "Peristaltic charge-coupled device : a new type of charge-transfer devices.", Electronics Letters, Vol. 8, 1972, pp. 620-621.

Esser 73 : L.J.M.Esser : "The peristaltic charge-coupled device", Charge-coupled device application conference, San Diego, pp. 269-277, 1973.

Esser 81 : L.J.M.Esser : "Charge-coupled devices : physics, technology and applications.", Proceedings of International Workshop on Physics of Semiconductor Devices, pp. 235-249, New Delhi, Nov 23-28, 1980.

Esser 88 : L.J.M.Esser, H.C.G.van Kuijk, J.G.C.Bakker, C.H.L.Weijtens, A.J.P.Theuwissen : "A smearfree accordion CCD imager.", Digest of Technical Papers ICCE, pp. 10-11, Chicago, June 8-10, 1988.

Esser 93 : L.J.M.Esser, A.J.P.Theuwissen : "Charge-coupled devices : physics, technology and imaging.", in "Handbook on Semiconductors", ed. C.Hilsum, North-Holland, Amsterdam, pp. 389-473, 1993.

Farrier 80 : M.G.Farrier, R.H.Dyck : "A large area TDI image sensor for low light level imaging.", IEEE Journal Solid-State Circuits, Vol. SC-15, No. 4, 1980, pp. 753-758.

Fossum 91 : E.Fossum : "Wire transfer of charge packets using a CCD-BBD structure for charge-domain signal processing.", IEEE Transaction on Electron Devices, Vol. ED-38, No. 2, 1991, pp. 291-298.

Furukawa 92 : J.Furukawa, I.Hiroto, Y.Takamura, T.Wada, Y.Keigo, A.Izumi, K.Nishibori, R.Tatebe, S.Kitayama, M.Shimura, H.Matsui : "A 1/3-inch 380k pixel (effective) IT-CCD image sensor.", IEEE Transaction on Consumer Electronics, Vol. CE-38, No. 3, 1992, pp. 595-600.

Goto 84 : H.Goto, N.Suzuki, K.Matsushima : "A 5000-element CCD linear image sensor.", Toshiba Review, No. 150, Winter 1984, pp. 39-43.

Hagiwara 78 : Y.D.Hagiwara, M.Abe, C.Okada : "A 380Hx488V CCD imager with narrow channel transfer gates.", Proceedigns 10th Conference on Solid State Devices, pp. 335-340, Tokyo, 1978.

Hanma 83 : K.Hanma, Y.Saito : "Fully automatic solid-state color video camera for consumer use.", Hitachi Review, Vol. 32, No. 3, 1983, pp. 117-120.

Hanneman 72 : H.W.Hanneman, L.J.M.Esser : "Field and potential distributions in charge transfer devices.", Philips Research Reports, Vol. 30, No. 1, pp. 56-72, 1972.

Harada 80 : N.Harada, N.Suzuki, O.Yoshida, K.Yano, H.Morita : "Frame transfer CCD imager with transparent electrodes", Japanese Journal Applied Physics, Vol. 19, 1980, pp. L177-L180.

Harada 92 : K.Harada, M.Negishi, T.Ohgishi, S.Kubota, T.Oda, M.Yamagishi : "A 2/3-inch 2M pixel FIT-CCD HDTV image sensor.", Digest Technical Papers ISSCC92, pp. 170-171, San Francisco, Febr 19-21, 1992.

Heidtmann 87 : D.L.Heidtmann, F.Yang : "The effects of FET dimensions and bias on the noise performance of CCD output amplifier.", Electronic Imaging 87, pp. 248-253, Boston, Nov 2-5, 1987.

Heijns 78 : H.Heijns, J.G.van Santen : "The resistive gate CTD area image sensor." IEEE Transaction on Electron Devices, Vol. ED-25, No. 2, 1978, pp. 135-139.

Herbst 76 : H.Herbst : "The quadrilinear CCD imager - A new device for high resolution imaging.", Conference on Charge-Coupled Device Technology and Applications, Washington DC, Nov 11-Dec 2, 1976.

Hirama 90 : M.Hirama, Y.Watanabe, S.Koike, T.Kodake, K.Tsuchiya, T.Narabu : "A 5000-pixel linear image sensor with on-chip clock drivers.", IEEE Transaction on Consumer Electronics, Vol. CE-36, No. 3, 1990, pp. 473-478.

Hoenk 92 : M.E.Hoenk, P.J.Grunthaner, F.J.Grunthamer, R.W.Terhune, M.Fattahi, H.-F.Tseng : "Growth of a delta-doped silicon layer by molecular beam epitaxy on a charge-coupled device for reflection-limited ultraviolet quantum efficiency", Applied Physic Letters, Vol. 61, No. 9, 1992, pp. 1084-1086.

Horii 81 : K.Horii, T.Kuroda, T.Kunii : "A new configuration of CCD imager with a very low smear level.", IEEE Electron Devices Letters, Vol. EDL-2, No. 12, 1981, pp. 319-320.

Horii 84 : K.Horii, T.Kuroda, S.Matsumoto : "A new configuration of CCD imager with a very low smear level - FIT-CCD imager.", IEEE Electron Devices, Vol. ED-31, No. 7, 1984, pp. 904-909.

Hojo 91 : J.Hojo, Y.Naito, H.Mori, K.Fujikawa, N.Kato, T.Wakayama, E.Komatsu, M.Itasaka : "A 1/3-in 510(H) x 492(V) CCD image sensor with mirror image function.", IEEE Transaction on Electron Devices, Vol. ED-38, No. 5, 1991, pp. 954-959.

Hsu 86 : S.C.Hu : "The Kell factor : past and present.", SMPTE Journal, February 1986, Vol. 95, pp. 206-214.

Hynecek 81 : J.Hynecek : "Virtual phase technology : a new approch to fabrication of large-area CCD's.", IEEE Transaction on Electron Devices, Vol. ED-28, No. 5, 1981, pp. 483-489.

Hynecek 85 : J.Hynecek : "Design and performance of a high-resolution image sensor for color TV applications.", IEEE Transaction on Electron Devices, Vol. ED-32, No. 8, 1985, pp. 1421-1429.

Hynecek 86 : J.Hynecek : "High-resolution 8-mm CCD image sensor with correlated clamp sample and hold charge detection circuit.", IEEE Transaction on Electron Devices, Vol. ED-33, No. 6, 1986, pp. 850-862.

<u>Hynecek 88</u> : J.Hynecek : "A new device architecture suitalbe for high-resolution and high-performance image sensors.", IEEE Transaction on Electron Devices, Vol. ED-35, No. 5, 1988, pp. 646-652.

<u>Hynecek 89</u> : J.Hynecek : "A new high-resolution 11-mm diagonal image sensor for still-picture photography.", IEEE Transaction on Electron Devices, Vol. ED-36, No. 11, 1989, pp. 2466-2474.

<u>Iesaka 88</u> : M.Iesaka, S.Osawa, S.Uya, Y.Egawa, Y.Matsunaga, S.Manabe, N.Harada : "Analysis of charge transfer loss in dual read-out registers used for HDTV CCD image sensor.", Proceedings 20th Conference on Solid State Devices, pp. 359-362, Tokyo, 1988.

<u>Ikeda 85</u> : S.Ikeda, T.Tanahashi, A.Kono : "NEC improves flexibility of color ENG cameras with CCD image sensors.", JEE, March 1985, pp.28-32.

<u>Ishihara 82</u> : Y.Ishihara, E.Oda, H.Tanigawa, N.Teranishi, E.Takeuchi, I.Akiyama, K.Arai, M.Nishimura, T.Kamata : "Interline CCD image sensor with an anti blooming structure.", Digest Technical Papers ISSCC82, pp. 168-169, San Francisco, Febr 10-12, 1978.

<u>Ishihara 83</u> : Y.Ishihara, K.Tanigaki : "A high photosensitivity IL-CCD image sensor with monolithic resin lens array.", Technical Digest IEDM83, pp. 497-500, Washington DC, Dec 6-8, 1983.

<u>Ishikawa 81</u> : K.Ishikawa, S.Hashimoto, Y.Sone, T.Kunii : "Color reproduction of a single chip color camera with a frame transfer CCD.", IEEE Journal Solid-State Circuits, Vol. SC-16, No. 2, 1981, pp. 101-103.

<u>Itakura 93</u> : K.Itakura, T.Nobusada, Y.Toyoda, Y.Saitoh, N.Kokusenya, R.Nagayoshi, M.Ozaki : "A multiple frame-interline-transfer (M-FIT) CCD for progressive-scan camera systems.", Digest Technical Papers ISSCC93, pp. 190-191, San Francisco, Febr 24-26, 1993.

<u>Janesick 89</u> : J.Janesick : "Open pinned-phase CCD technology.", Proceedings SPIE, Vol. 1159, San Diego, 1989.

<u>Janesick 90</u> : J.Janesick, T.Elliott, A.Dingizian, R.Bredthauer, C.Chandler, J.Westphal, J.Gunn : "New advancements in charge-coupled device technolgy - sub-electron noise and 4096x4096 pixel CCD's.", Proceedings SPIE, Vol. 1242, pp. 223-237, Santa Clara, Feb 12-14, 1990.

<u>Jastrzebski 87</u> : L.Jastrzebski, R.Soydan, G.W.Cullen, W.N.Henry, S.Vecrumba : "Silicon wafers for CCD imagers.", Journal Electrochemical Society, Vol. 134, No. 1, 1987, pp. 212-221.

<u>Jastrzebski 90</u> : L.Jastrzebski, R.Soydan, H.elabd, W.Henry, E.Savoye : "The effect of heavy metal contamination on defects in CCD imagers.", Journal Electrochemical Society, Vol. 137, No. 1, 1990, pp. 242-249.

<u>Kaneko 82</u> : S.Kaneko, M.Sakamoto, F.Okumura, T.Itano, H.Kataniwa, Y.Kajiwara, M.Kanamori, M.Yasumoto, T.Saito, T.Ohkubo : "Amorphous Si:H contact linear image sensor with Si_3N_4 blocking layer.", Technical Digest IEDM82, pp. 328-331, San Francisco, Dec 13-15, 1982.

<u>Kanoh 81</u> : Y.Kanoh, S.Usui, A.Sawada, M.Kikuchi : "A contact-type linear sensor with a GDa-Si:H photodetector array.", Technical Digest IEDM81, pp. 313-316, Washington DC, Dec 7-9, 1981.

<u>Kawamoto 91</u> : S.Kawamoto, Y.Watanabe, Y.Otsuka, T.Narabu : "A CCD color linear image sensor employing new transfer method.", Digest of Technical Papers ICCE, pp. 150-151, Rosemont, June 5-7, 1991.

Kimata 85 : M.Kimata, M.Denda, N.Yutani, N.Tsubouchi : "A 400x400 element image sensor with a charge sweep device.", Digest Technical Papers ISSCC85, pp. 100-101, New York, Febr 13-15, 1985.

Kioi 88 : K.Kioi, S.Toyoyama, M.Koba : "Monolithic character recognition system implementated as proto-type intelligent image sensor by 3D integration technology.", Technical Digest IEDM88, pp. 66-69, San Francisco, Dec 11-14, 1988.

Kioi 90 : K.Kioi, T.Shinozaki, S.Toyoyama, K.Shirakawa, K.Ohtake, S.Tsuchimoto : "Design and implementation of a 3D-LSI character recognition image sensor.", Technical Digest IEDM90, pp. 279-282, San Francisco, Dec 9-12, 1990.

Knop 85 : K.Knop : "A new class of mosaic color encoding patterns for single-chip cameras.", IEEE Transaction on Electron Devices, Vol. ED-32, No. 8, 1985, pp. 1390-1395.

Kobayashi 93 : A.Kobayashi, Y.Naito, T.Ishigami, A.Izumi, T.Hanagata, K.Nakashima : "A 1/2-in 380k-pixel progressive scan CCD image sensor." Digest Technical Papers ISSCC93, pp. 192-193, San Francisco, Febr 24-26, 1993.

Koch 83 : R.Koch : "Flachenhafte halbleiter-bildsensoren." Rundfunktechnische Mitteilungen, Jahrgang 27, Heft 5, 1983, pp. 213-224.

Kohno 85 : A.Kohno, T.Tanaka, N.Teranishi, T.Kitagawa : "A single chip CCD color video camera using a line sequential complete color difference signal method.", Digest of Technical Papers ICCE, pp. 284-285, Chicago, June 5-7, 1985.

Koike 80 : N.Koike, I.Takemoto, K.Satoh, S.Hanamura, S.Nagahara, M.Kubo : "MOS area sensor : Part 1 - Design consideration and performance of an n-p-n structrure 484x384 element color MOS imager.", IEEF Journal Solid-State Circuits, Vol. SC-15, No. 4, 1980, pp. 741-746.

Komiya 81 : K.Komiya, M.Kanzaki, T.Yamashita : "A 2048-element contact type linear image sensor for facsimile.", Technical Dig. IEDM81, pp. 309-312, Washington DC, Dec 7-9, 1981.

Konuma 91 : K.Konuma, S.Tohyama, A.Tanabe, K.Masubuchi, N.Teranishi, T.Saito, T.Muramatsu : "A 648x487 pixel Schottky-barrier infrared CCD image sensor.", Digest Technical Papers ISSCC91, pp. 216-217, San Francisco, Febr 13-15, 1991.

Kosman 90 : S.L.Kosman, E.G.Stevens, J.C.Cassidy, W.C.Chang, P.Roselle, W.A.Miller, M.Mehra, B.C.Burkey, T.H.Lee, G.A.Hawkins, R.P.Khosla : "A large area 1.3-megapixel full-frame CCD image sensor with a lateral-overflow drain and a transparent gate electrode.", Technical Digest IEDM90, pp. 287-290, San Francisco, Dec 9-12, 1990.

Kosonocky 87 : W.F.Kosonocky : "Infrared image sensors with Schottky-barrier detectors." Proceedings SPIE, Vol. 869, pp. 1-18, Cannes, Nov 16-20, 1987.

Kuriyama 91 : T.Kuriyama, H.Kodama, T.Kozono, Y.Kitahama, Y.Morita, Y.Hiroshima : "A 1/3-in 270000 pixel CCD image sensor.", IEEE Transaction on Electron Devices, Vol. ED-38, No. 5, 1991, pp. 949-953.

Kuroda 86 : T.Kuroda, T.Kuriyama, Y.Matsuda, T.Kozono, S.Matsumoto, Y.Hiroshima, K.Horii : "A smear-suppressing CCD imager.", Digest Technical Papers ISSCC86, pp. 94-95, San Francisco, Febr 19-21, 1986.

Lee 81 : T.H.Lee, T.J.Tredwell, B.C.Burkey, J.S.Hayward, T.M.Kelly, R.P.Khosla, D.L.Losee : "A novel solid-state image sensor for image recording at 2,000 frames per second.", Technical Digest IEDM81, pp. 475-478, Washington DC, Dec 7-9, 1981.

Lee 83 : T.H.Lee, T.J.Tredwell, B.C.Burkey, T.M.Kelly, R.P.Khosla, D.L.Losee, F.C.Lo, R.L.Nielsen, W.C.McColgin : "A 360,000 pixel color image sensor for imaging photographic negatives.", Technical Digest IEDM83, pp. 492-496, Wahington DC, Dec 6-7, 1983.

Lee 90 : T.H.Lee, W-C.Chang, W.A.Miller, G.R.Torok, K.Y.Wong, B.C.Burkey, R.P.Khosla : "A four million pixel CCD image sensor.", Proceedings SPIE, Vol. 1242, pp. 10-16, Santa Clara, Feb 12-14, 1990.

Lee 90 : T.H.Lee, B.C.Burkey, R.P.Koshla : "A four million pixel CCD image sensor.", Proceedings ESSDERC 90, pp. 599-602, Nottingham, Sept 10-13, 1990.

Lees 90 : R.Lees, L.Bernstein, H.Erhardt, R.Godden, G.Kennel, D.Kessler, A.Kurtz, J.Loveridge, L.Moore, R.Sharman : "High-Performance CCD Telecine for HDTV.", SMPTE Journal, October 1990, pp. 837-843.

Lemonier 87 : M.Lemonier, C.Piaget : "Solid-state image sensors for electronic readout of image tubes.", L'Onde Electrique, Vol. 67, No. 6, 1987, pp. 99-107.

Levine 84 : P.A.Levine, W.F.Kosonocky, E.D.Savoye, D.F.Battson : "High performance frame transfer CCD imager for television applications.", Electronic Imaging 84, pp. 1-4, Boston, Sept 10-13, 1984.

Losee 89 : D.L.Losee, J.C.Cassidy, M.Mehra, E.T.Nelson, B.C.Burkey, G.Geisbuesch, G.A.Hawkins, R.P.Khosla, J.P.Lavine, W.C.McColgin, E.A.Trabka, A.K.Weiss : "A 1/3" format image sensor with refractory metal light shield for color video applications.", Digest Technical Papers ISSCC89, pp. 90-91, New York, Febr 15-17, 1989.

Manabe 83 : D.Manabe, T.Ohta, Y.Shimidzu : "Color filter array for IC image sensor.", Proceedings CICC83, pp. 451-455, Rochester, May 23-25, 1983.

Manabe 88 : S.Manabe, Y.Matsunaga, M.Iesaka, S.Uya, A.Furukawa, K.Yano, H.Nozaki, Y.Endo, Y.Egawa, Y.Endo, Y.Ide, M.Kimura, N.Harada : "A 2-million pixel CCD imager overlaid with an amorphous silicon photoconversion layer.", Digest Technical Papers ISSCC88, pp. 50-51, San Francisco, Febr 17-19, 1988.

Manabe 91 : S.Manabe, Y.Matsunaga, A.Furukawa, K.Yano, Y.Endo, R.Miyagawa, Y.Iida, Y.Egawa, H.Shibata, H.Nozaki, N.Sakuma, N.Harada : "A 2-million-pixel CCD image sensor overlaid with an amorphous silicon photoconversion layer.", IEEE Transaction on Electron Devices, Vol. ED-38, No. 8, 1991, pp. 1765-1771.

Matsunaga 91 : Y.Matsunaga, S.Ohsawa : "A 1/3-in interline transfer CCD image sensor with a negative-feedback-type charge detector.", IEEE Journal Solid-State Circuits, Vol. SC-26, No. 12, 1991, pp. 1902-1906.

McGrath 83 : R.D. McGrath, J.W.Freeman : "An 8 megapixel/sec 800x800 virtual phase CCD imager for scientific applications.", Technical Digest IEDM83, pp. 489-491, Washington DC, Dec 6-7, 1983.

R.D.McGrath, J.W.Freeman, W.F.Keenan : "A 1024x1024 virtual phase CCD imager.", Technical Digest IEDM83, pp. 749, Washington DC, Dec 6-7, 1983.

Mitani 84 : N.Mitani, T.Furrusawa, Y.Tsuchihashi, Y.Kitamura, Y.Kiriyama, Y.Rai : "A single chip 1/2" frame transfer CCD color image sensor.", Technical Digest IEDM84, pp. 44-47, San Francisco, Dec 9-12, 1984.

Morimoto 92 : M.Morimoto, K.Orihara, N.Mutoh, A.Toyoda, M.Ohba, Y.Kawakami, T.Nakano, K.Chiba, K.Hatano, K.Arai, M.Nishimura, Y.Nakashiba, A.Kohno, I.Akiyama, N.Teranishi, Y.Hokari : "A 2M pixel HDTV CCD image sensor with tungsten photo-shield and H-CCD shunt wiring.", Digest Technical Papers ISSCC92, pp. 172-173, San Francisco, Febr 19-21, 1992.

Murata 83 : N. Murata, C.Hirano, M.Ohoba, S.Nagahara : "Development of a 3-MOS color camera.", SMPTE Journal, December 1983, pp. 1270-1273.

Mutoh 89 : N.Mutoh, M.Morimoto, M.Nishimura, N.Teranishi, E.Oda : "New low noise output amplifier for high definition CCD image sensor.", Technical Digest IEDM89, pp. 173-176, Washington DC, Dec 3-6, 1989.

Nakamura 86 : T.Nakamura, K.Matsumoto, R.Hyuga, A.Yusa : "A new MOS image sensor operating in a non-destructive readout mode.", Technical Digest IEDM86, pp. 353-356, Los Angeles, Dec 7-10, 1986.

Nakamura 91 : Y.Nakamura, H.Ohzu, M.Miyawaki, N.Tanaka, T.Ohmi : "Design of bipolar imaging device (BASIS).", IEEE Transaction on Electron Devices, Vol. ED-38, No. 5, 1991, pp. 1028-1036.

Negishi 91 : M.Negishi, H.Yamada, K.Harada, M.Yamagishi, K.Yonemoto : "A low smear structure for 2M-pixel CCD image sensors.", Digest of Technical Papers ICCE, pp. 154-155, Rosemont, June 5-7, 1991.

Nichols 87 : D.N.Nichols, W.-C.Chang, B.C.Burkey, E.G.Stevens, E.A.Trabka, D.L.Losee, T.J.Tredwell, C.V.Stancampiano, T.M.Kelly, R.P.Khosla, T.H.Lee : "A 1.4 million element, full frame CCD image sensor with a vertical overflow drain for anti-blooming and low color crosstalk.", Technical Digest IEDM87, pp. 120-123, Washington DC, Dec 6-9, 1987.

Nicollian 82 : E.H.Nicollian, J.R.Brews : "Field Effect" in "MOS Physics and Technology", John Wiley & Sons, 1982, pp. 26-70.

Nishizawa 87 : S.Nishizawa, T.Izawa, K.Furuichi, H.Sokei, I.Takemoto, M.Ashikawa : "Solid state imager implementing sensitivity control function on chip." Proceedings SPIE, Vol. 765, pp. 47-52, Los Angeles, Jan 13-14, 1987.

Nobusada 89 : T.Nobusada, M.Azuma, H.Toyoda, T.Kuroda, K.Horii, T.Otsuki, G.Kano : "Frame interline transfer CCD sensor for HDTV camera.", Digest Technical Papers ISSCC89, pp. 88-89, New York, Febr 15-17, 1989.

Noda 86 : M.Noda, T.Imaide, T.Kinugasa, R.Nishimura : "A solid-state color video camera with a horizontal read out MOS imager.", IEEE Transaction on Consumer Electronics, Vol. CE-32, No. 3, 1986, pp. 329-336.

REFERENCES

<u>Nohmi 90</u> : N.Nohmi, I.Inoue, Y.Endo, Y.Matsunaga, S.Manabe, N.Harada : "Capacitive lag-free CCD imager overlaid with an amorphous silicon photoconversion layer.", Proceedings 22th Conference on Solid State Devices and Materials, pp. 713-716, Sendai, 1990.

<u>Nomoto 93</u> : T.Nomoto, R.Hyuga, S.Nakajima, I.Takayanagi, T.Isokawa, R.Ohta, K.Matsumoto, T.Nakamura : "A 2/3-inch 2M-pixel CMD image sensor with multi-scanning function.", Digest Technical Papers ISSCC93, pp. 196-197, San Francisco, Febr 24-26, 1993.

<u>Oda 83</u> : E.Oda, Y.Ishihara, N.Teranishi : "Blooming suppression mechanism for an interline CCD image sensor with a vertical overflow drain.", Technical Digest IEDM83, pp. 501-504, Wahington DC, Dec 6-7, 1983.

<u>Oda 85</u> : E.Oda, K.Orihara, T.Kitagawa, K.Arai, T.Kamata, Y.Ishihara : "1/2-inch 768(H)x492(V) pixel CCD image sensor.", Technical Digest IEDM85, pp. 444-447, Washington DC, Dec 1-4, 1985.

<u>Oda 89</u> : E.Oda, K.Nagano, T.Tanaka, N.Mutoh, K.Orihara : "A 1920(H) x 1035(V) pixel high-definition CCD image sensor.", IEEE Journal Solid-State Circuits, Vol. SC-24, No. 3, 1989, pp. 711-717.

<u>Ogino 83</u> : M.Ogino, T.Usami, M.Watanabe, H.Sekine, T.Kawaguchi : "Two-step thermal anneal and its application to a CCD sensor and CMOS LSI.", Journal Electrochemical Society, Vol. 130, No. 6, 1983, pp. 1397-1402.

<u>Ohbo 88</u> : M.Ohbo, S.Yamamoto : "Integral delayed and differential noise suppression method for CCD camera.", Digest of Technical Papers ICCE, pp. 58-59, Chicago, June 8-10, 1988.

<u>Okada 91</u> : Y.Okada, M.Okigawa, K.Kazui, Y.Kitamura, T.Furusawa : "New FPN suppression method and flickerless mosaic color filter array for a frame transfer CCD image sensor.", Optoelectronics- Devices and Technologies, Vol. 6, No. 2, 1991, pp. 231-243.

<u>Okano 91</u> : F.Okano, Y.Fujita, K.Mitani, K.Kawashiri, T.Toma : "The dual green pickup experiment for a compact HDTV color camera with three-2/3 inch CCDs.", IEEE Transactions Broadcasting, Vol. 37, No. 1, 1991, pp. 17-22.

<u>Orihara 86</u> : K.Orihara, E.Oda : "New technologies in dual channel read-out registers for high density CCD image sensor.", Technical Digest IEDM86, pp. 365-368, Los Angeles, Dec 7-10, 1986.

<u>Ozaki 91</u> : T.Ozaki, H.Ono, H.Tanaka, K.Tokumasu : "A proposal of a highly sensitive interline CCD for HDTV.", Extended Abstracts of the 1991 Internal Conference on Solid State Devices and Materials, pp. 666-668, Yokohama, 1991.

<u>Peckerar 79</u> : M.C.Peckerar, D.McCann, F.Blaha, W.Mend, R.Fulton : "Deep depletion charge-coupled device for X-ray and IR sensing applications.", Technical Digest IEDM79, pp. 144-146, Washington DC, Dec 3-5, 1979.

<u>Pohlmann 87</u> : K.C.Pohlmann : "Fundamentals of Digital Audio" in "Principles of Digital Audio", Howard W. Sams & Co., 1987, pp. 37-64.

<u>Renshaw 90</u> : D.Renshaw, P.B.Denyer, G.Wang, M.Lu : "ASIC Vision.", Proceedings CICC 90, p. 7.3.1., Boston, May 13-16, 1990.

Roks 92 : E.Roks, P.Centen, L. Sankaranarayanan, J.Slotboom, J.Bosiers, W.Huinink : "A bipolar floating base detector (FBD) for CCD image sensors.", Technical Digest IEDM92, pp. 109-112, San Francisco, Dec 13-16, 1992.

Sahai 83 : R.Sahai, R.L.Pierson, R.J.Anderson, E.H.Martin, E.A.Sovero, J.A.Higgons : "GaAs CCD's with transparent (ITO) gates for imaging and optical signal processing.", IEEE Electron Devices Letters, Vol. EDL-4, No. 12, 1983, pp. 463-464.

Sakakibara 91 : K.Sakakibara, H.Yamamoto, S.Maegawa, H.Kawashima, Y.Nishioka, M.Yamawaki, S.Asai, N.Tsubouchi, T.Okumura, J.Fujino : "A 1" format 1.5M pixel IT-CCD image sensor for an HDTV camera system.", IEEE Transaction on Consumer Electronics, Vol. CE-37, No. 3, 1991, pp. 487-493.

Saks 80 : N.Saks : "A technique for suppressing dark current generated by interface states in buried channel CCD imagers.", IEEE Electron Devices Letters, Vol. EDL-1, No. 7, 1980, pp. 131-133.

Sangster 69 : F.L.J.Sangster, K.Teer : "Bucket-brigade electronics.", IEEE Journal Solid-State Circuits, Vol. SC-4, 1969, p. 131.

Sankara 91 : L.Sankaranarayanan, W.Hoekstra, L.G.M.Heldens, A.Kokshoorn : "1 GHz CCD transient detector.", Technical Digest IEDM91, pp. 179-182, Washington DC, Dec 8-11, 1991.

Sano 90 : Y.Sano, T.Nomura, H.Aoki, S.Terakawa, H.Kodama, T.Aoki, Y.Hiroshima : "Submicron spaced lens array process technology for a high photosensitivity CCD image sensor.", Technical Digest IEDM90, pp. 283-286, San Francisco, Dec 9-12, 1990.

Sano 91 : Y.Sano, T.Nomura, H.Aoki, S.Terakwa : "Submicron spaced lens array process technology for a high photo-sensitivity CCD image sensor.", Optoelectronics - Devices and Technologies, Vol. 6, No. 2, 1991, pp. 219-229.

Schlig 86 : E.S.Schlig : "A TDI charge-coupled imaging device for page scanning.", IEEE Journal Solid-State Circuits, Vol. SC-21, No. 1, 1986, pp. 182-186.

Schroder 78 : D.K.Schroder : "Transparent gate silicon photodetectors.", IEEE Transaction on Electron Devices, Vol. ED-25, No. 2, 1978, pp. 90-97.

Senda 85 : K.Senda, Y.Hiroshima, S.Matsumoto, T.Kuriyama, M.Susa, S.Terakawa, T.Kunii : "Fixed pattern noise in the solid-state imagers due to the striations in Czochralski silicon crystals.", Journal Applied Physics, Vol. 57, No. 4, 15 Febr 1985, pp. 1369-1372.

Sevenhans 83 : J.Sevenhans, J.Bosiers, E.Laes, G.J.Declerck : "A novel quadrilinear CCD for high resolution line arrays." IEEE Transaction on Electron Devices, Vol. ED-30, No. 2, 1983, pp. 1776-1779.

Shibata 92 : H.Shibata, I.Inoue, R.Miyagawa, H.Yanashitia, N.Nohmi, A.Furukawa, Y.Iida, T.Yamaguchi, Y.Endo, Y.Matsunaga, S.Manabe : "A 2M-pixel two-level vertically-integrated HDTV image sensor.", Digest Technical Papers ISSCC92, pp. 166-167, San Francisco, Febr 19-21, 1992.

Shinoda 87 : K.Shinoda, F.Nagumo, H.Terakawa : "Broadcast charge-coupled device imagers.", Internal Broadcast Engineer, Sept 1987, pp. 28-47.

Slotboom 91 : J.W.Slotboom, G.Streutker : "Physical aspects of charge-coupled devices.", Physica Scripta, Vol. T35, 1991, pp. 281-286.

Stevens 89 : E.Stevens, B.C.Burkey, D.N.Nichols, Y.Lee, D.L.Losee, T.H.Lee, T.J.Tredwell, R.P.Khosla : "A lag-free 1024 x 1024 progressive scan interline CCD image sensor with antiblooming and exposure control.", Technical Digest IEDM89, pp. 169-172, Washington DC, Dec 3-6, 1989.

Tabei 91 : M.Tabei, K.Kobayashi, M.Shizukuishi : "A new CCD architecture of high-resolution and sensitivity for color digital still picture.", IEEE Transaction on Electron Devices, Vol. ED-38, No. 5, 1991, pp. 1052-1058.

Takizawa 83 : Y.Takizawa, H.Kotaki, K.Saito, T.Sugiki, Y.Takemura : "Field integration mode CCD color television camera using a frequency interleaving method.", IEEE Transaction on Consumer Electronics, Vol. CE-29, No. 3, 1983, pp. 358-364.

Tanaka 90 : T.Tanaka, S.Katoh, I.Akiyama, N.Teranishi, K.Orihara, E.Oda : "HDTV single-chip CCD color camera.", IEEE Transaction on Consumer Electronics, Vol. CE-36, No. 3, 1990, p. 479-485.

Terakawa 80 : S.Terakawa, T.Yamada, K.Horii, T.Takamura, I.Teramoto : "A new organization area image sensor with CCD readout through charge priming transfer.", IEEE Electron Devices Letters, Vol. EDL-1, No. 5, 1980, pp. 86-87.

Teranishi 82 : N.Teranishi, A.Kohono, Y.Ishihara, E.Oda, K.Arai : "No image lag photodiode structure in the interline CCD image sensor.", Technical Digest IEDM82, pp. 324-327, San Francisco, Dec 13-15, 1982.

Teranishi 87 : N.Teranishi, Y.Ishihara : "Smear reduction in the interline CCD image sensor.", IEEE Transaction on Electron Devices, Vol. ED-34, No. 5, 1987, pp. 1052-1056.

Theuwissen 84 : A.J.P.Theuwissen, C.H.L.Weijtens, L.J.M.Esser, J.N.G.Cox, H.T.A.R.Duyvelaar, W.C.Keur : "The accordion imager : an ultra high density frame transfer CCD.", Technical Digest IEDM84, pp. 40-43, San Francisco, Dec 9-12, 1984.

Theuwissen 84a : A.J.P.Theuwissen, G.J.Declerck : "Optical and electrical properties of reactively DC-magnetron sputtered In_2O_3:Sn films." Thin Solid Films, Vol. 121, pp. 109-119, 1984.

Theuwissen 88 : A.J.P.Theuwissen, J.G.C.Bakker, H.J.N.G.Cox, A.L.Kokshoorn, P.A.C.van Loon, B.C.J.O'Dwyer, J.M.A.M.Oppers, C.H.L.Weijtens : "A 400k pixels 1/2" accordion CCD-imager.", Digest Technical Papers ISSCC88, pp. 48-49, San Francisco, Febr 17-19, 1988.

Theuwissen 90 : A.J.P.Theuwissen : "Charge-coupled devices als Bildaufnehmer.", Elektronik, Vol. 23, Oct 1990, pp. 76-84.

Theuwissen 91 : A.J.P.Theuwissen, H.L.Peek, P.Centen, R.G.M.Boesten, J.N.G.Cox, P.Hartog, A.L.Kokshoorn, H.J.C.van Kuijk, B.O'Dwyer, J.M.A.M.Oppers, F.Vledder : "A 2.2Mpixel FT-CCD imager, according to the Eureka HDTV-standard.", Technical Digest IEDM91, pp. 167-170, Washington DC, Dec 8-11, 1991.

Tompsett 73 : M.F.Tompsett : "The quantitative effects of interface states on the performance of charge-coupled devices.", IEEE Transaction on Electron Devices, Vol. ED-20, No. 1, 1973, pp. 45-55.

Toyoda 94 : Y.Toyoda, K.Itakura, T.Nobusada, Y.Saitoh, N.Kokusenya, R.Nagayoshi, H.Tanaka, M.Ozaki, M.Sugawara, K.Mitani, Y.Fujita : "A 2/3-inch 2.0 Mpixel M-FIT CCD with a single channel HCCD for HDTV camera.", Digest Technical Papers ISSCC94, pp. 220-221, San Francisco, Febr 16-18, 1994.

Tsaur 90 : B-Y.Tsaur, C.K.Chen, J-P.Mattia : "PtSi Schottky-barrier focal plane arrays for multispectral imaging in ultraviolet, visible and infrared spectral bands.", IEEE Electron Devices Letters, Vol. EDL-11, No. 11, 1990, pp. 162-164.

Tsukada 81 : T.Tsukada, T.Baji, Y.Shimomoto, A.Sasano, Y.Tanaka, H.Matsumaru, Y.Takasaki, N.Koike, T.Akiyama : "Solid-state color imager using an a-Si:H photoconductor film.", Technical Digest IEDM81, pp. 479-482, Washington DC, Dec 7-9, 1981.

Van der Spiegel 89 : J.Van der Spiegel, G.Kreider, C.Claeys, I.Debusschere, G.Sandini, P.Dario, F.Fantini, P.Bellutti, G.Soncini : "A foveated retina-like sensor using CCD technology." in "Analog VLSI implementation of neural systems", C.Mead and M.Ismail, eds., Kluwer Academic Publishers, Boston, 1989, Chapter 8, pp. 189-210.

Van der Spiegel 84 : J.Van der Spiegel, J.Sevenhans, A.Theuwissen, J.Bosiers, I.Debusschere, G.Declerck : "Study of different sensor types for high resolution linear CCD imagers.", Sensors and Actuators, Vol. 6, 1984, pp. 51-64.

Van de Steeg 85 : M.J.H.van de Steeg, H.L.Peek, J.G.C.Bakker, J.A.Pals, B.G.M.H.Dillen, J.M.A.M.Oppers : "A frame-transfer CCD color imager with vertical anti-blooming.", IEEE Transaction on Electron Devices, Vol. ED-32, No. 8, 1985, pp. 1430-1438.

Van de Wiele 76 : F.Van de Wiele : "Photodiode quantum-efficiency." in "Solid-State Imaging", NATO Advanced Study Institute on Solid-State Imaging, Eds. P.Jespers, F.Van de Wiele, M.White, Noordhoff, Leyden, 1976, pp.47-90.

Veendrick 90 : H.J.M.Veendrick in "Integrated MOS circuits", Delta Press, Overberg (NL), 1990, pp.1-45.

Wadsworth 84 : M.Wadsworth, R.D.McGrath : "Charge coupled device for X-ray imaging.", Technical Digest IEDM84, pp. 20-23, San Francisco, Dec 9-12, 1984.

Wadsworth 89 : M.Wadsworth : "Image windowing with CID cameras.", Electronic Imaging 89 West, pp. 949-953, Pasadena, April 10-13, 1989.

Walden 72 : R.H.Walden, R.H.Krambeck, R.J.Strain, J.McKenna, N.L.Schyer, G.E.Smith : "The buried-channel charge coupled device.", Bell Systems Technical Journal, Vol.51, 1972, pp. 1635-1640.

Watanabe 84 : T.Watanabe, K.Hashiguchi, T.Yamano, J-I.Nakai, S.Miyatake, O.Matsui, K.Awana : "A CCD color signal separation IC for single-chip color imagers.", IEEE Transaction on Electron Devices, Vol. ED-31, No. 2, 1984, pp. 183-188.

Weijtens 85 : C.H.L.Weijtens, W.C.Keur : "Reduction of reflection losses in solid-state image sensors.", Proceedings SPIE, Vol. 591, pp. 75-79, Cannes, Nov 26-27, 1985.

Weijtens 92 : C.K.L. Weijtens : "Reduction of oxide charge and interface-trap density in MOS capacitors with ITO gates.", IEEE Transaction on Electron Devices, Vol. ED-39, No. 8, 1992, pp. 1889-1894.

White 74 : M.H. White, D.R.Lampe, F.C.Blaha, I.A.Mack : "Characterization of surface channel CCD image arrays at low light levels.", IEEE Journal Solid-State Circuits, Vol. SC-9, No. 1, 1974, pp. 1-13.

White 76 : M.H. White : "Design of solid-state imaging arrays." in "Solid-State Imaging", NATO Adv. Study Inst. on Solid-State Imaging, Eds. P.Jespers, F.Van de Wiele, M.White, Noordhoff, Leyden, 1976, pp.485-522.

Wong 92 : H.-S.Wong, Y.L.Yao, E.S.Schlig : "TDI charge-coupled devices : design and applications.", IBM Journal Research Development, Vol. 36, No. 1, 1992, pp. 83-106.

Woody 91 : T.W.Woody : "High-performance CCDs : cameras and applications.", Photonics Spectra, Sept 1991, pp. 167-172.

Yamada 89 : T.Yamada, A.Fukumoto : "Trench CCD image sensor.", IEEE Transaction on Consumer Electronics, Vol. CE-35, No. 3, 1989, pp. 360-367.

Yonemoto 90 : K.Yonemoto, T.Iizuka, S.Nakamura, K.Harada, K.Wada, M.Negishi, H.Yamada, T.Tsunakawa, K.Shinohara, T.Ishimaru, Y.Kamide, T.Yamasaki, M.Yamagishi : "A 2 million pixel FIT-CCD image sensor for HDTV camera system.", Digest Technical Papers ISSCC90, pp. 214-215, San Francisco, Febr 14-16, 1990.

Yusa 85 : A.Yusa, J.Nishizawa, M.Imai, H.Yamada, J.Nakamura, T.Mizoguchi, Y.Ohta, M.Takayama : "The operating characterisitcs of a static induction transistor (SIT) image sensor.", Technical Digest IEDM85, pp. 440-443, Washington DC, Dec 1-4, 1985.

Yutani 91 : N.Yutani, H.Yagi, M.Kimata, J.Nakanishi, S.Nagayoshi, N.Tsubouchi : "1030 x 1040 element PtSi Schottky-barrier IR image sensor.", Technical Digest IEDM91, pp. 175-178, Washington DC, Dec 8-11, 1991.

INDEX

A

absorption coefficient	131
absorption of photons	131
accordion CCD	259
accumulation mode	8
alloying	338
AMI	214
amplification function	211
amplified MOS intelligent imager	214
amplifier noise	99
anisotropic etch	317
annealing	317
antiblooming	176
antiparallel pulses	54
aliasing	100,152
aperture ratio	195
array imager	114
ASIC vision	293

B

back-end	336
back-side illumination	211
base-stored image sensor	215
BASIS	215
BCCD	40
BCMD	213
bilinear imager	112
binning	66
blooming	176
body factor	69
bulk charge-modulation device	213
buried-channel (CCD	39
buttable devices	281

C

carrier collection	134
CCIR standard	159
center of gravity	48
channel potential	40

channel definition 66
channel-stopper implantation 66
charge binning 66
charge-coupled devices
    working principle 7
charge-handling capability 45
charge-injection device 123
charge-modulation device 213
charge multiplexing 248
charge-priming CCD 120
charge-pumped antiblooming 178
charge reset 183
charge-sweep device 202,262
charge transfer 25
    efficiency 36
    inefficiency 36
CID 123
clocked antiblooming 178
CMD 213
collection of carriers 134
collection efficiency 134
color filter 165
color imaging 165
color-splitting prism 171
column defects 222
complementary colors 165
compound channel 257
configurations 112
contact hole 337
contact type imager 311
continuity equation 25
conversion factor 76,234
convolution 148
correlated double sampling 228
cost price 365
cover glass 354
cross-gate CCD 207
current-density equation 36

D
dark current 92
dark-current nonuniformities 94
dark-current shot noise 223
deep-depletion CCD 304

deep depletion mode	10
delay line	85
denuded zone	320
depletion capacitance	31
depletion region	10
deposition technique	317
dichroic prism	171
diffusion constant	
effective	28
of electrons	28
of holes	28
diffusion length	134
diffusion MTF	143
diode cutoff	70
drive-in	324
dynamic CMOS logic	259
dynamic pinning	290
dynamic pixel management	264

E

EIA standard	159
electronic shutter	183
electronic still picture	267
electrostatic discharge	355
etch-back	346
etching	317

F

field-integration mode	161
field shield	66
fill and spill	73
FIT	119
flat-band voltage	18
floating diffusion (with reset)	76
floating gate (without reset)	79
Fourier transform	148
four-phase system	54
foveated-retina sensor	296
frame-integration mode	161
frame-interline-transfer CCD	119
frame-transfer CCD	114
fringing field	25
FT CCD	114
full-frame CCD	114

G

GaAs imagers	311
gate tapering	253
gate voltage	7
generation velocity	131
geometric MTF	148

H

HDTV imagers	262
hemispherical lens	196
high-speed clocks	262
high-speed devices	310
hole-accumulation device	223
horizontal anti-blooming	177

I

ideal MOS capacitance	7
IL CCD	117
imager configurations	112
implantation	317
indium tin oxide	208
infrared imaging	298
input section	53
input structure	66
integration time	94
interlaced scanning	161
interline-transfer CCD	117
inversion mode	13
IrSi detector	298
isotropic etch	317

J

junction	
induced	109
metallurgical	109
one-sided diffused	10

K

kTC noise	99

L

large-area devices	280
lateral anti-blooming	177
lift-off	342,344

linear imager   112
lithography   317
LOCOS   66

M
M-FIT   267
microlens   196,347
mobility of electrons   25
modulation transfer function   87,142
    diffusion   143
    geometric   148
    horizontal   360
    transport   145
    vertical   360
Moiré effects   152
mosaic filters   168
MOS capacitance
    ideal   7
    in accumulation   8
    in deep depletion   10
    in inversion   13
    in weak inversion   16
    real   18
MOS-XY imager   120
multilayered structure   205
multi-phase pinned CCD   288
multiple-frame-interline-transfer CCD   267
multiple outputs   311

N
noise
    amplifier   224
    elimination   228
    reset   227
    thermal   225
    1/f   99,227
noise electron density (NED)   225
nonvisible imaging   297
notch CCD   282
Nyquist frequency   145
Nyquist theorem   100

O
one-and-a-half-phase system   62

one-dimensional potential analysis   48
open-phase pinned CCD   287
optical cavity   298
optical thickness   205
output amplifier architectures   231
output amplifier noise   224
output amplifier sensitivity   234
output section   53
output structure   73
oxide charge   18

P
partitioning noise   227
Peltier element   94
penetration depth   131
photoconductor   311
photoconversion layer   201
photon absorption   131
photon sensing   109
pinned-phase CCD   285
pixel pitch   145
pixel nonuniformities   222
planarization layer   345
point defects   221
Poisson distribution   97
Poisson's equation   29
potential analysis   48
primary colors   166
progressive scanning   162
PtSi detectors   298

Q
quantum efficiency   139
quadrilinear imager   112
quasi-frame-interline-transfer imager   119

R
rectangular dome lens   196
reflow   348
reset   76
reset noise   227
reset pulse   76
reset transistor   76
resolution   136

ripple clock 63

S
sampling 100
saturation level 18
scan circuit 120
scanning modes 159
SCCD 40
scientific imagers 280
scratch protection 340
self-aligned implantation 335
self-induced drift 29
self-induced (electrical) fields 29
semicylindrical lens 196
shot noise 97
signal-to-noise ratio 193
SIT 212
skipper CCD 283
smart image sensors 293
smear 236
space-charge density 29
spatial domain 142
spectral response 136
sputtering technique 338
static-induction transistor 212
step coverage 338
strapping technique 262
striations 222
stripe filters 166
surface states 36

T
TDI 274
technology related noise 221
thermal diffusion 28
thermal noise 99,225
three-dimensional integrated image sensor 294
three-phase system 57
threshold voltage 13
time-delay and integrating CCD 274
transfer time 42
transfer function 87
transfer noise 222
transparent conductive gates 208

transport
    efficiency 31
    depth 42
    inefficiency 87
    MTF 145
    section 2
    systems 54
trapping noise 98
two-phase system 58

U
underdiffusion 322
UV imaging 302

V
vertical anti-blooming 179
virtual-phase system 63,207

W
weak inversion 16
work function difference 18

X
x-ray imaging 306

Y

Z
z-domain 87

Printed in the United Kingdom
by Lightning Source UK Ltd.
93049